Natural Gas Markets in India

Sanjay Kumar Kar · Ayush Gupta
Editors

Natural Gas Markets in India

Opportunities and Challenges

Springer

Editors
Sanjay Kumar Kar
Department of Management Studies
Rajiv Gandhi Institute of Petroleum Technology
Jais, Uttar Pradesh
India

Ayush Gupta
GAIL Training Institute
Noida, Uttar Pradesh
India

ISBN 978-981-10-3116-8 ISBN 978-981-10-3118-2 (eBook)
DOI 10.1007/978-981-10-3118-2

Library of Congress Control Number: 2016957862

Printed on acid-free paper

This Springer imprint is published by Springer Nature
The registered company is Springer Nature Singapore Pte Ltd.
The registered company address is: 152 Beach Road, #22-06/08 Gateway East, Singapore 189721, Singapore

Foreword

It is my pleasure to know that Dr. Sanjay Kumar Kar and Mr. Ayush Gupta have come up with a comprehensive book bringing out the opportunities and challenges before India in the natural gas sector.

Continuous efforts are being made to reduce the gap between the demand and supply of natural gas in the country through domestic production and import of gas in the form of liquefied natural gas (LNG). The natural gas pipeline is being continuously expanded, and gas is now reaching to thousands of new domestic consumers through the city gas distribution networks being established in new cities/geographical areas. With natural gas just contributing only about 9% to the India's energy basket as against the world average of 25%, there lies a huge opportunity for the growth of natural gas sector in India. However, with limited domestic production and increasing volatility in the international gas markets, the natural gas sector in India is presently passing through a challenging phase. The steep fall in the crude oil prices over last one year is already having a deep impact on the natural gas market in India.

The book is an excellent compilation of issues and challenges in the Indian natural gas sector as brought out by the industry professionals and academicians in their respective domain areas. The contributing authors have attempted to provide a comprehensive view on the latest developments and advances in the various processes involved in the natural gas value chain which includes areas such as exploration, production, transmission distribution, business development, marketing, and other relevant areas with a special focus on India.

The different chapters in the book cover all important aspects of natural gas business including natural gas demand and supply scenario, gas sourcing, contract management, and pricing, opportunities and challenges of building infrastructure for gas business, city gas distribution, regulatory developments and drivers for gas market, capability building for natural gas business, taxation, and health, safety, environment issues in natural gas value chain.

Dr. Sanjay and Mr. Ayush have been actively working and publishing in the field of green energy for past several years. Dr. Sanjay's coedited book "Energy Sustainability Through Green Energy" serves as an interesting reference material for the readers. I am sure that the industry professionals, academic researchers, and students having an interest in the natural gas sector in India will find this book to be of immense use. I wish Dr. Sanjay Kumar Kar and Mr. Ayush Gupta, all the best for their endeavours made.

Prof. K.B. Akhilesh
President, Board of Governors
Rajiv Gandhi Institute of Petroleum Technology
Jais, India

Preface

Overview

Environment and climate change continue to attract attention of developed, developing, and least developed countries. Today, global leaders, policymakers, think tank, and consuming class are more concerned about climate change and global warming. Considerable efforts are being made to build a consensus on controlling rising temperature and retaining mother earth's natural beauty. Energy remains the biggest factor behind any climate change discussions or negotiations because growth, development, and industrialization are largely dependent on energy. Even today, the global community heavily relies on fossil fuels, notably more polluting fuels such as coal and oil. As far as India is concerned, its primary energy consumption basket is too skewed towards fossil fuels. In 2014, coal contributed 56.47% followed by oil (28.33%), natural gas (7.14%), hydro (4.64%), renewables (2.18%), and the rest contributed by nuclear energy. India's share of natural gas in the primary energy basket is much lower than global average of 23.71%.

There are plenty of reasons for low-consumption share of natural gas in the country. India being an energy-deficit country, it faces serious challenges related to natural gas demand–supply gap. Continuous efforts are being made to reduce the gap between the demand and supply of natural gas in the country through domestic production and import of gas in the form of liquefied natural gas (LNG). Further, efforts are being made to import natural gas through transnational natural gas pipelines. The government has been implementing many policy measures to increase the use of natural gas in various segments such as domestic and transport.

This book shares the latest market developments and advances in natural gas demand, supply, transmission, distribution, and consumption. Chapters are written by researchers and industry professionals working in the field of natural gas and energy. The book covers entire natural gas value chain, and the contributing authors provide insightful analyses in the areas of natural gas demand–supply, exploration and production policy, downstream regulatory developments, city gas distribution, pipeline, pricing, and taxation policy natural gas market development in India. We hope the readers of this book would enjoy thoughts, ideas, practices, and policy analyses shared by the authors.

Part I: Natural Gas Market Scenario in India

India's Active Engagement with Natural Gas: Imperatives and Challenges
In this chapter, Manish Vaid and Sanjay Kumar Kar discussed about post-COP 21 scenario and India's reconfigured energy goals—which are linked with climate policy goals. Their submission is that India's Intended Nationally Determined Commitments (INDC) submitted to the UNFCCC signals that India wishes to strive for green growth. By 2030, India's INDC commits for 40% share of electricity that would come from non-fossil fuel sources. This calls for a paradigm shift in India's energy strategy by reshuffling its energy basket in a way that it not only continues and expedites with its renewable capacity expansion programmes, which is largest in the world, but also sticks to its active engagement with natural gas as a base fuel for electricity generation and city gas distribution. The chapter discusses the imperatives for India's continued engagement with natural gas and the challenges it would face in increasing its share in its natural gas basket amidst stiff emission targets set by India in the Paris Summit.

Impact of Slashing Oil Prices on the Natural Gas Market
Husain Ahmad discusses the challenges faced by natural gas due to proverbial free fall in the prices of crude oil during the year 2014–15. The chapter also includes a recapitulation of the key drivers that have slumped the global crude oil prices. Further, it discusses about the macroeconomic and financial propositions and impact on Indian economy due to plummeting oil prices thereof.

Part II: Gas Sourcing, Contract, and Pricing

Sourcing of Natural Gas Through Cross Border Pipelines
Coauthors Ayush Gupta and Bhavesh Gupta present how in the era of economic modernization and globalization, India has unfolded itself as central economic and trading zone in emerging Asian market and has focused a lot towards its foreign policies impacting its accelerating economic performance. Private sector involvement in power sector and rationalization of prices has forced India to look for efficient energy management system. Natural gas, a highly promising commodity in India, is presently supplied from only two sources, i.e. domestic production and liquefied natural gas (LNG). But now instead of importing costly LNG, India has to examine and start importing through low-cost cross-border pipelines. Technological

development in the field of deep-sea pipelines has made it possible for India to consider importing natural gas from Iran and other Middle East countries, bypassing countries such as Pakistan and Afghanistan having serious security issues. Construction of Turkmenistan–Afghanistan–Pakistan–Indian (TAPI) pipeline has already started, and Indian government has also initiated discussions with Oman, Myanmar, Kazakhstan, and Russia for new pipeline projects. Indian Prime Minister Narendra Modi can imagine energy has beating heart of accelerating India and pipelines as its veins; hence, to deal with energy diplomacy and to have sufficient amount of fuel, he is focusing on developing strong and long-term strategic relationships with resource rich countries. This will give a big boost to the natural gas industry in India.

Issues and Challenges of Gas Contracts

Manas Das discusses important issues and challenges involved in natural gas contracting. Contracts are agreements entered into by two parties with the intention of creating a legal obligation. As the gas sector in India has evolved from pre-NELP regime to a market open to international LNG contracts, the domestic gas contracts also evolved with newer dimensions, newer clauses, newer obligations, or bindings. Gas contracts are entered into various forms with different objectives across the gas value chain. In the upstream sector, production sharing contract (PSC) is signed between Government of India and the producer consortium. Joint Operating Agreement (JOA) is signed among the consortium members. Gas producer also signs gas supply contract (GSC)/Gas Supply Agreement (GSA) with the transporter/trader in the midstream segment. In the downstream, the transporter/trader signs agreements with the customers/consumers of gas. The supply gets governed by the GSC, and the transmission is governed by Gas Transmission Agreement (GTA). In some cases, the transmission and the supply get clubbed in a single contract termed as GSTA or Gas Supply and Transmission Contracts. The downstream contracts may be named differently, but structurally, they still remain same with minor modifications. This first section of this chapter covers basic features of production sharing contracts, how profit/cost petroleum is arrived at, contractual obligations under PSCs, issues and concerns in PSCs, and changes proposed by Rangarajan Committee and Kelkar Committee. The second section brings in illustration on different types of midstream and downstream contracts, the process description of execution of contracts, how the contracts have evolved, and futuristic changes expected in these contracts.

Natural Gas Pricing

In this chapter, the author Pramod Paliwal simplified the concept of pricing of natural gas. Consumption of natural gas has grown rapidly over the last few decades and currently accounts for nearly a quarter of the world's primary energy basket. Natural gas prices are mainly a function of market supply and demand. Natural Gas pricing–like any other commodity—is not immune to the forces of demand–supply. However, it would be interesting to understand the uniqueness of natural gas and the subsequent context with its pricing. The physical properties of crude oil and the fact that it is relatively uncomplicated to transport and to store facilitated the emergence of commodity pricing mechanisms in the oil sector. However, these considerations do not apply in the same way to natural gas. Globally, one of the biggest challenges in the market is the pricing techniques. As known, natural gas does not have a globalized market; rather, there are regional markets with substantial differences. The overriding mechanism for the international gas trade nevertheless remains oil indexation. The phenomenon of international natural gas trade and pricing is under evolution also because once-isolated regional gas markets are now practically interconnected through the increasing trade in liquefied natural gas (LNG). Starting with Natural Gas International Demand Supply Matrix, the chapter discusses distinctiveness of natural gas pricing followed by international context of natural gas pricing including the conventional oil indexation-based pricing. Natural gas pricing in India has later been discussed in detail with a brief mention of the concept of natural gas price pooling in India. The chapter concludes with a reality check on the prospects of evolution of global natural gas trading hub.

Feasibility of Price Pooling for Gas-Based Power Generation in India: A Reality Check!

In this chapter, the author RK Tripathy throws lights on gas price pooling and its feasibility. Gas-based power projects are on the verge of becoming bad assets for the country. While nearly 14 GW of gas-based projects are non-functional, 9 GW of projects operates at a suboptimal level of 25% or less. Domestic gas supply to power sector seems a distant dream with no major gas discovery in sight. In a bid to kick start the plants and to enable the project developers to service their debts and cover up the operational expense, the government has started a scheme of reverse e-auctioning for supplying sport RLNG to power plants. This plan will help revive some of stressed gas projects in the country with the help of government subsidy out of power system development fund. While this is a welcome step from the government and would function well in constantly depreciating spot RLNG price, the scheme provides only short-term visibility for operations and servicing of debt. The long-term operations and viability of these plants would remain dependent upon ramping up production of gas from domestic sources. The current situation where long-term RLNG price hovers around USD 7 to 8 per MMBTU, the government has to rethink its strategy and should move away from indirectly bailing out of gas-based power projects and open the sector for competition with necessary policy support. Gas price pooling is long debated in India, and finally, it was made effective for fertilizer sector from 1 July 2015. While the decision to introduce pooling price concept would go a long way for fertilizer unit revival, the debate is still on its implementation for the power sector. The feasibility of such an option for the power sector would be accessed in this chapter along with government's preparedness in addressing the concerns of various stakeholders.

Part III: City Gas Distribution in India

A Comprehensive Review of City Gas Distribution in India
Authors S.K. Kudaisya and Sanjay Kumar Kar provide a comprehensive review and analysis of city gas distribution in India. City gas distribution is going through transformation in India. A sincere attempt is made to offer a comprehensive review of city gas distribution in the country. The chapter provides critical insights into developmental gas market and evolution of city gas distribution business in India. The authors discuss about causes of city gas distribution evolution. Important linkage between regulatory developments and city gas distribution has been discussed. The authors highlight the progress of city gas distribution system after constitution of Petroleum and Natural Gas Regulatory Board. The readers can find a comprehensive review of various bidding rounds as per the regulations developed by the Petroleum and Natural Gas Regulatory Board. The chapter provides meaningful insights on opportunities and challenges available in the city gas distribution business. It points out the critical factors responsible for achievement or failure of commitments made by the entities to the regulator. The authors suggest with the support of the government, constant vigil and involvement of the regulator, and professional management of entities; the city gas distribution business would gain greater penetration in India.

Building and Sustaining Natural Gas Business in India
Authors Sanjay Kumar Kar, Piyush Kumar Sinha, and Bhashit Dholakia offer insights on building and sustaining natural business in India. Natural gas is a green fuel, and the Government of India is highly committed to promote green fuel like natural gas to replace more polluting fuels. The chapter provides critical insights on demand–supply scenarios, enabling factors, challenges, and opportunities to build sustainable gas business in the country. It highlights role of global factors such as demand, supply, pricing, and geopolitics in natural gas market in India. Author presents insightful analysis on natural gas business environment, industry growth, role of competition, new entrants, opportunities available, challenges experienced, and suggestions for sustainable natural gas market growth in the country.

Issues and Challenges in the Development of Efficient Gas Market
In this chapter, author Akhil Mehrotra discusses critical issues and challenges in the development of efficient natural gas market developments in India. India needs energy, whereby its population can access and afford. Government of India intends to increase energy availability by increasing the share of gas, in the energy mix from 7.1 to 20% by the year 2030. However, this is not possible unless there is a thriving gas market in the country. The current Indian gas market has distortions such as monopoly power, inadequate infrastructure, and lack of transparency. Gas market evolution curve helps segment various phases of gas market, starting from government controlling the gas chain, negotiated market, early wholesale, and fully

developed gas market. Study of major countries reveal India at negotiated market phase, with aspirations similar to other countries, to move to a fully developed market phase such as the USA and UK. The learning from the USA and UK can help to understand factors for such a transition; however, this can only be achieved if challenges such as issue relating to large population below poverty line, lack of integrated planning, continued distortion in all energy markets, unstable fiscal regime, and misalignment between federal and state governments are addressed. Government needs to calibrate its approach to develop gas market in order to increase the share of gas in the energy mix.

Demand Assessment and Design Aspect of City Gas Distribution Network
Author Husain Ahmad discusses about design aspects of city gas distribution network in India. Natural gas is a promising policy element in achieving equitable, balanced, and sustainable economic growth by widening its user base beyond conventional industries. The city gas distribution is what we call the "last mile" in entire gas value chain. Gas supplies usually are at low/medium pressure to the residential/domestic, commercial, and industrial consumers; this is called as pipe natural gas and is compressed to very high pressure to increase the volumetric efficiency in order to make it compatible for the usage in automobile sectors as compressed natural gas.

Part IV: Legal, Regulatory Developments: Impact on Natural Gas Market in India

Petroleum Exploration and Licensing Policy in India
T.P. Rao and Sanjay Kumar Kar discuss progress of Petroleum Exploration and Licensing Policy in India. This chapter lucidly brings out policy developments in the area of petroleum exploration in India. Authors discuss historical perspectives, existing policy frameworks, recent developments, and set futuristic directions. Findings suggest that the current policy fails to attract desired level of investment and retaining confidence of investors. Therefore, to bring back petroleum exploration and production on track, the government should implement investor-friendly E&P policies at the earliest.

Downstream Petroleum and Natural Gas Regulatory Developments in India
The downstream regulatory environment is passing through developmental stages in India. Sanjay Kumar Kar covers historical and recent downstream regulatory developments in India in this chapter. A sincere attempt is made to discuss about vision, functions, and operations of the Petroleum and Natural Gas Regulatory Board (PNGRB). The chapter highlights all important regulations developed since

the inception of the PNGRB and authorizations granted to entities as per the regulations. The chapter discusses challenges faced by the regulator to develop efficient natural gas market in India. It is suggested that the regulator should be empowered to regulate the entire petroleum sector.

Taxation Issues in Natural Gas Industry in India
Author Pankaj Gupta discusses important taxation issues relevant for natural gas business in India. The author points out that the entire natural gas supply chain is plagued by plethora of taxes in India. These taxes are levied through separate statutes administered by different authorities from central and state government. Each statue levies tax at different stages in the value chain and therefore provides separate valuation methodology. Most of these taxes are non-vatable having huge cascading effect. The customers mainly from power and fertilizer sectors are also not able to take credit of most part of taxes paid on purchase of natural gas. This inefficiency is further compounded by administrative cost resulting from compliance of taxation statutes which are administered by different government authorities. More often than not, divergent position taken on same issue by various wings of the taxation administrative authorities results in further complication. Consequent litigation takes anywhere between 5 to 10 years for final position to emerge. Any adverse tax ruling results in huge unrecoverable tax burden as indirect taxes are largely passed through, and it becomes difficult to recover past tax dues from the customers after such a long lag. Therefore, a reform in the current tax regime is required for increasing the share of natural gas in the energy basket of India, being a cleaner fuel compared to other fossil fuels. But the expectation is being belied, as natural gas is being kept out of the proposed goods and services Tax.

Part V: Other Important Areas

Human Resource Challenges for Gas Sector in India
Authors Ayush Gupta, Sanjay Kumar Kar, Amit Ray, and DV Shastry highlight how natural gas industry across the world is facing a severe crunch of talented, skilled, and experienced manpower. The industry is losing experience by way of superannuation of experienced employees and loosing talent to other industries as oil and gas sector is no longer a preferred sector among younger generation. The chapter discusses critical issues and emerging challenges related to human resource requirements for natural gas sector in India. Further, the authors throw some lights on human resource demand, skilled manpower availability, academic and professional institutions offering specialized courses/training programmes, and skill sets required to succeed in the sector. The authors discuss about ways to address prominent HR challenges and meet existing gap in the natural gas value chain in India.

Health, Safety, Environment: Issues and Challenges for Gas Value Chain in India

In this chapter, author Jitin Saxena highlights health, safety, and environment issues and challenges across natural gas value chain. With growing demand of energy in country, activities in sector are increasing due to which managing health, safety, and environment are becoming major and critical concern for gas companies. Any laxity in upkeep/handling of highly inflammable gas products can be a reason of accident taking place which possess high risk to life, property, and environment. Over the period, accident rates in sector have gradually reduced which shows that companies are becoming more vigilant and proactive on safety issues. Still, there are number of safety-related issues and challenges across gas value chain in India which needs to be addressed by the companies with the support of government. This chapter describes the various safety issues and challenges emerging these days in sector and highlights the needs for developing safety culture and well-established HSE management system to address those issues.

Reviewing Europe's Transition from a Gas Market to Renewables

In this chapter, Gerard Kreeft presents an interesting review of Europe's transition from a gas market to renewables. While this chapter is focused on the European gas market, it is also an overview/summary of trends and innovations that have started in Europe and helped shape renewable energy. Trends can indeed influence various parts of the global energy market. Much of this has been done within the confines of the global energy village, an annual storage event conducted by EnergyWise, an event which brings together the best minds so that new trends and innovations can be explained, compared, and scenarios developed. Key issues include Europe's diminished gas production and its increased dependency on gas imports, in particular from Russia. While gas imports will probably stay at their current level, in long term, this can change. There is the question whether Gazprom can finance its Yamal's assets to ensure a continued supply of gas to Europe. And Europeans are increasingly turning to renewable fuels. How this will play out remains to be seen. Europe's energy politics is in the short-medium term a muddle. The rise of shale gas in the USA has given a sharp rise to coal imports to Europe. CO_2 emission trading prices are so low that they are irrelevant. Continuous subsidies for green energy continue to distort the marketplace. And in Germany, "brown coal" continues to be popular as an industrial fuel. Also, an analysis is given of the oil and gas industry. The basic tenant of a hydrocarbon base, known in the industry as RRR (Reserve Replacement Ratio), is critically examined. Instead, it is argued that the industry should introduce an "Energy Unit", based on renewables, and be CO_2 neutral. This can be the basis of a new energy road map.

Jais, India Sanjay Kumar Kar
Noida, India Ayush Gupta

Acknowledgements

Before the beginning of this project in late 2014, we were reasonably aware of investment of time and effort required for editing a book. We thought of bringing an edited book to share knowledge, market developments, and policy advancements in the field of natural gas business with large group of potential readers. So we felt we can simply get connected with the researchers and industry experts concerned about natural gas business in India by inviting them to share their work in the form of book chapters for an edited volume.

To us, it sounded quite simple at the beginning. But to our surprise, we were proven absolutely wrong on many occasions. Now, we understand better! Since the authors contributed to this book are from the industry and the academic community with their prior commitments and unforeseen engagements, our deadlines for submitting book chapter proposals, first drafts of the chapters, peer-reviewed manuscripts, and submission of final revisions of the contributions received serious challenges. However, our contributors and reviewers extended all kinds of possible support to take the project further. We are really honoured to have all the distinguished contributors. We are very thankful to all our passionate contributors and reviewers!

This book never would have taken off without the generous support of Dr. J.P. Gupta, former Director RGIPT, Prof. P.K. Bhattacharya, the current Director RGIPT, and our colleagues at RGIPT and GAIL. We are extremely thankful to Prof. Abhinadan Kumar Jain, Indian Institute of Management, Ahmedabad, for constantly encouraging to complet this project. It would never have been completed without the cooperation, administrative, and editorial support of Swati Meherishi, Aparajita Singh, and the rest of the team at Springer. Our heartfelt thanks to all these dedicated and extremely cooperative individuals!

Finally, this project consumed quite some of our leisure time that should have been dedicated to our families. There is no doubt that we are very excited, enthusiastic, and proud of the final outcome of our work; we also feel sorry about unwittingly neglecting them on many occasions, specially weekends. For all their patience and moral support, we dedicate this book to Tapaswini, Kashyapi, Monika, and Tavishi.

Sanjay Kumar Kar
Ayush Gupta

Contents

Editors and Contributors

About the Editors

Dr. Sanjay Kumar Kar is an Associate Professor and the Head of the Department of Management Studies at Rajiv Gandhi Institute of Petroleum Technology (RGIPT), Jais, India. Prior to joining RGIPT, Dr. Kar taught at the Institute of Petroleum Management, Gandhi Nagar. As a visiting faculty, he had been associated with the Chonnal National University (South Korea), National Institute of Fashion Technology, India, and other national level management institutes. He has more than 14 years of research and teaching experience. He regularly contributes articles to refereed journals and books. He is the co-editor of *Energy Sustainability Through Green Energy* published by Springer in 2015. He has contributed numerous book chapters to several books published by reputed international publishers. To his credit, he has several case studies, research papers, book chapters, conference papers, and presentations. Sanjay regularly writes columns for Economic Times (Energy World), Financial Express, Daily News Analysis (DNA), Iran Review, HansIndia, Modern Diplomacy and South Asia Monitor. His current research interests are in the area of retailing, distribution of petroleum and natural gas, energy policy, and green energy.

Ayush Gupta is a bachelor's degree holder in electrical engineering from IIT Roorkee and done MBA in operations management, and he is a Research Scholar. Ayush has a rich experience of over 24 years in process and oil and gas industry. He has served the industry in various functions and capacities encompassing project management, operations and maintenance, and training and currently working as deputy general manager (HRD) with largest public sector natural gas transmission and distribution company in india, GAIL India Limited, New Delhi. A Chevening Scholar, Mr. Gupta, has presented several papers at national and international forums and has been instrumental in design, development, and delivery of a number of courses related to natural gas pipeline projects management, O&M, cathodic protection, and skill development for natural gas sector.

Contributors

Husain Ahmad EPIC (Engineering, Procurement, Installation and Commissioning) Company, Doha, Qatar

Manas Das Strategic Planning, Kuwait Petroleum Corporation, Kuwait City, Kuwait

Bhashit Dholakia Business Development, Adani Gas, Ahmedabad, India

Ayush Gupta GAIL India Limited, New Delhi, India

Bhavesh Gupta IM & I-BPM Utilities, Accenture Services Pvt. Ltd., Gurgaon, India

Pankaj Gupta GAIL India Limited, New Delhi, India

Sanjay Kumar Kar Department of Management Studies, Rajiv Gandhi Institute of Petroleum Technology, Jais, India

Gerard Kreeft EnergyWise, Arnhem, The Netherlands

S.K. Kudaisya Sabarmati Gas Limited, Gandhinagar, Gujarat, India

Akhil Mehrotra Mumbai, India

Pramod Paliwal School of Petroleum Management, PDPU, Gandhinagar, India

T.P. Rao NELP, DGH, Noida, India

Amit Ray GAIL India Limited, New Delhi, India

Jitin Saxena GAIL India Limited, Noida, India

D.V. Shastry GAIL India Limited, New Delhi, India

Piyush Kumar Sinha Marketing and Retailing, Indian Institute of Management, Ahmedabad, India

Radha Krishna Tripathy Athena Infonomics, Chennai, India

Manish Vaid Observer Research Foundation, New Delhi, India

Part I
Natural Gas Market Scenario in India

India's Active Engagement with Natural Gas: Imperatives and Challenges

Manish Vaid and Sanjay Kumar Kar

Abstract Post-Conference of Parties (COP 21) scenario has reconfigured India's energy goals and has intrinsically linked it to its climate policy goals. India's Intended Nationally Determined Commitments (INDC) submitted to the United Nations Framework Convention on Climate Change (UNFCCC) signals that India wishes to strive for green growth. By 2030, India's INDC commits for 40% of electricity that would come from non-fossil fuel sources. This calls for a paradigm shift in India's energy strategy by reshuffling its energy basket in a way that it not only continues and expedites with its renewable capacity expansion programs, which is largest in the world, but also sticks to its active engagement with natural gas as a base fuel for electricity generation, city gas distribution and all other possible options of use. This chapter discusses the imperatives for India's continued engagement with natural gas and the challenges it would face in expanding consumer base and increasing share of natural gas in the energy basket amidst stiff emission targets set by India in the Paris Summit.

Keywords INDC · Energy · Natural gas · LNG · Import

1 India's Energy Scenario

According to the World Energy Outlook Special Report 2015, India is in early stages of a major transformation, and with current economic growth rate of around 7% and policy priority of making India as world's manufacturing hub, its energy sector is set to expand further. Moreover, meeting Sustainable Development Goals

M. Vaid (✉)
Observer Research Foundation, New Delhi, India
e-mail: manishv@orfonline.org

S.K. Kar
Department of Management Studies, Rajiv Gandhi Institute of Petroleum Technology, Jais, India
e-mail: skar@rgipt.ac.in

S.K. Kar and A. Gupta (eds.), *Natural Gas Markets in India*,
DOI 10.1007/978-981-10-3118-2_1

(SDGs) will always keep the trajectory of energy demand on the higher side. All this would result in a scenario wherein India would continue to face the challenge in improving the standard of living of their folks with the focus on eliminating the poverty and providing electricity to its people. Notably, 237 million people which translates to about 19% of its population still under electricity starvation (Table 1). The numbers are staggering, and a comparison indicates that people without electricity in India could have constituted 31% of the entire population of Europe.

Since the start of this century, increased economic activities and the pace of urbanization have resulted in tremendous growth in India's energy sector. During the period from 2000–2013, India's share of global energy demand has been around 10%. In this period, India's energy demand grew by 46% but remained only around one-third of the world's average, primarily due to its large part of population remaining without access to the modern fuels.[1]

India's current energy situation is characterized by strained supplies increasingly being met through imports due to insufficient domestic production. While India imports roughly 35% of its primary energy needs, its imports of oil, natural gas and coal stand at 70, 30 and 14%, respectively.[2]

According to India Energy Outlook 2015, India's total primary energy demand rose by almost 75% from 441 million tonnes of oil equivalent (Mtoe) in 2000 to 775 Mtoe in 2013 (Fig. 1).

Fossil fuels account for almost three-quarters of India's energy demand attributed largely to increased coal consumption, which as compared to other energy resources is more affordable and easily available. This fuel is the main source of India's power sector. This is evident from Fig. 1, where the share of coal demand has increased by 10% in 2013 since 2000. But during this period, the traditional biomass fuel registered 24% of the total energy demand in 2013. This was due to major shift from solid biomass, such as fuel wood, straw, charcoal and dung, to modern fuels such as liquefied petroleum gas (LPG) and piped natural gas (PNG).

According to the Census 2011, 87% of rural household and 26% of urban household still depend on the biomass. Natural gas, being regarded as a bridge fuel, has so far failed to make a mark during this period. While the share of natural gas grew by just 1% during 2000–2013, the overall share confined to only 6%, compared to the global average of 21% (Fig. 1). Oil demand increased by 57% in 2013 since the year 2000 from 112 Mtoe in 2000 to 176 Mtoe in 2013 (Table 2).

If one observes the trend of fossil fuel growth during the last three years from 2012 to 2014, BP Statistical Review of World Energy 2015 suggests, while both consumption of oil and coal witnessed a significant increase, the consumption of natural gas showed a reversed trend. During this period (2012–2014), oil consumption grew from 173.6 to 180.7 Mtoe, while consumption of coal increased

---

[1]International Energy Agency, "India Energy Outlook", *World Energy Outlook Special Report, 2015.*

[2]Ministry of Petroleum and Natural Gas, "Report of the Committee on Roadmap for Reduction in Import Dependency in Hydrocarbon Sector by 2030—Part I", December, 2013. http://petroleum.nic.in/docs/kelkar.pdf. Accessed on 17 February 2016.

Table 1 Number and share of people without access to electricity by state in India, 2013

	Population without access (million)			Share of population without access		
	Rural	Urban	Total	Rural (%)	Urban (%)	Total (%)
Uttar Pradesh	80	5	85	54	10	44
Bihar	62	2	64	69	19	64
West Bengal	17	2	19	30	7	22
Assam	11	0	12	45	9	40
Rajasthan	10	0	11	22	2	17
Odisha	10	0	11	32	4	27
Jharkhand	8	0	9	35	4	27
Madhya Pradesh	7	1	8	16	3	12
Maharashtra	6	1	6	11	2	7
Gujarat	2	2	3	7	6	6
Chhattisgarh	2	0	3	14	6	12
Karnataka	1	0	1	5	1	3
Other states	3	2	6	2	2	2
Total	221	16	237	26	4	19

Source India's Energy Outlook, World Energy Outlook Special Report 2015, IEA

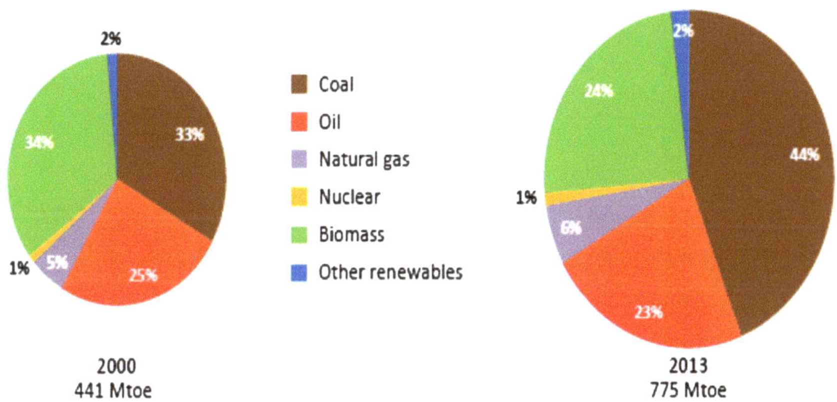

Fig. 1 Primary energy demand in India by fuel. *Source* India Energy Outlook—World Energy Outlook Special Report 2015, IEA

from 302.3 to 360.2 Mtoe, registering 4.1 and 19.15% growth, respectively. Natural gas consumption, on the other hand, falls sharply by 7.3%.

Therefore, the trends of natural gas demand and consumption, as shown in Figs. 1 and 2, respectively, suggest that natural gas, despite being dubbed as a cleaner fossil fuel, has actually failed to increase its share and rather registered consistent fall from 2012 to 2013.

Table 2 Primary energy demand by fuel in India in the New Policies Scenario (Mtoe)

	2000	2013	2020	2030	2040	Shares		2013–2040	
						2013 (%)	2040 (%)	Change	CAAGR[a] (%)
Oil	**112**	**176**	**229**	**329**	**458**	**23**	**24**	**282**	**3.6**
Natural gas	**23**	**45**	**58**	**103**	**149**	**6**	**8**	**104**	**4.6**
Coal	**146**	**341**	**476**	**690**	**934**	**44**	**49**	**592**	**3.8**
Nuclear	**4**	**9**	**17**	**43**	**70**	**1**	**4**	**61**	**7.9**
Renewables	**155**	**204**	**237**	**274**	**297**	**26**	**16**	**93**	**1.4**
Hydropower	***6***	***12***	*15*	*22*	*29*	***2***	***1***	*16*	*3.2*
Bioenergy	***149***	***188***	*209*	*217*	*209*	***24***	***11***	*20*	*0.4*
Other renewables	***0***	***4***	*13*	*35*	*60*	***0***	***3***	*56*	*11.0*
Fossil fuel share (%)	64	72	75	78	81	72	81	8	n.a.
Total	441	775	1018	1440	1908	100	100	1133	3.4

[a]Compound average annual growth rate
Source International Energy Agency, World Energy Outlook 2015

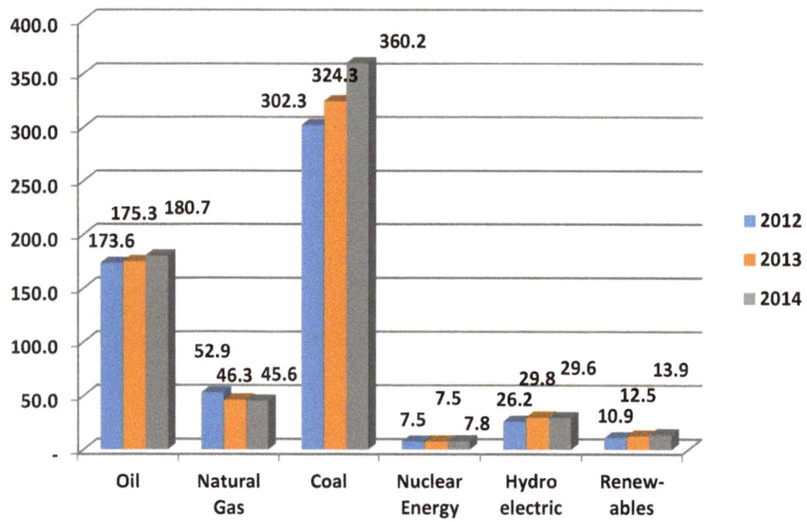

Fig. 2 Primary energy: consumption by fuel type (Mtoe). *Source* BP Statistical Review of World Energy (2013–2015)

Fall in natural gas demand and consumption was due to the sharp fall in domestic gas production coupled by no new gas resources addition which resulted in increased import of liquefied natural gas (LNG). The share of clean energy sources such as nuclear energy, hydroelectric and renewables has played a small role in total energy mix despite their consistent rise in their demand and consumption.

As far as India's projected energy outlook is concerned, 2016 edition of BP Energy Outlook 2035[3] has estimated India to have the fastest energy consumption growth among all major economies with continued dependence on energy imports, notwithstanding its ever-increasing share of non-fossil fuel production.

India's energy imports would continue to rise by 153% by 2035 because its energy production as a share of consumption declines from 57% in 2014 to 54% by 2035.[4]

While registering 8.7% of global energy demand by 2035, India would have the second largest share in BRICS countries only after China at 25%. The share of Russia and Brazil would be at 4 and 2%, respectively.[5]

As regards fossil fuel imports by 2035, the increase in imports of oil and coal would be to the tune of 161 and 122%, respectively, while that for natural gas would be the highest at 301% in volumetric terms.[6]

The above trends highlight India's rise in energy consumption with corresponding fall in their domestic production, creating increasing gap between domestic demand and supply which would be filled largely through imports. The case in point is the natural gas, where by 2035 its demand is set to increase by 155%,[7] which would be largely met though imports with 301% rise, which is the highest among all the fossil fuels.

2 Why We Need Natural Gas?

Natural gas is one of the three fossil fuels which can energize modern economies. This is the cleaner fuel among all other fuel alternatives, such as wood, coal and liquids derived from crude oil. When compared to these alternative fuels, natural gas scores quite well on the criteria such as transportability, storability, combustion efficiency, convenience, cleanliness and flexibility of use. It also scores superior to its counterpart when it comes to emission factor (Table 3), reliability and durability of supply. All these characteristics combined make natural gas a preferred fuel for any economy.

According to the Central Statistical Office (CSO), India's gross domestic product (GDP) grew at 7.3% in 2014–2015. The projections for GDP growth is on the higher side for the year 2016 and 2017. According to the Organization for Economic Cooperation and Development (OECD), India's GDP is expected to

[3]BP Energy Outlook, Country and regional insights—India, 2016. Available at http://www.bp.com/content/dam/bp/pdf/energy-economics/energy-outlook-2016/bp-energy-outlook-2016-country-insights-india.pdf.

[4]Ibid.

[5]Ibid.

[6]Ibid.

[7]Ibid.

Table 3 Emission factors

Coal and lignite	1.614 tonne of CO_2 per tonne of Indian coal and lignite mix
Crude oil	0.175 tonne of CO_2 per tonne of crude oil refined
Petroleum products	Oil refined petroleum products
Natural gas	0.002 tonne of CO_2 per cubic metre

Source The Final Report of the Expert Group on Low Carbon Strategies for Inclusive Growth, Planning Commission, Government of India, April 2014

Table 4 OECD's growth forecast

	2016	Difference from November forecast
World	3.0	−0.3
US	2.0	−0.5
Euro Area	1.4	−0.4
China	6.5	0.0
India	7.4	0.1
Brazil	−4.0	−2.8

Source The Economic Times Bureau (2016)

grow at 7.4% in 2016 (Table 4), while Moody projects the growth to be around 7.5% each in 2016 and 2017.[8] The reason is India's domestic demand growth which would make India insulated from global financial turmoil.

Therefore, to maintain the current growth trajectory without compromising on sustainable growth path, India would require natural gas as a bridge fuel while slowly reducing its dependency on dirtier fuels such as coal and crude oil, alongside striving for renewable energy goals. The Government of India has set an ambitious target of renewable energy capacity addition of 175 GW by 2021–2022 from the current 30 GW by 2016–2017.[9] Besides in its submission of Intended Nationally Determined Contributions (INDC) in Conference of Parties (COP—21) Summit, India declared a voluntary goal of reducing the emission intensity of its GDP by 20–25% over 2005 levels, by 2020. It has also proposed to reduce the emission intensity of its GDP by 33–35% from 2005 level.[10]

Moreover, in its INDC, India has also decided to enhance its share of non-fossil fuel in the cumulative electric power installed capacity from 30% in 2015 to 40% by 2030, with the help of transfer of technology and low-cost international financing including from Green Climate Fund.[11] In India, there is a space of around 60% to support the base load of power generation, wherein it can use coal and natural gas as a main fuel efficiently and more environment friendly way. Despite

[8]Mahesh (2016).

[9]Press Information Bureau, Government of India. Ministry of Environment and Forests. http://pib.nic.in/newsite/PrintRelease.aspx?relid=128403. Accessed on 14 February 2016.

[10]Ibid.

[11]Ibid.

public posturing for renewable energy in an international arena, India would continue to depend on coal and natural gas for power generation in the long run. The primary reason behind this argument is that the renewables such as solar and wind have relatively lower capacity utilization factor compared to coal- and natural gas-based power plants. That means 160 GW of solar and wind installed capacity could translate to actual production of 32–40 GW at current average capacity utilization rate. Therefore, natural gas becomes the next best alternative for cleaner and large-scale electricity production in India. Cleaner coal technologies with carbon sequestering are another option. Kar et al. (2015) and Kar (2015) suggest that natural gas could play complementary role in driving green growth in India.

3 Oil Indexation Playing a Crucial Role

The most important transformation in the energy sector since mid-June 2014 has been the steep fall in global oil prices to the tune of around 70% to around $30 billion (bn) a barrel. The natural gas prices also fell as a result of this but to a lesser extent. Notably, the international gas market is dominated by long-term contracts linked to the price of oil, for gas delivered through pipeline and as LNG. Mirroring the crude oil price drop, the price of LNG went down by 58% from $18.3 per million British thermal unit (MMBTU) in April 2014 to $7.7 per MMBTU. Lower gas price helped natural gas to become more competitive fuel than coal in markets like Asia. In India, for September 2015, the gas-fired power plants operated at 25.4%, an improvement over 20% during 2014–2015, though fell fairly short to coal and nuclear fuel power plants, which operate at 60–80% plant load factor (Table 5).

The low oil price regime, wherein the gas prices have also plummeted, can resurrect India's natural gas-fired plants, many of which were either idle or operating at a fraction of their capacity. This has also given relief to the banks which were struggling after providing loans in billions to these power plant projects. The 1980 megawatt (MW) Dabhol power project, for instance, lying idle since 2013, owes Rs. 8500 crore to a group of banks including IDBI Bank, State Bank of India, ICICI Bank and Canara Bank.[12] This plant has once again started generating electricity for India Railways on 26 November 2015.

Similarly, private sector power plants such as Torrent Power, GMR, GVK and Lanco owning 9773 MW of gas fuelled power plants are also recovering well. During October 2015, these plants generated 2.5 times the electricity they produced in October 2014, increasing electricity generation by 76% from April 2015 to October 2015.[13]

[12]Bhandari (2015a).

[13]Bhandari (2015b).

Table 5 Plant load factor for gas, coal and nuclear power plants

	September 15 (%)	April–September 15 (%)	April–March 2105 (%)
Gas	25.4	21.8	20.8
Coal	63.4	60.4	64.8
Nuclear	66.9	75.1	80.5

Source Central Electricity Authority, India

Notwithstanding the low gas prices, the biggest benefit from switching the coal to gas is the reduction in carbon intensity from centralized power by as much as 50%.[14]

Further, oil indexation continues to play an important role on European and Asian gas prices. The case in point is the signing of revised contract by India's Petronet LNG (PLL) with RasGas of Qatar, where the former have to pay the lower price at \$6–6 per mmBtu, down from \$12–13 per mmBtu, to purchase LNG in a long-term contract, which would help PLL to save Rs. 4000 crore.[15]

With shortfall in domestic gas production, India has already increased LNG imports, which now has surpassed 15 million tonnes per annum (MMTPA) in 2015, an increase of 5% from 2014.[16] Research and ratings agency Moody's stated that India could double its LNG imports to 24 MMTPA by 2020.[17] Currently following LNG terminals are operational with installed capacity of 25 MMTPA:

- Dabhol LNG (5MMTPA), Maharashtra;
- Dahej LNG (10 MMTPA), Gujarat;
- Hazira LNG (5 MMTPA), Gujarat; and
- Kochi LNG (5 MMTPA), Kerala.

4 Natural Gas and INDC

Interestingly, India's INDC had no expressed mention of the role of natural gas in enabling outcomes for the COP 21 Paris Summit, but it surely mentioned about revisiting its National Action Plan on Climate Change (NAPCC). Though government is proposing to set up four new missions such as wind energy, health,

[14]US. Energy Information Administration. Frequently Asked Questions. "How much carbon dioxide is produced when different fuels are burned?". https://www.eia.gov/tools/faqs/faq.cfm?id=73&t=11. Accessed on 2 February 2016.

[15]Sharad Raghavan (2016).

[16]ICIS. "Indian LNG imports surpass 15 m tonnes in record 2015", 13 January 2016, http://www.icis.com/resources/news/2016/01/13/9960246/indian-lng-imports-surpass-15m-tonnes-in-record-2015/. Accessed on 15 February 2016.

[17]BS Reporter. "India to double LNG imports to 24 mtpa by 2020: Moody's", Business Standard, 6 May 2015. http://www.business-standard.com/article/economy-policy/india-to-double-lng-imports-to-24-mtpa-by-2020-moody-s-115050601048_1.html. Accessed on 18 February 2016.

waste to energy, coastal areas and redesigning the National Water Mission and National Mission on Sustainable Agriculture, the scope of the use of natural gas as a mean to mitigate greenhouse gas emissions (GHG) always remains there.

In its actions relating to GHG mitigation, NAPCC has mentioned that to clean the urban air pollution, compressed natural gas (CNG) would be introduced as a transport fuels along with the fuel switching option for industrial sectors. For this, efforts would be made to augment domestic availability of natural gas by both increased domestic production and greater LNG imports.

Moreover, Intergovernmental Panel on Climate Change (IPCC), in its Fifth Assessment of the Working Group III report,[18] titled ***"Climate Change 2014: Mitigation of Climate Change"*** in its chapter, "Energy Systems", stated the role of natural gas as a means for mitigating fugitive emissions. This report thus states that

> GHG emissions from energy supply can be reduced significantly by replacing current world average coal-fired power plants with modern, highly efficient natural gas combined-cycle (NGCC) power plants or combined heat and power (CHP) plants, provided that natural gas is available and the fugitive emissions associated with its extraction and supply are low or mitigated (robust evidence, high agreement).

Therefore, even though India has pushed hard for renewable energy resources in its INDC Scenario so as to diversify its energy supply, fossil fuel would remain a major part of it, with natural gas expected to play a greater role, particularly in power, fertilizer and city gas distribution sectors. This is also evident from 2016 edition of BP Energy Outlook of India, which mentions that despite efforts being pushed for renewable energy resources, India would by and large remain fossil fuel-dependent, wherein natural gas demand and its import would be on the higher side. This is aptly put by IEA's Executive Director Dr. Faith Birol,[19] who stated that "To push for renewables, to push for energy efficiency with technologies doesn't mean that we abandon gas and other fuels."

According IEA's World Energy Outlook Special Report, titled "Energy and Climate Change", under INDC Scenario, though natural gas acting as a substitute to renewables and nuclear would contribute to emissions, but when acting as a substitute for other fossil fuels, it will help in suppressing the emissions, as world's natural gas demand increases by around 30% by 2030. This would result in natural gas accounting for 25% of energy-related CO_2 emissions, which would be significantly low compared to coal and oil which would account for 41 and 34% of energy-related CO_2 emissions by 2030 (pp. 38, Energy and Climate Change[20]).

[18]IPCC (2014).

[19]Clemente (2015).

[20]IEA. "Energy and Climate Change", World Energy Outlook Special Report, 2015. https://www.iea.org/publications/freepublications/publication/WEO2015SpecialReportonEnergyandClimateChange.pdf. Accessed on 17 February 2016.

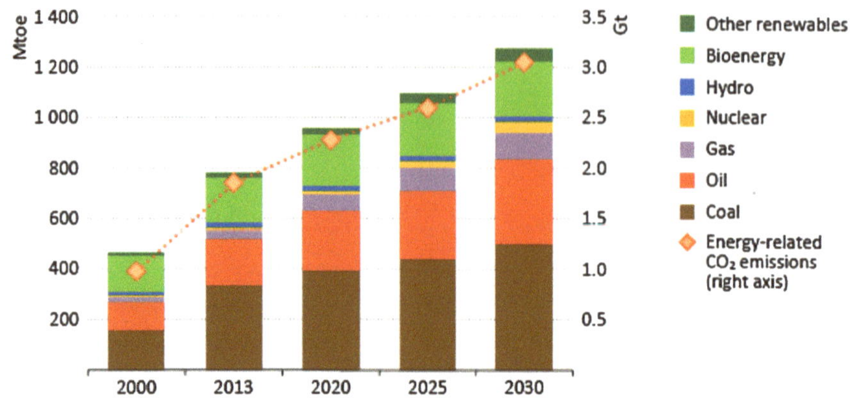

Fig. 3 India primary energy demand by fuel type and related CO_2 emissions in the INDC Scenario. *Source* International Energy Agency, World Energy Outlook Special Report 2015, Energy and Climate Change

In the INDC Scenario, India's major shift would be towards renewables to diversify its energy supply. But higher energy demand will not eliminate its demand for fossil fuels, which together provides for three-quarters of total energy demand. In this scenario, by 2022, India's energy-related CO_2 emissions are around 30% higher than 2013, at 2.4 gigatone (Gt), which will cross 3 Gt by 2030 (Fig. 3).

5 Natural Gas Scenario of India

5.1 Natural Gas Reserves

According to IEA's India Energy Outlook 2015, natural gas constitutes for only 6% of the domestic energy mix. According to the BP Statistical Review of World Energy 2015, India's proven reserves of natural gas at the end of December 2014 was 1.4 trillion cubic metres (TCM) which is about 0.74% of global natural gas reserves of 187.1 TCM.

As per the Ministry of Petroleum and Natural Gas, the total natural gas reserves in India as on April 2015 stood at 1488.73 billion cubic metres (BCM) of which the offshore gas reserves stood at 988.88 BCM, whereas the onshore reserves stood at 499.59 BCM. India's reserve-to-production ratio for natural gas reserves stood at 45 years compared to 54.1 years for the world.

The main sources of onshore gas-producing fields are in the states of Assam, Gujarat, Tamil Nadu and Andhra Pradesh, while the offshore-producing areas include the Krishna Godavari Basin off the east coast.

5.2 Domestic Production of Natural Gas

The production of natural gas in the country is mainly from the nominated fields operated by the National Oil Companies (NOCs), namely Oil and Natural Gas Corporation Limited (ONGC) and Oil India Limited (OIL), Panna-Mukta-Tapti and from the blocks of New Exploration and Licensing Policy (NELP) blocks such as KG D6 and from few small fields. Recently, the production from Krishna Godavari Basin (KG-D6) offshore field (Fig. 4) impacted the overall gas production due to the reasons such as water loading/sand ingress in D1 and D3 fields and MA fields and non-drilling of the required number of gas producers well in D1 and D3 fields by the contractor in the line with the Addendum to initial Development Plan (AIDP),[21] resulting in an overall fall in India's gas production since 2011.

KG-D6 basin which started producing hydrocarbons in 2009 has seen a steep decline to 11.8 million metric standard cubic metres per day (MMSCMD) in 2014 from 69.42 MMSCMD in March 2010. Consequently, India total gas production, after peaking at 50.8 BCM in 2010, started to fall sharply and consistently, registering 31.7 BCM in 2014.[22] The shortfall in natural gas demand was met through LNG imports.

When we observe the historical trends of natural gas production since 2005–2006 to 2014–2015 and compare the same with the production of NOCs and private/joint ventures (JVs), the gas production by the former remained steady throughout this period, while that of the later picked up sharply from 2009 to 10, exceeding that of ONGC and OIL combined. Thereafter, total gas production, which peaked in the year 2010–2011 with 50,847.17 MMSCMD[23], came down sharply to 32,693.22 MMSCMD during 2014–2015 (Fig. 5).

The state-wise trends in natural gas production from 2005–2006 to 2014–2015 is shown in Table 6.

According to the Working Group Report of the Petroleum and Natural Gas Sector (2012–2017), the natural gas production by the end of 12th Five-Year Plan was estimated at 63,876 MMSCM, wherein oil public sector undertakings, namely ONGC and OIL, expected to contribute at 42,876 MMSCM, while private companies/JVs at 21,000 MMSCM (Table 7).

---

[21]Standing Committee on Petroleum and Natural Gas (2013–14), 15th Lok Sabha, MoPNG. "Allocation and Pricing of Gas", pp. 14. http://164.100.47.134/lsscommittee/Petroleum%20&%20Natural%20Gas/15_Petroleum_And_Natural_Gas_19.pdf. Accessed on 4 March 2016.

[22]BP. "BP Statistical Review of World Energy June 2015". https://www.bp.com/content/dam/bp/pdf/energy-economics/statistical-review-2015/bp-statistical-review-of-world-energy-2015-full-report.pdf. Accessed on 10 February 2016.

[23]Petroleum Planning and Analysis Cell, Ministry of Petroleum and Natural Gas, MoPNG. http://ppac.org.in/content/151_1_ProductionNaturalGas.aspx. Accessed on 22 February 2016.

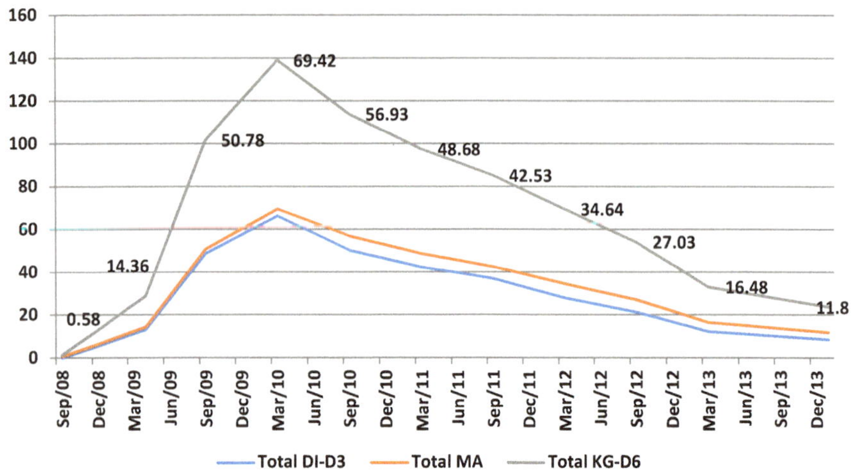

Fig. 4 Cumulative production (MMSCMD) from active wells from Indian gas fields (KG-D6). *Source* Parliament of India, Rajya Sabha Unstarred Question No. 296, answered on 26 February 2013, Updated to January 2014 (Vaid and Maini 2014)

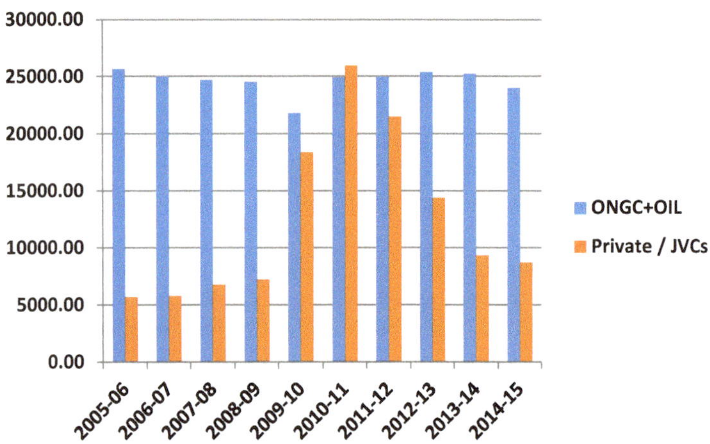

Fig. 5 Trends in natural gas production (BCM) (year 2005–2006 to 2014–2015). *Source* MoPNG, ONGC, OIL and Directorate General of Hydrocarbons (DGH)

5.3 *Natural Gas Demand*

Normally, the role of demand and supply influences the energy consumption, but in the case of India, it is the role of supply which has played a significant role in shaping the natural gas consumption pattern. India's natural gas demand, like in any other country, is based on the cost economics, availability of the alternative fuels and actual availability of natural gas itself. In addition, natural gas prices,

Table 6 Trend of natural gas production in India (year 2005–2006 to 2014–2015) [MMSCM]

State	2005–06	2006–07	2007–08	2008–09	2009–10	2010–11	2011–12	2012–13	2013–14	2014–15
A. Onshore										
Assam and Arunachal Pradesh										
Gross production (GP)	2698	2803.04	2883.00	2819.00	2742.67	2729.38	2943.99	2950.84	2909.47	2991.77
Net production	2493	2570.11	2666.00	2623.00	2529.53	2516.73	2744.29	2724.34	2703.91	2786.78
Gujarat										
Gross production (GP)	3831	3293.72	2931.00	2605.00	2444.11	2261.09	2169.40	2032.10	1656.94	1526.83
Net production	3810	3277.39	2913.00	2574.00	2404.23	2230.24	2099.97	1993.50	1584.59	1404.95
Tamil Nadu										
Gross production (GP)	906	1130.13	1169.00	1243.00	1177.76	1119.18	1284.76	1206.04	1304.02	1191.57
Net production	867	1112.95	1152.00	1233.00	1163.62	1105.95	1276.57	1199.15	1298.28	1174.55
Andhra Pradesh										
Gross production (GP)	1663	1525.07	1567	1524	1479.06	1384.41	1363.79	1248.17	1171.34	540.59
Net production	1655	1515.78	1557.00	1517.00	1478.13	1382.62	1361.56	1245.21	1167.69	518.76
Tripura										
Gross production (GP)	480	519.78	534.00	553.00	563.78	609.98	643.90	646.74	822.23	1140.04
Net production	480	519.78	534.00	553.00	562.35	609.87	643.90	646.68	815.91	1139.87
West Bengal (CBM)										
Gross production (GP)	0	0.00	14.97	19.79	38.40	41.46	79.11	100.85	156.49	223.68
Net production	0	0.00	0.57	3.02	5.22	36.38	68.79	95.99	140.94	203.36
Jharkhand (CBM)										
Gross production (GP)	0	0	0	0	0	0	3.56	2.95	3.38	2.48
Net production	0	0	0	0	0	0	3.56	2.95	3.38	2.48

(continued)

Table 6 (continued)

State	2005–06	2006–07	2007–08	2008–09	2009–10	2010–11	2011–12	2012–13	2013–14	2014–15
Madhya Pradesh(CBM)										
Gross production (GP)	0	0	0	0	0	0	1.53	3.44	5.65	2.08
Net production	0	0	0	0	0	0	0.63	1.14	1.57	0.79
Rajasthan										
Gross production (GP)	0	0.00	0.00	0.00	238.59	430.79	590.18	685.42	982.20	1177.77
Net production	0	0.00	0.00	0.00	221.58	382.11	556.87	633.67	919.90	1079.72
A. Onshore total										
Gross production (GP)	9578	9271.74	9098.57	8763.79	8684.37	8576.29	9080.21	8876.55	9011.73	8796.80
Net production	9305	8996.01	8822.57	8503.02	8364.66	8263.90	8756.15	8542.62	8636.17	8311.27
B. Offshore										
Mumbai high + Eastern offshore										
Gross production (GP)	16,823	16,566.62	16,457	16,737	17,475.99	17,590.64	17,564.96	18,102.42	17,967.54	17,271.76
Net Production	16,351	16,010.6	15,879	16,027	13,452.23	17,035.24	16,872.86	17,572.76	17,574.35	16,862.80
Private/JVCs										
Gross production (GP)	5801.00	5908.49	6861.00	7348.00	19,359.72	26,054.46	20,909.84	13,699.93	8427.66	7588.88
Net production	5669.00	5784.83	6777.00	7221.00	18,371.89	25,930.18	20,824.27	13,637.56	8363.54	7519.16
Total (A&B)										
Gross production (GP)	32,202.00	31,746.85	32,416.97	32,848.79	47,512.00	52,221.40	47,555.00	40,678.90	35,406.93	33,657.44
Net production	31,325.00	30,791.44	31,478.57	31,751.02	46,485.88	51,229.29	46,453.29	39,752.94	34,574.06	32,693.22
Net production (BCM)	31.32	30.79	31.48	31.75	46.48	51.23	46.45	39.75	34.57	32.69

Source Petroleum Planning and Analysis Cell, MoPNG, Government of India

Table 7 Estimates for natural gas production in 12th Plan (2012–2017)

Company	Natural gas production				
	2012–2013[b]	2013–2014[b]	2014–2015[a]	2015–2016	2016–2017
ONGC	23,549	23,284	22,022	28,215	38,676
OIL	2639	2626	2722	4100	4200
Pvt/JV	14,491	9497	8912	18,500	21,000
Total	40,679	35,407	33,656	50,815	63,876
Total (MMSCM)	111	97	92	137	172

[a]Provisional
[b]Actual
Note Natural gas production estimates were revised on 30 April 2012
Source Working Group Report of the Petroleum and Natural Gas Sector (2012–2017)

particularly in Asian markets, are indexed to crude oil. Therefore, the decline in the crude oil prices since mid-June 2014 has also been reflected in the natural gas prices in this region, including India. The corresponding demand supply dynamics and LNG imports from 2005 to 2014 are shown in Fig. 6. The sharp fall in domestic production of natural gas resulted in sharp increase in LNG imports, particularly during the period 2010–2014.

According to the report by Industry Group for Petroleum and Natural Gas Regulatory Board (PNGRB), titled "Vision 2030—Natural Gas Infrastructure in India", India's "realistic natural gas demand[24]" is set to grow at significantly at compound annual growth rate (CAGR) of 6.8% from 242.6 MMSCMD in 2012–2013 to 746 MMSCMD in 2029–2030.[25]

Table 8 presents sector-wise natural gas consumption in India. It is evident from Table 8 that the fertilizer sector has been one of the largest beneficiary of natural gas, particularly from 2013–2014 to 2014–2015, followed by power, CGD, refinery and petrochemical industry.

According to the report of the Working Group on Petroleum and Natural Gas Sector for the 12th Five-Year Plan (2012–2017), estimated demand and natural gas availability during 12th and 13th Five-Year Plan is shown in Tables 9 and 10.[26]

As on January 2016, the natural gas production, consumption and imports have been shown in Table 11.

---

[24]Realistic demand means the demand estimated after considering limiting factors that are likely to restrict growth in demand.

[25]PNGRB, 'Vision 2030': Natural Gas Infrastructure in India, Report by Industry Group for Petroleum and Natural Gas Regulatory Board, 6 September 2013, at http://www.pngrb.gov.in/newsite/pdf/vision/vision-NGPV-2030-06092013.pdf. Accessed on 29 February 2016.

[26]PPAC, MoPNG. "Ready Reckoner—Oil Industry Information at a Glance". http://ppac.org.in/WriteReadData/Reports/201601050441524912702ReadyReckoner,November2015.pdf. Accessed on 24 February 2016.

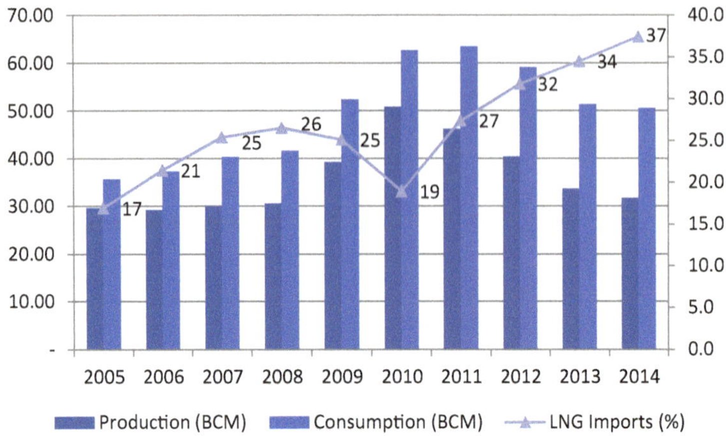

Fig. 6 Demand–supply dynamics of natural gas (year 2005–2014). *Source* BP Statistical Review of World Energy June 2015

According to PPAC, while net natural gas production witnessed a rise by around 4.98% during 2015–2016 compared to the corresponding figures of the previous year (2014–2015), the LNG imports during the same period rose by around 18% thereby suggesting that the increasing demand was been met through LNG imports.

5.4 *Natural Gas Trade*

Therefore, with the rise in LNG imports, particularly from 2010 (Fig. 6), the LNG sector of India looks brighter, necessitating sufficient additions to its current infrastructure to support the existing as well as futuristic natural gas demand, as shown in Tables 9 and 10, showing projected natural gas demand and availability for 12th and 13th Plan, respectively.

According to the BP Statistical Review of World Energy June 2015, Qatar remains India's biggest exporter of LNG. In addition, during 2014, India also imported LNG from countries such as Trinidad and Tobago, United Arab Emirates, Oman, Yemen, Algeria, Nigeria, Malaysia, South Korea, Norway and other European countries (Fig. 7). Accordingly, India imported 18.9 BCM of LNG taking India natural gas consumption to 50.6 BCM. The total production during the year 2014 (calendar year) was 31.7 BCM.

Attributed to oil indexed gas price and low global crude oil prices, LNG prices are staying low, allowing India to broader its natural gas base in its energy mix. This would help end use sectors such as power and fertilizer to increase the usage of natural gas, through increased LNG imports at cheaper rates.

Table 8 Sector-wise consumption of natural gas (MMSCM)

Sector	2008–2009	2009–2010	2010–2011	2011–2012	2012–2013	2013–2014	2014–2015[a]
1	2	3	4	5	6	7	8
(a) Energy purpose							
Power	12,603	21,365	25,787	22,628	16,078	11,284	10,720
Industrial	[c]	[c]	[c]	284	269	156	395
Manufacture	[c]	[c]	[c]	30	0	105	138
Road transport	[c]	[c]	[c]	0	0	66	0
City or local natural gas							
Distribution network	6014	2568	3427	5599	5780	5838	5415
Tea plantation	154	167	193	175	182	196	188
LPG shrinkage	[b]	[b]	[b]	1068	1027	982	1005
Int. con. for pipeline system	[b]	[b]	[b]	385	387	372	351
Miscellaneous	3420	7272	7546	9064	7976	7479	5934
Total (a)	22,191	31,372	36,953	39,233	31,699	26,478	24,146
(b) Non-energy purpose							
Fertilizer Industry	9082	13,168	11,464	14,003	14,733	15,869	15,190
Petrochemical	1105	1264	1309	1858	2486	2405	2890
Refinery	611	703	1529	4257	3891	3968	4575
Sponge iron	1535	1838	765	1333	1106	274	154
Total (b)	12,333	16,973	15,067	21,451	22,216	22,516	22,809
Grand total (a + b)	34,524	48,345	52,020	60,684	53,915	48,994	46,955
Total in MMSCMD	94.59	132.45	142.52	166.26	147.71	134.23	128.64

Includes consumption of LNG in 2011–2012 and onwards
[a]Provisional
[b]Includes in Miscellaneous sector
[c]Includes Industrial, Manufacture and Road Transport sectors covered under CGD sector
Source PPAC and Oil Companies

At this end, Government of India has already planned to increase the LNG regasification terminals[27] from existing four in numbers to 8–10 in the coming years. New LNG terminals are coming up in Dhamra (Odisha) Ennore (Tamil Nadu) and Mundra (Gujarat). Further, Floating Storage and Regasification Units (FSRUs) are coming up in Jafrabad (Gujarat), Kakinada (Andhra Pradesh) and off-coast of Haldia (West Bengal). LNG terminals coming up in the eastern India would certainly improve associated natural gas infrastructure which includes natural gas pipeline in the region.

Recently, India has tied up with several countries to import LNG. Notably, Gas Authority of India Limited (GAIL) has signed around 25 sale purchase agreement with global LNG suppliers such as Total, Shell and BG Group. GAIL has also

[27]Ibid.

Table 9 Demand of natural gas and projected natural gas availability 12th Five-Year Plan

Figures in MMSCMD					
	2012–2013	2013–2014	2014–2015	2015–2016	2016–2017
Power	135	153	171	189	207
Fertilizer	62	110	113	113	113
City gas	15	19	24	39	46
Industrial	20	20	22	25	27
Petrochemicals/refineries/internal consumption	54	61	67	72	72
Sponge iron and steel	7	8	8	8	8
Grand total demand	293	371	405	446	473
Domestic availability	124	149	170	177	209
Imports—LNG	63	87	87	129	150
Total availability	187	236	257	306	359

Source Report of the Working Group on Petroleum and Natural Gas Sector for the 12th Five Year Plan (2012–17)

Table 10 Demand of natural gas and projected natural gas availability 13th Five-Year Plan

Figures in MMSCMD					
	2017–2018	2018–2019	2019–2020	2020–2021	2021–2022
Power	225	243	261	289	307
Fertilizer	113	113	113	113	113
City gas	47	50	53	55	57
Industrial	28	32	35	37	37
Petrochemicals/refineries/internal consumption	72	76	80	82	82
Sponge iron and steel	9	9	10	10	10
Grand total demand	494	523	552	586	606
Domestic availability	216	222	229	236	243
Imports—LNG	172	193	215	236	258
Imports—trans border pipelines	30	30	30	30	30
Total availability	418	445	474	502	531

Source Report of the Working Group on Petroleum and Natural Gas Sector for the 12th Five Year Plan (2012–17)

secured LNG under long-term agreements from suppliers such as Sabina Pass for 3.5 million metric tonnes per annum (MMTPA), U.S. Dominion Cove Point (2.3 MMTPA) along with Russia's Gazprom (2.5 MMTPA).[28] Supplies from these sources are expected to commence during 2018–2020.

[28]Gas Authority of India Limited. http://www.gailonline.com/final_site/lng.html.

Table 11 Natural gas production, consumption and imports

Qty. in MMSCM					
	January 2016		April to January 2016 (cumulative)		
	2014–2015	2015–2016 (P)	2014–2015	2015–2016 (P)	Change (%)
Net natural gas production for sale (excluding flare, loss and internal consumption)	2279.55	1892.91	20,300.02	21,311.56	4.98
LNG import	1889.72	2131.29	14,679.30	17,298.25	17.84
Total natural gas consumption (Net NG production + LNG import)	4169.27	4024.20	34,979.32	38,609.81	10.37

Note Current financial year figures are provisional
Source PPAC, MoPNG, GoI

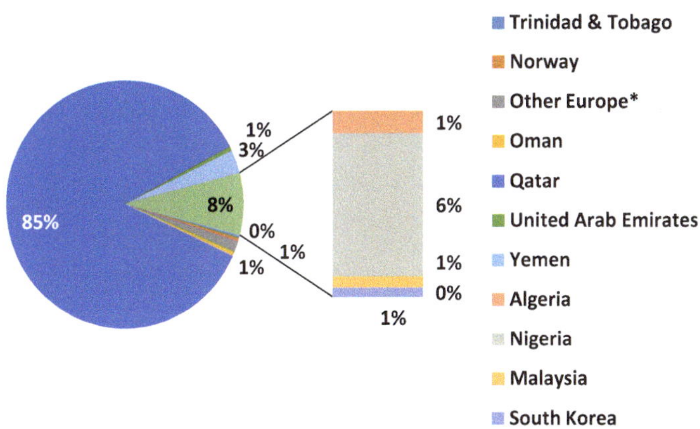

Fig. 7 India's LNG imports (BCM) in 2014. *Source* BP Statistical Review of World Energy June 2015

Moreover, Qatar's RasGas, which has slashed the price of the gas supplied to India by almost half, removing the floor and the cap of the pricing formula,[29] could help PLL to increase the volumes of purchase.

Drop in LNG prices becomes one of the key drivers for hunting LNG from multiple destinations, providing options for importing cheaper energy sources from overseas. India is also eyeing LNG imports from Australia and the USA to restart the stranded gas-based power plants.

[29]PTI. "Qatar scraps floor, cap in gas price formula with India", The Economic Times, 3 January 2016. http://articles.economictimes.indiatimes.com/2016-01-03/news/69486436_1_petronet-lng-rasgas-oil-price. Accessed on 26 January 2015.

In order to source LNG on a long-term basis, availability of import and regasification infrastructure plays a critical role. However, lack of adequate regasification capacity (Table 12) has been the constraint in making sufficient quantity of LNG through long-term contract to meet additional requirement for substituting costlier feedstock/fuel.[30]

High LNG prices coupled with highly capital-intensive nature of LNG regasification terminal acted as a barrier for India seeking more LNG contracts, making it an unviable proposition at that time, particularly to the sectors such as power and fertilizer, which are very sensitive to the price of natural gas but also the main source of natural gas demands. Henceforth, sourcing of LNG was left only under single long-term contract, signed between PLL and RasGas in July 1999 for import of 7.5 MMTPA LNG for 2004–2028.

But now with lower LNG prices, India should strive for enhancing both the capacity of existing LNG regasification terminals and the numbers of these terminals to cater to the needs of increasing natural gas demand.

Despite these plans for the expansion of LNG, the future LNG still depends on the volatility of the global oil price. Therefore, given this dynamics and the sensitivity of power and fertilizer sectors, the long-term contract, having take or pay clause, needs to be planned on the basis of proper study of the market and demand along with the affordability of LNG among different sectors.

5.5 *Gas Utilization Policy*

Natural gas is a part of Union List, and the power of allocation of domestic gas vests with Union of India. In India, since natural gas is a scarce commodity, the Central Government has been following the policy of sectoral priority in allocation of natural gas in the following order:

(1) gas-based fertilizer plants producing subsidized fertilizers;
(2) gas-based LPG plants;
(3) gas-based power plants supplying power to distribution companies at regulated tariff; and
(4) City Gas Distribution (CGD) entities for supply to domestic (PNG) and transport sectors (CNG).

The government placed CGD (domestic PNG and CNG) under no cut category. In line with the ruling of the honourable High Court of Gujarat, the government currently supplies at uniform base price to each CGD entity across the country. The government allocated 6.4 MMSCMD of domestic (APM and PMT) gas to GAIL

[30]Comptroller and Auditor General of India. "Report of the Comptroller and Auditor General of India on Supply and Infrastructure Development for Natural Gas for the year ended March 2014". Performance Audit No. 6 of 2015. http://www.cag.gov.in/content/report-no-6-2015-performace-audit-supply-and-infrastructure-development-natural-gas-union. Accessed on 15 February 2016.

Table 12 Statement of requirement of R-LNG and actual regasification capacity

							Quantity in MMSCMD	
Year	LNG import through long-term contract	LNG import (spot)	Requirement of gas for proposed schemes under fertilizer sector	Requirement of gas to avoid use of costlier feedstock in fertilizer sector	Requirement of gas to avoid use of costlier fuel in power sector	Total requirement of R-LNG	Actual regasification capacity	Minimum additional requirement for regasification capacity
	1	2	3	4	5	6 (1–5)	7	8 (6–7)
2010–2011	27.00	8.05	12.37	6.33	1.75	55.50	48.96	6.54
2011–2012	27.00	12.62	12.37	6.81	1.02	59.82	48.96	10.86
2012–2013	27.00	13.07	20.37	2.88	1.58	64.90	61.20	3.70

Source Performance Audit Report of CAG on Supply and Infrastructure Development for Natural Gas for the year ended March 2014 (CAG Report)

for supplying to CGD entities without discrimination. Requirement beyond 6.4 MMSCMD may be sourced by CGD entities through RLNG or any other alternative source, as may be available.

The above four sectors taken together are regarded as "core sector", while the remaining sectors, namely steel, petrochemicals, refineries, industrial and commercial consumers, being placed better to respond to the market price of inputs, put them under "non-core sectors".

According to this policy, the allocation of natural gas from different denominated fields such as APM, Non-APM, KG-D6 and PMT is done sector-wise as above as decided by Empowered Group of Ministers (EGoM).[31] Notably, the allocation mechanism of gas produced from different fields has evolved over a period of time, which has been pursuant to recommendations made by committees/GoMs and court directives.

However, with the falling gas production in the KG-D6 basin, to which priority of allocation for gas has been set, the MoPNG initially introduced pro-rata cuts across sectors between July 2010 and March 2011.

But the consistent fall in its production prompted the MoPNG to enforce a priority cut in the reverse order starting with the non-core sector, CGD, power, LPG and fertilizer with effect from 9 May 2011. The primary reason for such a pro-rata cut was to insulate sectors such as fertilizer plants, power and LPG from the subsidy burden.

However, this policy of the government has been an obstacle to the development of greater market orientation within the gas sector.

5.6 Regional Imbalance in Natural Gas Infrastructure

According to the Vision 2030 report, India can be divided into six major regional natural gas markets (Table 13) namely northern, western, central, southern, eastern and north-eastern market, out of which the western and northern markets currently have the highest consumption due to better pipeline connectivity.

Natural gas pipeline infrastructure connects various gas sources to gas demand centres. These demand centres may have customers belonging to powers, fertilizer, CGD and other industries in the country. It has been observed that new demand centres emerge due to the availability of natural gas. The gas pipeline infrastructure

[31]Standing Committee on Petroleum and Natural Gas (2014–15), 16th Lok Sabha, MoPNG. "Allocation and Pricing of Gas", [Action Taken by the Government on the recommendations contained in the Nineteenth Report (Fifteenth Lok Sabha) of the Standing Committee on Petroleum and Natural Gas (2013–14)], pp. 30. http://164.100.47.134/lsscommittee/Petroleum%20&%20Natural%20Gas/16_Petroleum_And_Natural_Gas_3.pdf. Accessed on 4 March 2016.

Table 13 Regional gas markets in India

Region	% of consumption	States with infrastructure and consuming gas	States lacking pipeline infrastructure
Western	53	Gujarat, Maharashtra	Goa
Northern	26	Delhi, UP, Haryana, Rajasthan	Punjab, J&K, Himachal Pradesh, Uttarakhand
Central	3	Madhya Pradesh	Chhattisgarh
Southern	14	Tamil Nadu, Andhra Pradesh	Kerala, Karnataka
Eastern	Nil	–	Bihar, West Bengal, Jharkhand, Orissa
North-eastern	4	Assam, Tripura	Meghalaya, Sikkim, Arunachal Pradesh, Mizoram, Manipur, Nagaland

Source Saumitra Chaudhuri report on Policy for pooling of natural gas prices and pool operating guidelines, Vision 2030—Natural Gas Infrastructure in India

has not only developed the much needed industrialization, but all brought socio-economic changes in the major regions from where the gas pipeline passes.[32]

The eastern market accounts for the lowest consumption of natural gas in the country, while it is the western market which consumes even more than 50% of the total gas consumption, followed by northern and southern markets which has around 26 and 14% of gas consumption, primarily because of either being closer to the gas resources or having a pipeline infrastructure. States consuming more gas includes Gujarat, Maharashtra and Andhra Pradesh.

States lacking gas pipeline infrastructure includes, Goa, Punjab, J&K, Bihar, Chhatisgarh and Kerala have not been able to utilize the benefits of gas, either due to lower availability of gas or poor pipeline infrastructure.

5.7 Gas Pipeline Grid

Natural gas infrastructure map suggests that natural gas pipelines are concentrated (Fig. 8) in the western (40%), northern (20%), southern (16%) and central (13%) part of India. Therefore, largely the consumption centres are located in the western and northern India.

As on 30 September 2015, around 16,065 km of natural gas pipelines were operational. Still the eastern part of the country lacks natural gas pipeline connectivity. To address the pipeline infrastructure disparity and enhance the connectivity, new pipelines of about 11,302 km (Table 14) are under various stages of construction/execution in the country. Surat–Paradip (2112 km) pipeline with a

[32]http://www.gailonline.com/final_site/naturalgas_transmission.html accessed on 25 February 2016.

Fig. 8 Map of pipeline network in India. *Source* PPAC Ready Reckoner June 2016

designed capacity of 77.8 MMSCMD would connect the west with the ast. Similarly, Jagdishpur–Haldia pipeline (1860 km) with a designed capacity of 32 MMSCMD ensure that the East gets pipeline connectivity with the North.

Pipeline transportation of gas offers a safe, economic and environmentally sound alternative to most other modes of energy transport. Therefore, till this, point firms or customers from the eastern belt are expected to be deprived of benefits of natural gas. However, the government and regulator are constantly pushing to develop Pan-India natural gas distribution network. The National Gas Grid would be a reality within a decade, and the gas starved customers in various parts of the country would have better access to natural gas.

Table 14 Natural gas pipeline under construction

Network/region	Entity	Length in km	Design capacity (mmscmd)	Pipeline size	Status of pipeline laid (km)
Kochi–Kottanad–Bengaluru–Mangalore	GAIL (India) Ltd.	1063	16.00	30′/24′/18′/8′/4′	35
Dabhol–Bengaluru (DBPL)	GAIL (India) Ltd.	315	16.00	8′/4′	65
Surat–Paradip[a]	GAIL (India) Ltd.	2112	76.81	36′/24′/18′	0
Jagdishpur–Haldia Phase-1 (755) km, 7.44 MMSCMD capacity	GAIL (India) Ltd.	1860	32.00	30′/24′/18′/12′/8′/4′	0
Mallavaram–Bhilwada[a]	GSPL India Transco Ltd.	2042	78.25	36′/30′/24′/18′	0
Mehsana–Bhatinda[a]	GSPL India Gasnet Ltd.	2052	77.11		0
Bhatinda–Srinagar[a]	GSPL India Gasnet Ltd	725	42.42		0
Kakinada–Srikakulam[a]	A P Gas Distribution Corporation	391	90.00		0
Shadol–Phulpur[a]	Reliance Gas Pipelines Ltd.	312	3.50	16′	0
Ennore–Nellore[a]	Gas Transmission India Pvt. Ltd.	430	36.00		0
Total		11,302			

Source Ready Reckoner, PPAC, November 2015
[a]Competitive Bidding

5.8 Gas Pricing

Efforts are on to enhance LNG import infrastructure and regasification facilities in the country. Appreciating the need to develop natural gas market and enhancing domestic gas production, the government is now considering deregulation of the natural gas prices. This is important at the time when gas producers both the PSUs and private companies are seeking independent pricing so as to make current investments viable.[33] India would do well to introduce the gas pricing freedom when it comes up with the new exploration regime. The government has approved the Marginal Field Policy which allows the contractor to have arm's length price for gas produced from marginal fields.

[33]Chakraborty and Singh (2015).

Many argue that market-driven gas pricing should be implemented to encourage companies to explore and produce all kinds of conventional and non-conventional hydrocarbons such as shale gas, coal bed methane and gas hydrates. This non-conventional hydrocarbon provides further potential if tapped properly to increase domestic gas production. The government is thinking in this line, and in the Union Budget 2016–2017, the finance minister hinted about the possible adoption of market-driven gas pricing.

6 Discussions

Indian is an energy-deficit country and primarily depends on energy import, especially on petroleum and natural gas. In the recent past, the Government of India shown high level of commitment to reduce carbon emission and build green infrastructure for green growth. Without any doubt, the current NDA government puts environment on top of its agenda. In the changing scenario, economic growth back on track and energy demand bound to increase in the country. Over the next decade, millions of Indian would be moving out of energy poverty.

'Make in India' initiative of the prime minister increases hopes of thousands of new enterprises to have access to energy, thereby creating opportunity for additional demand. Amidst global concern for environment and increased orientation towards cleaner energy, India's continued thrust on natural gas is noteworthy. The future growth and development of natural gas business in India would depend on global factors such as supply–demand, pricing, and geopolitics. To a larger extent, the global factors could be either beyond control or least controlled by a country like India. However, the domestic factors which are largely under the control of the country could play very important role to transform market developments. Figure 9 shows some of the very important factors. Authors believe the government and regulators would play their part in the future gas market development in the country.

The domestic exploration and production (E&P) policy plays an important role in ensuring adequate production of domestic natural gas, thereby ensuring stable supply at affordable price to the domestic buyers. Investor friendly E&P policy would enable domestic and global investors to commit their capital in risky projects. Domestic gas pricing is very tricky and sensitive issue, which would continue to be loaded in favour of people under energy poverty. It is equally challenging for the government to find balance between free market price and subsidized price. Scholars with social prosperity view argue that economist view of free market price ignore hardship of poor people. Possibly both economists and sociologists are right at their position. But government is free to devise its strategy to address concerns of all the stakeholders. Considering socio-economic condition of the large section of the society, the authors find merit in the government's existing pricing policy.

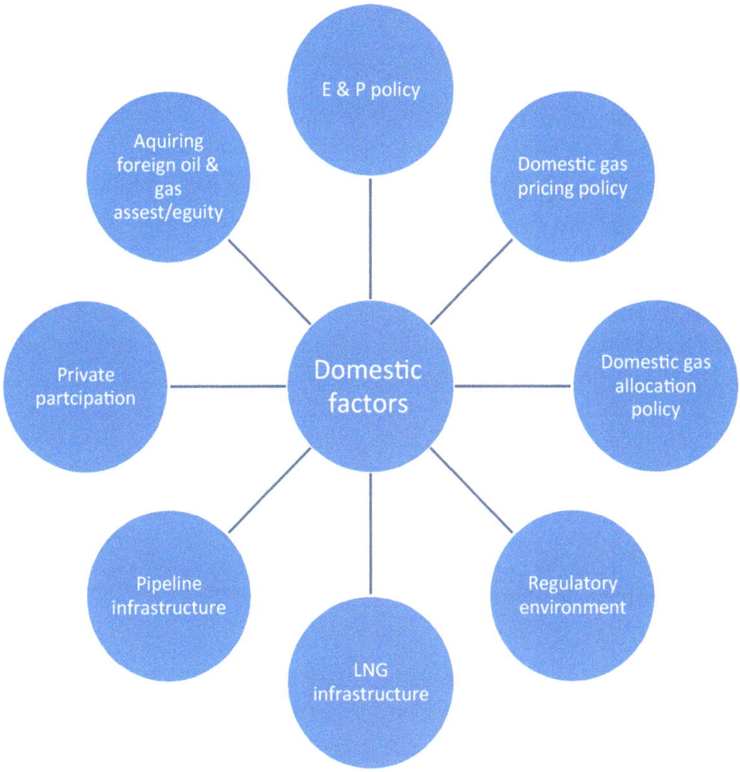

Fig. 9 Driving domestic factors for developing natural gas market in India

The areas where government needs to progressively think are as follows:

- exploration and production policy, especially for non-conventional sources such as gas hydrates and shale gas;
- encouraging active private participation in gas market developments; and
- encouraging Indian companies acquiring foreign oil and gas assets/equity.

Future growth would depend on following natural gas infrastructure developments:

- LNG ship building,
- LNG terminal,
- Cross-country natural gas pipeline and
- Transnational natural gas pipeline.

Role of regulatory environment is paramount in building a developed natural gas market. At this stage, the regulator would play the role of guide than that of a watchdog. Progressive regulations would serve the market well.

7 Conclusion

India's commitment to natural gas is well documented. The current government is absolutely committed to improve share of natural in the energy basket of the country. The government took many initiatives improving supply to meet rising demand of natural gas. The domestic pricing and gas allocation policies suggest that government committed to ensure optimal use of natural gas in the most need segments. The government has been extremely proactive in developing desired natural gas infrastructure to support growth and development gas market in the country. Within a decade, number of LNG terminal would double and the existing terminals double their capacity. As far as evacuating and transporting regasified LNG, natural gas pipeline across the country would be a reality. No longer would the eastern and central belt of the country be deprived of natural gas supply. Many contemporary economists appreciate Gujarat model of economic growth and prosperity. Drawing parallel, authors suggest that the Gujarat model of natural gas market development is quite robust and replicable in other states. The authors are reasonably sure that Pan-India natural gas market transformation would be realized through adequate LNG infrastructure along with extensive cross-country gas pipeline and a fully functional national gas grid across India.

References

Bhandari A (2015a) India gets power boost from 58% drop in gas prices. Business Standard, 14 Oct 2015. http://www.business-standard.com/article/specials/india-gets-power-boost-from-58-drop-in-gas-prices-115101400207_1.html. Accessed on 1 Feb 2016

Bhandari A (2015b) Falling global gas prices bonanza for India. Business Standard, 14 Dec 2015. http://www.business-standard.com/article/specials/falling-global-gas-prices-bonanza-for-india-115121400147_1.html. Accessed on 1 Feb 2016

Bureau ET (2016) Moody's pegs India's growth at 7.5% for next two years, says India insulated from turmoil. The Economic Times, 19 Feb 2016. http://economictimes.indiatimes.com/articleshow/51047456.cms?utm_source=contentofinterest&utm_medium=text&utm_campaign=cppst. Accessed on 19 Feb 2016

Chakraborty D, Singh RK (2015) Govt said to be considering deregulation of natural gas prices. Live Mint, 14 Oct 2015. http://www.livemint.com/Politics/zAc26QWsWV3dpcJUWG4IUJ/Govt-said-to-be-considering-deregulation-of-natural-gas-pric.html. Accessed on 24 Feb 2016

Clemente J (2015) COP21 means more natural gas and the U.S. must help. Forbes, 29 Nov 2015. http://www.forbes.com/sites/judeclemente/2015/11/29/cop21-means-more-natural-gas-and-the-u-s-must-help/#15361f0072eb. Accessed on 28 Jan 2016

IPCC (2014) Climate change 2014: mitigation of climate change. In: Edenhofer O, Pichs-Madruga R, Sokona Y, Farahani E, Kadner S, Seyboth K, Adler A, Baum I, Brunner S, Eickemeier P, Kriemann B, Savolainen J, Schlömer S, von Stechow C, Zwickel T, Minx JC (eds) Contribution of working group III to the fifth assessment report of the intergovernmental panel on climate change. Cambridge University Press, Cambridge. http://mitigation2014.org/report/publication/. Accessed on 16 Feb 2016

Kar SK (2015) Natural gas to drive green and sustainable developments in India. In: Sharma A, Kar SK (eds) Energy sustainability through green energy. Green Energy and Technology, Springer India. Doi:10.1007/978-81-322-2337-5_15

Kar SK, Mishra SK, Bansal R (2015) Drivers of green economy: an Indian perspective. In: Thangavel P, Sridevi G (eds) Environmental sustainability: role of green technologies. Springer, Berlin

Mahesh P (2016) Local growth can shield India from global crisis: Didier Borowski, Amundi Asset Management. The Economic Times, 19 Feb 2016. http://economictimes.indiatimes.com/articleshow/51048007.cms?utm_source=contentofinterest&utm_medium=text&utm_campaign=cppst. Accessed on 19 Feb 2016

Sharad Raghavan TCA (2015) Qatar sashes gas price for India, waives penalty. The Hindu, 31 Dec 2015. http://www.thehindu.com/business/qatar-slashes-gas-price-for-india-waives-penalty/article8049955.ece. Accessed on 1 Feb 2016

Vaid M, Maini TS (2014) Shale gas development—another avenue for India-US Co-operation. In: Global Asia, vol 9(1). Spring 2014. https://www.globalasia.org/wp-content/uploads/2014/03/547.pdf. Accessed on 15 Feb 2016

Impact of Slashing Oil Prices on the Natural Gas Market

Husain Ahmad

Abstract The epitome of this chapter is to discuss the challenges faced by natural gas due to proverbial freefall in the prices of crude oil during the year 2014–15. This chapter also includes a recapitulation of the key drivers that has slumped the global crude oil prices. Further, it discusses about the macroeconomic and financial propositions and impact on Indian economy due to plummeting oil prices and thereof.

Keywords Natural gas · Crude oil slump · Factors affecting gas prices · Challenges ahead

1 Natural Gas: Background in Indian Perspective

The global energy landscape is presently undergoing considerable changes. Nearly all countries are significantly affected by growth in the oil market, either as a producer or consumer or may be both. Efficient, unswerving and competitively determined price of energy supplies which are rudiments for boosting fiscal development. For any emerging nation, the policy to obtain and meet the energy supplies and energy developments are the crucial ingredients of the overall economic strategy. Proficient use of resources and long-standing sustainability in its utilization is of key significance for economic development. Sustainability would take into consideration not only existing natural resources but also to ensure that the related ecological and social aspects meets the priority requirements of the economy. Concurrent and synchronized action is, thus, needed to make certain that the interim concerns do not detract the nation away from the long-term goals.

H. Ahmad (✉)
EPIC (Engineering, Procurement,
Installation and Commissioning) Company, Doha, Qatar
e-mail: husainsinbox@gmail.com

S.K. Kar and A. Gupta (eds.), *Natural Gas Markets in India*,
DOI 10.1007/978-981-10-3118-2_2

Given India's growth, its fuel requirements are likely to expand at a substantial rate. The requirement for energy in India, the economic augmentation, and the resulting prosperity of the country have amplified substantially in the recent years. For India, crude oil forms an important proportion of the country's energy basket, next only to coal and natural gas. Given 17% of the world's population and 2.4% of globe's geographical area, India has meager known liquid fuel reserves. The per capita energy consumption for India is anticipated to be of a very small amount, i.e. ~530 kg of oil equivalent (Kgoe) in contrast of the world's average of 1800 Kgoe (Ahmad 2011).

Natural gas is cleanest burning fuel among the competitive fuels, enhancing its desirability as a fuel choice in a carbon-constrained environment. Natural gas is rising as the most significant energy source having huge potentials for the future. The profusion of natural gas, globally as well as nationally, coupled with its green soundness and manifold applications across varied industries, entails that natural gas is going to register and mark a gradually more vital role in meeting the demand for energy across the globe (Ahmad 2011).

In India, natural gas has picked up as a fuel of significance in the last few decades when Indian Government started to set up separate public sector, undertaking to be in charge the gas distribution business. Until that time, ONGC and OIL were up keeping production as well as natural gas distribution. Since last ten years, natural gas started discovering various usages as environmental friendly fuel and as a rattling efficient fuel. The demand for transport fuels in India is growing by leaps and bounds owning to the surge in vehicular population. Transport sector accounts for almost all the petrol, and about 60% of the diesel is consumed in India (Ahmad 2011). Hence, the deployment of natural gas in the form of compressed natural gas (natural gas compressed to high pressure to make it compatible for the usage in automotive sector) would serve as a much cheaper and highly cost-effective fuel for automotive purpose, resulting in huge saving and thereof, thus helping India to bridge the fiscal deficit.

2 Environmental and Safety Aspects

Demystifying, the conundrum arising out of the usage of natural gas particularly on safety aspect, Natural gas is a safer fuel owning to its inherent characteristics which takes into account the low-specific gravity of fuel. In other words, it means natural gas is lighter than air and in the event of leakage, natural gas will rise and escape into the atmosphere. Hence, it does not pose any threat of fire or explosion. Another vital innate feature of natural gas is its high autoignition temperature which further minimizes the hazards (Ahmad 2011).

Furthermore, global warming has intensified the debate on various energy sources and their usage across the globe. Natural gas is the cleanest of all the fossil fuels which mainly comprises of methane. The key end products produced by burning natural gas are carbon dioxide and water vapor. Coal and oil comprises of much more intricate molecules, with a higher carbon ratio and higher nitrogen and sulfur contents. This implies that when burnt, coal and oil release higher levels of hazardous emissions, together with a higher ratio of carbon emissions, nitrogen oxides and sulfur dioxide all being catastrophic to health and environment (Ahmad 2011).

Harmful coal and fuel oil used for various industrial applications also discharges ash particles into the atmosphere, substance which do not burn up but instead are carried into the environment and add up to effluence. The burning of natural gas, on the contrary, releases negligible quantity of sulfur dioxide and nitrogen oxides, almost no ash, lower levels of carbon dioxide, carbon monoxide, etc.

3 Key Factors Distressing the Prices of Natural Gas

Price of natural gas is primarily determined by market demand and supply. As there are restricted short-term substitutes available to natural gas as a fuel for heating and power generation during peak demand phases, changes in supply or demand over a short time period may upshot in large price variation. Price itself often act to stabilize demand and supply.

There are various factors on the supply side which influence the price of natural gas such as gas production, imports, international sanctions, geopolitical situations, and underground storage levels. Boost in supply is likely to reduce prices, while drop in supply is possibly going to increase prices. Increase in prices has a propensity to push production, imports, and sales from storage inventories. Decreasing prices are inclined to have the turnaround effects.

The demand side also has various factors which affect natural gas pricing such as temperatures, economic situations, and liquid fuel prices. Low temperatures increments the demand for heating, while high temperatures amplifies the demand for cooling, which in turn increases the gas demand by electric power plants. Economic situations influence the demand for natural gas, in particular by manufacturers. Demand may be moderated by liquid fuel prices, which may be a cheap alternative for natural gas for power generation, manufacturers, and others. Higher demand is likely to guide to increased prices, while lower demand can result in decreased prices. Increase and decrease in prices have a propensity to lessen or boost demand.

The major supply-side factors that may influence natural gas price are as follows:

- changes in the quantity of natural gas being produced;
- quantity of natural gas being imported and/or exported;
- international sanctions and geopolitical situations; and
- amount of gas available in storage facilities.

Increase in supply has a propensity to lead to lower prices and decrease in supply is likely to increase prices. The major demand-side factors which may have an effect on price are as follows:

- level of economic development;
- changes in weather conditions; and
- prices of other competing fuels.

Higher demand leads to demonstrate the way to escalated prices, whereas dropped demand has a tendency to direct to decreased prices.

4 Tale of Falling Crude Oil Prices

Oil prices chopped down stridently between second quarter and third quarter of financial year 2014, bringing the four-year episode of permanence around $105 per barrel to a conclusion. The turn down, which is much bigger than that of then other commodity price indices weighed against early 2011 crests, may point out an end to a price which is often referred as super cycle. The oil prices had fallen to about one third to $36 per barrel during December 2015 and even after almost a year, as of November 2016, the oil price remain only at $47 per barrel. The prices are expected to ascend slightly in the year 2017. The sources and repercussions of the spiky reduction in oil prices have led to serious debate.

Recapitulation:

Since 2010, the oil prices were high bouncing more or less $100 per barrel. Due to high oil consumption by nations such as India, China, and clash in chief oil nations like Iraq, oil production could not meet the demand which led to increase in prices. Interestingly, underneath the earth's surface, many of the dynamics kept swiftly changing. High prices urged firms in the USA and Canada to start exploring for new-fangled, shale reserves in North Dakota and oil sands in the area of Alberta. Simultaneously, demand for liquid fuels in nations of Europe (EU), Asia, and the USA started dwindling, credit to deteriorating economies and latest efficiency measures. On top of that, nations like Iraq and Russia began producing more oil (Ahmad 2015b).

By late 2014, global oil supply was heading in the right direction, increasing much higher than genuine demand, and from September onwards, the prices started declining piercingly.

Following are the key drivers that had led to the slump in the global crude oil prices:

The foremost reason is the Shale Oil Boom in North America. As the crude oil was hovering above $100 per barrel in between 2011 and 2014, many energy companies based in North America which found it lucrative began extracting oil from difficult hard to drill areas. In the USA, firms began using methods such as fracking, i.e. hydraulic fracturing and horizontal drilling to extract oil from shale structures in North Dakota and Texas. In Canada, firms were heating oil sands in

Alberta with vapors to take out usable crude which resulted in an increased "Light Tight Oil" yield (Ahmad 2015b). Since 2008, US itself has added 4 million barrels per day (mbpd). For the record in last 30 years, US imports from OPEC have been cut in more than half.

Light tight oil (LTO) is unconventional oil stored in shale and needs modern drilling and techniques for recovery to get it out. Extra requirement of water for fracking process adds to the costs. Unconventional oil reserves are located in geologically aloof environments, requiring sophisticated technologies adding further to exploration costs. This unconventional oil becomes profitable only if the crude oil stays above $60 a barrel. Also the refining cost of heavy and extra-heavy oils is high since it can contain up to three times more carbon and high concentrations of sulfur and heavy metals. Bitumen, being the petroleum component in unconventional oil requires it to be pre-processed into synthetic crude oil before refining (Ahmad 2015b).

Secondly, Geopolitical clashes were roaring up in main oil producing nations. A civil war broke in Libya. Iraq was a complete muddle. The US and EU clouted oil sanctions on Iran and strained its oil selling to other countries. Those clashes took more than 3 mbpd from the market. But a lot has changed by middle of 2014 (Ahmad 2015b). A lot of those disturbances started settling. In July, Libyan fought back and opened up two key export terminals, Es Sider and Ras Lanuf, which were shut down for almost more than year which rose Libyan's export out of the blue.

Thirdly and most important reason is due to OPEC internal strife, there have been several articles, tales as regards the conflict amid OPEC member nations themselves, resulting in to let the market to think that OPEC shall not be able to control production like it has done so in the earlier period. OPEC faced internal battles, since the member countries are petro-states heavily reliant on oil revenue. The proverbial "let us try to catch the falling dagger but I'm not going to try it," expounds the ongoing situation among the cartel members (Outlook India 2015). The cartel has for almost a third of global oil production with its authorization to synchronize and bring together the petroleum policies of its member nations and to make sure stabilization of oil markets so as to secure an well-organized, profitable, and standard supply of petroleum to customers, a stable income to producers, and a fair return on capital for those investing in the petroleum industry (Outlook India 2015). Although, the cartel's objective is now appearing less believable than earlier and insinuates to have been destroyed in the domestic tussle of defending their respective market shares regardless of prevailing low prices.

Though OPEC officially maintained its production target of 30 mbpd, the actual oil production has been over and above 31.5 mbpd in the preceding months. Consequently, global oil prices have been dwindling thus hemorrhaging the economies of the nations who are cartel members of OPEC which are counting heavy losses with massive budget deficits and anemic economic growth.

Fourthly, the Iranian Nuclear deal, Iran has reached a historic nuclear deal that is likely to result in easing of sanctions in return for limiting Tehran's nuclear program. The comprehensive agreement has clinched between Iran and P5+1—the US, Britain, France, China, Russia, and Germany. The agreement essentially prevents Iran from getting a nuclear bomb. In exchange, many other nations will remove a lot of the economic sanctions placed on Iran (Ahmad 2015a). Iran looks to double its crude exports to over 2 mbpd in the next six months or so. Subsequently, more oil production means prices of oil will plummet further. Cash-strapped Tehran may decide to offload this stock and run with whatever price it gets, even though the global market is oversupplied by 2.6 mbpd. Iran would also likely accept lower prices in return for income and as a means to rebuild its share in the global oil market (Ahmad 2015a).

The front runner of OPEC, Saudi Arab which produces approximately 30% of the total OPEC's production lowered its official selling prices to Asia and the US. Saudi oil minister made it clear that Saudi Arabia prefers keeping its market share rather than cut production, and that the low price of oil could restrain future drilling elsewhere including the production of shale oil in North America. Apart from it, other Middle East OPEC crude producers are competing with counterparts in Latin America, North Africa, and Russia for buyers in Asia. They have lowered price differentials and maintained output in order to defend market share. Also, the reason that why OPEC is uncertain to cut the production thus a creating the demand–supply gap so as to increase the price is that if they will still keep the prices of the crude on the higher side, it shall incentivize further the investment in shale oil technology and production which could sway away the remaining command of OPEC to govern the crude prices.

Even more significantly, slowing economic activity in regions such as Europe, Japan, and China has begun deteriorating, mainly credit to slowdowns in China and Germany. Largely, oil demand has been slugging in lot of other countries across the globe. Simultaneously, countries such as Indonesia and Iran have been reducing on fuel subsidies.

5 Low Oil Are Challenging Natural Gas Markets

After three years of stability, oil prices have fallen piercingly since middle of 2014. The effects of this price fall on a wide range of energy firms have been material wherein many of the firms have been pushed to rethink their investments, cost structures, and even business models. A high degree of uncertainty regarding oil prices still remains. Will prices bounce back, as historical records hint that they will? Or have we cross the threshold and entered a "new normal" comprising of a mostly different pricing milieu?

International natural gas markets have already felt some shock from the uncontrolled oil prices. But these markets shall be affected to a much greater extent, if the oil prices stayed amid $40–$50 per barrel range for an extended episode, given the relation that exists between the two fuels. Gas prices in some markets are still contractually linked to oil prices. There are also substitution effects, i.e. users switch between oil and gas as a result when fuel price changes (The Energy and Resources Institute 2014).

Asia has had the world's fastest growth in demand for natural gas in the recent times with demand increasing over 6% per annum for the past five years, exceeding 600 billion cubic meters (bcm) in 2014. In the past, Asia has been reliant on heavy imports to stabilize the gap the between the huge demand and comparatively inadequate local production. To make certain security of supply, Asian consumers have been dependent primarily on long-term contracts, essentially indexed to oil prices. The relation to oil prices also replicates the region's escalating switch to natural gas from liquid based fuels.

Given the relationship of Asian gas contracts to oil prices, the drop in oil prices will unswervingly bang Asian gas prices, pushing them considerably lower. This will be ever more apparent in the upcoming months ahead, as present long-term gas import contracts are indexed to oil with a time lag that ranges between 6 and 12 months.

It may be anticipated that if prices of oil stay at existing levels, Asian LNG import prices will decrease to approximately $5 per mmbtu by middle of 2016, signifying the fact that long-term contracts' oil indexation is usually derived from oil's average price over the last 6–12 months. Indeed, present future price is already discounting this probable price reduction.

Another important factor that will weigh down the Asian LNG market are planned LNG projects in Australia and Papua New Guinea which may introduce about 90 bcm per annum to the marketplace. This additional quantity will raise downward force on prices. It will also lessen Asian buyers' motivation to amalgamate upstream (Rogers 2015).

Over the past few years, Asian buyers have been looking to expand the basket of indices they use while securing LNG supply, with a particular push to back off from oil in support of Henry Hub—indexed quantity, whose prices became ever more striking, compared with prices for oil indexed quantity when oil was approx $100 per barrel. As oil prices currently are fine and with more ambiguity over the competitiveness of the US LNG, Asian buyers have to evaluate their diversification strategies. Over the medium to longer term, there appears possibility that if the current oil price situation persists, it will augment the sustainable development of the natural gas market in Asia (Rogers 2015).

6 Macroeconomic and Fiscal Propositions

Oil prices nosh into development and price rises mainly through three channels:

- Input costs: Reduced oil prices lessen energy outlay usually, as prices of alternate and competing energy resources are pushed down too. As oil is feedstock for various segments in the industry, alongside petrochemicals, paper, and aluminum, the decrease in price unswervingly affects an extensive array of processed or partially processed inputs. The automobile, petrochemicals, and agricultural industries and some manufacturing sector would be foremost recipient from stumpy prices.

- Real income shifts: Plummeting prices of oil causes changes in real income thus benefiting oil importers and hammering oil exporters. The shift in income from oil exporting nations with greater average saving rates to net importers with a larger proclivity to splurge should by and fat upshot in sturdy international demand over the medium term. Nevertheless, the effects could show a discrepancy considerably across various economies in due course. Some exporting nations may be pushed by economic constrictions to adjust both government expenses and imports hastily in the short duration, while payback for importing nations could scatter and equalize by greater defensive savings if poise in recovery stays low.
- Fiscal and economic policies: In oil importing nations where waning prices of oil may decrease medium-term inflation potential lower than target, central banks could counter with other economic and fiscal policy relaxing, which, consecutively, can sustain development. The blend of lower inflation and higher productivity entails a positive short run strategy as an end result. In oil exporting nations, nonetheless, reduced oil prices might set off compressing economic policy actions, if not shock absorbers are available to shield outlays from the turn down in tax revenues from the oil sector.

This channel functions with diverse potency and covers the whole nation. Though, it seems apparent that oil price turn downs usually have less significant yield effects on oil importing economies than oil price increases. This lopsidedness could be caused by the resistance and adjustment costs coupled with variations in oil prices.

The impacts of oil price changes on yield may also change amid emerging and developed nations. Output in developing economies may be moderately more energy intensive and therefore, may have an advantage more from reduction in energy input costs. Domestic inflation outlook in emerging nations may also be more reactive to changes in fuel prices than in developed nations, to a certain extent on account of a larger weight of fuel and food in consumption baskets. This is reflected in sturdier effects of commodity price shocks on inflation in emerging nations than in highly developed countries.

7 Impact on the Indian Economy

India is a major importer of oil in the world and meets about 70% of its crude oil demands through imports. This hefty reliance on oil-producing countries makes it extremely vulnerable to price inconsistencies. With oil prices tumbling to less than half in 2014, the Indian financial system has been subject to both constructive and depressing phases from the international oil market.

Positive effects of oil prices on the Indian economy:

- The reduction in oil prices leads to a direct savings. Hence the capital, which was earlier used for either working capital or raw material, can at present be used for spending in long term assets that have a fine market return. This could lead to increased income of a good amount.

- Oil is a major constituent of the Consumer Price Index, a gauge of inflation in India. The decrease in oil price has caused inflation to fall since January 2014. Lower inflation leads to lower interest rates as the holding cost of money decreases and lower interest rates consecutively lead to higher investments as debt servicing becomes more reasonable. The flexibility in investment boosts the gross domestic product (GDP) of the nation and induces higher foreign direct and indirect Investments as the economy becomes a feasible investing destination due to its growth pace. This gives rise to a multiplier effect for the growth of the country.
- Oil and gas subsidies had weighed down the Indian budget with debt to shield marginalized households from soaring crude oil and gas prices. With the reduction in prices of petrol, diesel and gas subsidies have either been decreased or removed totally. This has freed a good amount capital for investments in important industries like roads, power, education, health, and telecommunication. Thus, instead of being used as consumption expenditure, this amount can be set out to fuel growth in the economy.

Negative effects of oil prices on the Indian economy:

- Oil-producing countries are chief markets for India's investors and products. A decline in their imports or currency valuation may have a repulsive effect on Indian companies which have investments and market share in various foreign oil-producing countries.
- India is the sixth largest petroleum product exporter in the world. The suppressing demand will cut unit prices and emaciate margins hence negatively striking the petroleum product manufacturers.
- Whenever oil revenues take a hit, oil-producing nations in the Gulf emphasize on local manufacturing and production and cut back on their foreign labor force. Since huge parts of the labor force in the Gulf are Indians, which are greatly reliant on remittances. Downsizing of labor force in the Gulf countries may take a significant toll. As remittances contribute more to the Indian GDP than the flagship information sector exports, halving oil prices would be correspondent to downsizing the IT sector by half.

8 Conclusion

Subsequent to four years of firmness at about $105 per barrel, oil prices have chopped aggressively after mid of 2014. When compared to the early 2011 commodity price peaks, the turn down in oil prices was a lot bigger than that in other commodity price indices. The turn down in oil prices was rather noteworthy compared with the previous occurrences of oil price declines all through the earlier decades.

There have been numerous long- and short-term factors in the wake of the latest thrust in oil prices: Quite a few years of huge rising revelations in oil supply, some downward unexpected changes in demand; slowing down of geopolitical situations which had threatened production changes in policy objectives of OPEC and appreciation of the US dollar. Supply linked drivers have evidently played a chief role, with the new OPEC policy intended at market share eliciting a further prickly decline.

The decrease in oil prices has considerable macroeconomic, fiscal, and policy connotations. If continued, it will shore up commotion and decrease inflationary, external, and economic forces in oil-importing nations. But then, it would affect oil exporting economies detrimentally by abating fiscal and external situations and reducing economic activities. Stumpy oil prices influences investor's sentiments about oil exporting up-and-coming market economies and can cause sizeable unpredictability in financial markets, as previously occurred in some of the nations in 2014. Still, waning oil prices also introduces a major window of prospects to restructure energy taxes and fuel subsidies, which are important in quite a lot of developing nations, and reinvigorate reforms to diversify oil-dependent economies.

References

Ahmad H (2011) Use more natural gas, it can be an Instrument of sustainable economy and a way to save the environment. Guest column, Daily edn. Sakaaltimes

Ahmad H (2015a) Policy proposals for India, Iran nuclear deal: a sanguine and scintillating breakthrough for Indian oil and gas sector. New Delhi

Ahmad H (2015b) Policy proposals for India, an insight in to the falling crude oil price. New Delhi

Outlook India—Web Edition (2015) Husain Ahmad, business, opinion, good and bad news about crude oil. New Delhi

Rogers HV (2015) The impact of lower gas and oil prices on global gas and LNG markets. OIES Pap 99

The Energy and Resources Institute (2014) Global oil markets and india's vulnerability to oil shocks. New Delhi

Part II
Gas Sourcing, Contract, and Pricing

Sourcing of Natural Gas Through Cross Border Pipelines

Ayush Gupta and Bhavesh Gupta

Abstract In the era of economic modernization and globalization, India has unfolded itself as central economic and trading zone in emerging Asian market and has focused a lot towards its foreign policies impacting its accelerating economic performance. Private sector involvement in power sector and rationalization of prices has forced India to look for efficient energy management system. Natural gas, a highly promising commodity in India, is presently supplied from only two sources, i.e. domestic production and liquefied natural gas (LNG). But now instead of importing costly LNG, India has to examine and start importing through low-cost cross-border pipelines. Technological development in the field of deep-sea pipelines has made it possible for India to consider importing natural gas from Iran and other Middle East countries, bypassing countries such as Pakistan and Afghanistan having serious security issues. Construction of Turkmenistan–Afghanistan–Pakistan–Indian (TAPI) pipeline has already started and Indian government has also initiated discussions with Oman, Myanmar, Kazakhstan and Russia for new pipeline projects. Indian Prime Minister Narendra Modi can imagine energy has beating heart of accelerating India and pipelines as its veins, hence to deal with energy diplomacy and to have sufficient amount of fuel, he is focusing on developing strong and long-term strategic relationships with resource-rich countries. This will give a big boost to the natural gas industry in India.

Keywords TAPI · IPI · Asian Development Bank · Deep-sea pipeline · Onland pipeline · Geopolitics

---

A. Gupta (✉)
GAIL India Limited, New Delhi, India
e-mail: ayushgupta@gail.co.in

B. Gupta
IM & I-BPM Utilities, Accenture Services Pvt. Ltd., Gurgaon, India
e-mail: bgupta234@gmail.com

S.K. Kar and A. Gupta (eds.), *Natural Gas Markets in India*,
DOI 10.1007/978-981-10-3118-2_3

1 Introduction

As world's fourth largest energy consumer, India has been dabbling in plans for last two decades in order to reduce dependence on oil from unstable Middle East and West Asia, by sourcing natural gas from Central Asia. Due to stalled indigenous natural gas production in recent years and widening demand–supply gap, import through pipeline is becoming must for India. As importing LNG is a costly process, India has shown interest in sourcing natural gas through cross-border pipelines from countries such as Turkmenistan, Iran, Oman and Myanmar.

There are several projects in various stages of planning and discussions but till date India is not importing natural gas through pipeline. For over a decade now, proposals for the two pipelines such as Iran–Pakistan–India (IPI) pipeline and Turkmenistan–Afghanistan–Pakistan–India (TAPI) have remained unsteady due to tense relations between Iran and USA, and New Delhi's tumbling relationship with Islamabad (Kasturi 2015). Along with this, India has been trying to source gas from its gas-rich eastern neighbours such as Bangladesh and Myanmar.

MoPNG has predicted availability of 30 mmscmd natural gas through pipeline from 2017–18 to 2021–22 in its five-year plan documents. No other cross-border pipeline is under active consideration and so the same assumption of 30 mmscmd has been further assumed to continue till 2029–30 (Industry Group 2013) (Tables 1 and 2).

Table 1 Gas availability (cross-border pipelines)—12th and 13th five-year plans

MMSCMD	12th plan					13th plan				
	2012–13	2013–14	2014–15	2015–16	2016–17	2017–18	2018–19	2019–20	2020–21	2021–22
Gas imports	0	0	0	0	0	30	30	30	30	30

Source *"Vision 2030" Natural Gas Infrastructure in India*, Report by Industry Group for Petroleum and Natural Gas Regulatory Board (2013)

Table 2 Gas availability (cross-border pipeline)—2023–2030

MMSCMD	2022–23	2023–24	2024–25	2025–26	2026–27	2027–28	2028–29	2029–30
Gas import	30	30	30	30	30	30	30	30

Source *"Vision 2030" Natural Gas Infrastructure in India*, Report by Industry Group for Petroleum and Natural Gas Regulatory Board (2013)

There are two options for importing natural gas through pipelines—the land route and the sea route. Over the years, several routes for gas pipelines have been proposed and are at various levels of implementation.

1.1 Turkmenistan–Afghanistan–Pakistan–India (TAPI) Pipeline

South Asian countries are considering TAPI as a new and effective step towards the evolution of the modern architecture of global energy security, a robust driver of economic and social stability—that is the reason Afghanistan President Ashraf Ghani called the project a "new silk route".

Pipeline construction for the Turkmen portion has already started on 13 December 2015. This pipeline will be 1420 mm (56 in.) in diameters with a working pressure of 100 standard atmospheres. It will be supplying 33 billion cubic metres of natural gas per year over a period of 30 years, out of which 0.5–1.5 billion cubic metres will be taken by Afghanistan whereas India and Pakistan have contracted for 14 billion cubic metres each (Tanchum 2015). The pipeline is expected to be operational by 2019 (Reyaz 2015).

In 2015, Turkmengaz was selected as the pipeline consortium leader (51% stake) and the state gas companies of Turkmenistan, Afghanistan, Pakistan and India (GAIL) established a consortium that will build, own and operate the planned 1814 km pipeline.

Asian Development Bank (ADB) did initial feasibility studies for the project and originally the cost of pipeline was reportedly estimated at US$7.6 billion; however, a more recent estimate predicts the project cost in the range of US$10 billion.

During the Road Shows conducted in 2012 scouting an International Oil Company as the Consortium Leader, several international companies including Mobil, Chevron, Total SA, Exxon had shown interest in the project. However, they had requested stake in onshore upstream gas fields which was not in conformity with the Turkmen law.

The pipeline will carry gas from Turkmenistan Galkynysh field, also known by its previous name South Yolotan Osman that holds gas reserves of 16 tcft (trillion cubic feet). From there, pipeline will pass through Herat and Kandahar provinces of Afghanistan before entering Pakistan. In Pakistan, it will run through Multan via Quetta before ending at Fazilka (Punjab) in India (Chaudhary 2015) (Fig. 1).

Security issues in Pakistan and Afghanistan and financial closure are some major hurdles for this project's success. Institute of Defence Studies and Analysis (IDSA) has also shown some serious concerns regarding completion of this project but high stakes of Pakistan (42%) and Afghanistan (16%) reduce the risk of the project being sabotaged (Turkmenistan launches $10bn gas pipeline to South Asia 2015).

Fig. 1 Turkmenistan–Afghanistan–Pakistan–India (TAPI) pipeline route

Status of Agreement

Intergovernmental agreement (IGA)	Signed by the Head of the states of all the member countries on 11 December 2010
Gas pipeline framework agreement (GSFA)	Signed by respective petroleum ministers of four countries on 11 December 2010
Gas sales and purchase agreement (GSPA)	Turkmenistan–Pakistan and Turkmenistan–India bilateral GSPAs were signed by the heads of respective commercial entities on 23 May 2012. Turkmenistan–Afghanistan signed the bilateral GSPA on 9 July 2013
Operations agreement	Signed between all four TAPI countries on 9 July 2014
Transaction advisory services agreement (TASA)	TAPI parties and ADB signed TASA on 19 November 2013 at Ashgabat, Turkmenistan
Transit fee agreement	Parties agreed to a transit fee figure with consensus. Currently, the tripartite transit agreement negotiations between Afghanistan–Pakistan–India are continuing
Shareholders agreement	After resolution of all issues, Shareholder Agreement has been signed among the parties during the Ground Breaking Ceremony held on 13 December 2015 in Turkmenistan
Investment agreement	Negotiations are under way for the finalization of Investment Agreement

Current Status of the Project:

- **TAPI Pipeline Company Limited (TPCL)**: TPCL has been incorporated in Isle of Man on 11 November 2014. The first Board Meeting has been held in Ashgabat, Turkmenistan on 19 November 2014.

- **Completion of Technical Studies and Preparation of Information Memorandum**: ADB as Transaction Advisor has submitted different technical studies required for the preparation of Information Memorandum/ Bidding Documents; however, with the selection of Consortium Leader, the Information Memorandum is no longer required.
- **Selection of Consortium Leader**: Initially Turkmen Party was negotiating with TOTAL (France) who showed interest to be potential Consortium Leader. However, in the 22nd TAPI Steering Committee (SC) meeting held on 6 August 2015 in Ashgabat, breakthrough was achieved when Turkmenistan proposed itself as Consortium Leader and TAPI-SC unanimously endorsed State Concern "Turkmengaz" as the Consortium Leader of TAPI Pipeline Company Limited (Turkmenistan-Afghanistan-Pakistan-India Gas Pipeline (TAPI) 2015).

1.2 Iran–Pakistan–India Pipeline (IPI)

From the day in 1988 when Iran discovered natural gas in South Pars field, their government is trying to promote higher gas exports abroad. Iranian government can easily see high-growth prospects in South Asian countries such as India and Pakistan, where natural gas reserves are low and demand–supply gap is wide. Also potential for economic and developmental gain from this pipeline has forced governments of Iran, Pakistan and India to reassess their roles, policies and national security issues.

In 1995, Pakistan and Iran signed a preliminary agreement for construction of a $3 billion, 870 mile onshore pipeline linking South Pars with Karachi Pakistan (Iran to India Natural Gas Pipeline 2000). Later on, new proposals were introduced and pipeline was intended to be extended to India. Four major companies who have shown interest in constructing IPI gas pipeline are BHP of Australia, NIGC, Petronas of Malaysia and Total (Alexander's Gas and Oil Connections 2000). Over a period of last 20 years, Indian, Iranian and Pakistani government officials have continued to negotiate the possible routes, modes of transportation and geopolitics, but nothing good has happened.

The following map shows the main route for IPI pipeline. Starting from the left side of the map, pipeline emanates in Asaluyeh, Iran on the coast of Persian Gulf near South Pars field and stretches 1172 km through Iran (Iran to India Natural gas pipeline 2000).

In Pakistan, the length of pipeline is 785 km (Iran–Pakistan Gas pipeline). It will pass through Baluchistan and Sindh and then in Khuzdar, a branch would spur off to Karachi on the Arabian Sea coast and the main section will continue towards Multan (Iran to India Natural gas pipeline 2000) (Fig. 2).

The initial capacity of the pipeline was supposed to be 22 BCM (780 BCF) of natural gas per year, to be equally distributed to India and Pakistan. This pipeline

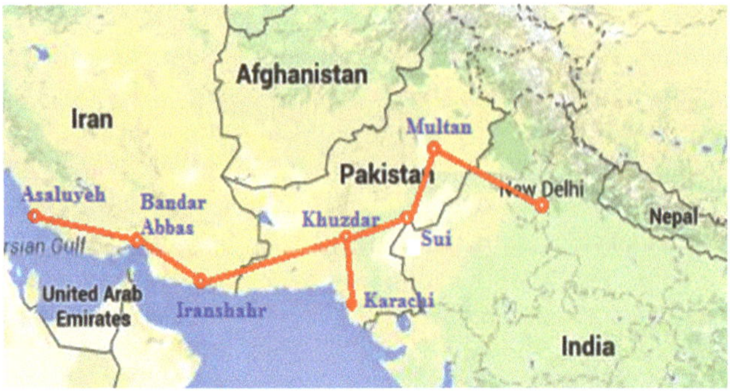

Fig. 2 Iran–Pakistan–India pipeline route

was expected to be extended to 55 BCM (1.9 TCF) later on with a diameter of 56 in (1400 mm) (Iran: No security snag in pipeline deal 2011).

Along with all the serious concerns of security with this pipeline project, another main reason of abandonment was US sanction on Iran for nuclear arm activities. But now with an agreement between Iran and six major powers (USA, UK, France, China, Russia and Germany) to limit Iran's nuclear arm capacity has restarted discussions to find all possible ways to evacuate gas from the field of Iran.

Although there is no more pressure by USA on India for this pipeline project, but serious security issues and geopolitical problems persist without any solution. However, on 16 March 2010 Iran and Pakistan signed an agreement on the pipeline in Ankara which stated that each country must complete its section of pipeline by 2014. In July 2011, Iran announced that it has finished construction of its section, but Pakistan fell behind the target to take delivery of gas in the winter of 2014 (Chaudhary 2015). This scenario of ever-rising hunger for energy and non-resulting pipeline projects has forced India to seriously consider an alternate deep-sea route which will eliminate all the long pertaining concerns.

1.3 Middle East to India Deep-Sea Pipeline (Oman–India Pipeline)

In the last few years, deep-sea gas pipeline technology has matured a lot and has been used globally to reduce on land security concern. Indian cross-border pipeline projects had been delayed in past due to security concern only with its neighbouring countries, so deep-sea pipeline is probably the most promising option to be considered (Ramakrishna 2014).

Oman–India deep-sea pipeline project is wilful to transport 8 tcft of natural gas to India over a period of 20 years. This pipeline is worked out to be about 1300 km long, laid at an extent of 3400 m below the seabed. The cost estimated by India's South Asia Gas Enterprise Pvt. Ltd. (SAGE) for this pipeline project is around

Fig. 3 Oman–India pipeline route

$4–5 billion and is supposed to be executed in near about next five years (Panda 2014). It will connect receiving terminal near Gujarat to Middle East Compression Station near Oman.

Comparatively more feasible and closeness of sources (less than 1500 km) of natural gas in Middle East are the major reasons Indian Government is quite serious about this pipeline project. Along with this another key reason in favour of this project is its landed cost, which will be lesser by $1.5–$2 per million BTU as compared to import through LNG. In future, this pipeline could be linked to other sources of natural gas in Middle East and nearby countries if required (Ramakrishna 2014) (Fig. 3).

1.4 Myanmar–Bangladesh–India (MBI) Pipeline

MBI pipeline was first time propounded in 1997 by Mohona Holdings Limited of Bangladesh. It was a 900-km pipeline which will source gas from Myanmar, pass through Bangladesh and will supply to India. It was meant to deliver around 5 billion cubic metres of gas from the Shwe field in Southern Myanmar.

The three countries reached an agreement in 2005 but the Khaleda Zia-led Bangladesh government withdrew from the project after India refused to accept its three additional conditions. On her return as Prime Minister in 2010, Sheikh Hasina gave her approval to pursue the project. By then, Myanmar was disinterested in the project as it had signed an MOU with China for a 2388-km pipeline from Kyaukphyu in Myanmar to China's Yunnan Province (Express News Service 2015). The Myanmar–China gas pipeline is already under operation from mid-2013 (Mukherjee 2015).

With the renewal of discussions and negotiations with Myanmar government, India is expecting availability of Myanmar's estimated gas reserves (90 TCF) for its

Fig. 4 Myanmar–Bangladesh–India pipeline route

energy needs. Also the project was re-evaluated during recent visit of our Prime Minister Narendra Modi to Bangladesh and some positive outputs are expected out of it (Express News Service 2015) (Fig. 4).

1.5 Russia–India Gas Pipeline

Energy supply from Russia to India, through a pipeline, has been on table since 2005, but until recently, real steps to implement this have not been taken. First definitive declaration at official level about this mega onland pipeline project came after meeting between Vladimir Putin and Manmohan Singh in October 2013 and now Narendra Modi-led government is pushing hard to lay down $40 billion, one of the India's biggest energy projects till date (Sanzhiev 2015).

Russia–India joint working group was established in Russia–India summit 2014 and is studying the possibility of building hydrocarbon pipeline. Along with this, Prime Minister's Office (PMO) and Ministry of Petroleum and Natural Gas (MoPNG) are also preparing blueprint to examine feasibility of this new project and evaluating various technical and economic aspects.

Two routes are being considered for the gas pipeline project. One is from Russia's southern border to India via the Himalayas, most likely through the

Xinjiang Province of China. Another one is from Russia–Astrakhan–Kazakhstan–Uzbekistan and then along the TAPI (Airy 2014).

Initially, the idea is to use the hydrocarbon pipeline for transportation of gas. This project is very important from Russian oil and gas industry perspective as it would not only diversify its sales markets, but also make exports more stable. Russian economy will receive an eminent boost if the Putin administration is able to set up network of oil and gas to major consumers in South East, i.e. India and China.

For India, although it is an expensive project as compared to TAPI and IPI, but it will ensure a reliable supply of oil and gas and further enhance India's energy security. Government of India has already approved the acquisition of a 15% stake in Vankor Field from Rosneft by Indian Oil and Gas Corporation. As soon as Indian Companies will develop their own resource base in Russia, pipeline will be the cheapest and most convenient way to proceed (Sanzhiev 2015).

Whether this pipeline passes through the Himalayas or through TAPI, security will remain a huge area of concern. Though China assures with a tight control over insurgency in its province, there is no guarantee that the pipeline will not be targeted by restive province. Also the topsy–turvy relations between India and China including historic border tensions, war and clash of geostrategic interests have created a wall of distrust for this project. The bitter truth with this project is that China is unavoidable if India wants to avoid dealing with Afghanistan and Pakistan (Mahapatra 2014).

Another route linking the pipeline with TAPI is also prone to some serious risks such as Taliban influence in Afghanistan and terrorism activities in Pakistan. India-specific terrorist groups such as Lashkar-e-Toiba in Pakistan will try to encash every opportunity to sabotage prospects of India–Pakistan Cooperation.

Beside security issues, pipeline construction in such a harsh terrain of Himalaya also presents several barriers. There are glaciers, ravines, mountains and landslide-prone areas which must be taken into account before putting pipeline in ground but India's record in effective implementation of projects in such terrain is not good and encouraging. In that case, Russian technology and experience in pipeline laying is going to play a very helpful role and later on can be extended to Indian hydrocarbon exploration field (Mahapatra 2014).

Russia has the ability to fulfil India's energy demand such as defence and Indo-Russian pipeline diplomacy, which would trigger a major reboot between the two strategic partners and weld their economies more cohesively.

1.6 Other Options

India is also working on the idea of a hydrocarbon pipeline with Kazakhstan that would bring fuel through a five-nation route. Our Ex-External Affairs Minister Salman Khurshid proposed a pipeline, roughly 940 miles long, to ship 3 billion cubic metres gas annually to India. The pipeline would start in the southern Kazakh city of Shymkent, close to the Uzbek border, then run southwards through Uzbekistan. After that it would enter Afghanistan to follow the route of the proposed TAPI gas pipeline (Daly 2013).

2 Conclusion

Looking at the graph of rising energy demand of India, government has to plan their efforts for everlasting solutions to the problem. The sooner the government takes a serious view on these pipeline project proposals, the better it will be in the interest of the country to ensure its energy security. For India, in the last two-to-three decades geopolitics has always replaced geoeconomics but now accelerated growth has forced India to synchronize its energy security issues with its security and foreign policy.

References

Airy A (2014) Hindustan times. Retrieved from India eyes $40-bn pipeline from Russia to import gas: http://www.hindustantimes.com/business/india-eyes-40-bn-pipeline-from-russia-to-import-gas/story-EzCMa3tghOKLjFSNO5V3qM.html

Alexander's Gas and Oil Connections (2000) Retrieved from http://www.gasandoil.com/news/south_east_asia/7cf6ad4d5de7917f89bc4418c83f2455. Accessed 9 March 2000

Chaudhary DR (2015). India poised to benefit as equal stakeholder in TAPI gas pipeline. Retrieved from The Economic Times: http://economictimes.indiatimes.com/industry/energy/oil-gas/India-poised-to-benefit-as-equal-stakeholder-in-tapi-gas-pipeline/articleshow/50020924.cms

Daly J (2013) Proposed Kazakh-Indian pipeline a long shot at best. Retrieved from www.OilPrice.com, http://oilprice.com/Geopolitics/Asia/Proposed-Kazakh-Indian-Pipeline-a-Long-Shot-at-Best.html

Express News Service (2015) Official talks on Myanmar-India-Bangla pipeline to start soon. Retrieved from Indian Express: http://indianexpress.com/article/business/business-others/official-talks-on-myanmar-india-bangla-pipeline-to-start-soon/

Industry Group (2013) "Vision 2030" natural gas infrastructure in India. Petroleum and Natural Gas Regulatory Board

Iran to India Natural Gas Pipeline (Ted Case Studies 2000). http://www1.american.edu/ted/iranpipeline.htm

Iran: No Security Snag in Pipeline Deal (2011) Retrieved from Tehran Times: http://www.tehrantimes.com/index.php/component/content/article/4304

Iran-Pakistan Gas pipeline (n.d.) Retrieved from Wikipedia: http://en.wikipedia.org/wiki/Iran%E2%80%93Pakistan_gas_pipeline

Kasturi CS (2015) Twin push for gas pipeline. Retrieved from The Telegraph: http://www.telegraphindia.com/1150704/jsp/nation/story_29513.jsp#.VnBN9krRKko

Mahapatra DA (2014) Russia-India gas pipeline faces several constraints. Russia and India report

Mukherjee S (2015) Petroleum geosciences: Indian contexts. Springer

Panda A (2014) India, Iran and Oman open talks on deep sea gas pipeline. Retrieved from The Diplomat: http://thediplomat.com/2014/03/india-iran-and-oman-open-talks-on-deep-sea-gas-pipeline/

Ramakrishna HS (2014) Oman-India gas pipeline a most promising option. Retrieved from The Economic Times: http://ecnomictimes.indiatimes.com/industry/energy/oil-gas/oman-india-gas-pipeline-a-most-promising-option/articleshow/45201546.cms
Reyaz M (2015) TAPI pipeline: a new silk route or a pipe dream. Retrieved from Aljazeera: http://www.aljazeera.com/news/2015/12/tapi-pipeline-silk-route-pipe-dream-151215211343976.html
Sanzhiev A (2015) Gas pipeline to India being considered. Russia and India report
Tanchum M (2015) A fillip of the TAPI pipeline. Retrieved from The Diplomat: http://thediplomat.com/2015/12/a-fillip-for-the-tapi-pipeline/
Turkmenistan launches $10bn Gas Pipeline to South Asia (2015) Retrieved from Aljazeera: http://www.aljazeera.com/news/2015/12/turkmenistan–launches-10bn-gas-pipeline-south-asia-151213165943882.html
Turkmenistan-Afghanistan-Pakistan-India Gas Pipeline (TAPI) (2015) Retrieved from Inter State Gas Systems (PVT) LTD.: http://www.isgs.pk/projects/turkmenistan-afghanistan-pakistan-india-gas-pipeline-tapi-4

Issues and Challenges of Gas Contracts

Manas Das

Abstract Contracts are agreements entered into by two parties with the intention of creating a legal obligation. As the gas sector in India has evolved from pre-NELP regime to a market open to international LNG contracts, the domestic gas contracts also evolved with newer dimensions, newer clauses, and newer obligations or bindings. Gas contracts are entered into in various forms with different objectives across the gas value chain. In the upstream sector, production sharing contract (PSC) is signed between Government of India and the Producer Consortium. Joint operating agreement (JOA) is signed among the consortium members. Gas producer also signs gas supply contracts (GSC)/gas supply agreement (GSA)/gas sales and purchase agreement/contract (GSPA) with the transporter/trader in the midstream segment. In the downstream, the transporter/trader signs agreements with the customers/consumers of gas. The supply gets governed by the GSC, and the transmission is governed by gas transmission agreement (GTA). In some cases, the transmission and the supply gets clubbed in a single contract termed as GSTA or gas supply and transmission contracts. The downstream contracts may be named differently but structurally they still remain same with minor modifications. This chapter brings in illustration on different types of midstream and downstream contracts, the process description of execution of contracts, how the contracts have evolved, and futuristic changes expected in these contracts.

Keywords Gas sales agreement · Gas sales and purchase agreement · Gas supply and transmission agreement · Gas transmission agreement · Capacity tranche · Gas sales and purchase notice

M. Das (✉)
Strategic Planning, Kuwait Petroleum Corporation, Kuwait City, Kuwait
e-mail: manasdas.iift@gmail.com

S.K. Kar and A. Gupta (eds.), *Natural Gas Markets in India*,
DOI 10.1007/978-981-10-3118-2_4

1 Introduction

Major developments that have occurred in the Indian gas industry in last two decades would include the introduction of NELP regime, import of liquefied natural gas (LNG), KG D6 gas find, formulation of Petroleum and Natural Gas Regulatory Board (PNGRB),[1] and Governmental control on domestic gas prices.

The evolution of gas sales contracts has accordingly evolved around these developments in terms of content and in terms of nomenclature as well. From the period of gas supply by couple of national companies like ONGC and Oil India Limited (OIL), today gas is sold in India not only from domestic sources but from sources in abroad in the form of regasified LNG (RLNG). In fact, last two decades have been very significant as far as the gas industry in India is concerned. There have been changes in the exploration aspects from pre-New Exploration Licensing Policy (NELP) regime to bidding process under NELP,[2] from unregulated transportation aspects to regulated services under PNGRB, from limited domestic gas sources to vast RLNG shipments from far flung countries. These all aspects have influenced the contractual arrangements for sale and purchase of gas from few pages of sale provisions to complex and voluminous contractual details.

In the pre-NELP regime, gas allocations were done by the Gas Linkage Committee (GLC)[3] pursuant to which contractual arrangement was agreed between the seller and buyer (mostly fertilizer and power customers). During this period, there was no elaborate gas sales and purchase framework in the upstream. Gas supply between ONGC (as seller) and GAIL (buyer) was pursuant to an MOU executed in early 1990 (Business Standard 2006). There was similar sort of arrangement for the gas supplies from OIL. The downstream contracts with the designated customers were accordingly in place wherein gas price applicability was as per the various gas pricing orders and allocation in the contract—daily contract quantity was in volumetric reference and would be based on the actual plant requirement, though actual supply of gas would be substantially lesser. These contracts provided for quarterly commitments from the buyers—referred to as MGO (minimum guaranteed offtake). These contracts only provided commitments from buyers for offtake, without any supply commitment from seller side. This is still prevalent in APM contracts (those contracts where allocation is made by GLC/MoP&NG), and seller (GAIL) could not extend flexible terms, as same would not be available to GAIL (as buyer) from seller (ONGC).

In due course, MOU between GAIL and ONGC was converted into a GSA during 2007; however, downstream contracts between GAIL and gas users did not undergo any major change. Significant change in the upstream contract (between

---

[1] http://ppac.org.in/WriteReadData/userfiles/file/THEPETROLEUM2006.pdf accessed on 12 July, 2016.

[2] Evolution of Indian Oil and Gas Industry, http://www.dghindia.org/index.php/page?pageId=56&name=INDIA'S%20E%20and%20P%20REGIME.

[3] Page 204, Handbook of Natural Gas Technology and Business by Parag Diwan.

ONGC and GAIL) was the billing cycle from monthly invoicing to fortnightly invoicing. Gas pricing has been in accordance with gas pricing circulars issued by MOP&NG from time to time.

On the other hand, gas supplies received from joint ventures were based on certain provisions as incorporated in the production sharing contracts itself. Here, GAIL was purchasing gas from consortium (ONGC, RIL and BG-earlier Enron), and the supply arrangement was in the form of interim sales and purchase agreement, wherein GAIL as Government of India nominee would receive gas as per the provisions of the PSC. Periodic extensions were agreed between GAIL and consortium till March 2005. Ravva and Ravva satellite were the only detailed contracts in late 1990 between GAIL and Ravva consortium.

The major change in the gas sales contracts can be attributed to commencement of RLNG in the country. The agreements of GAIL, IOCL, and BPCL with PLL are very elaborate contract. This is due to the fact that PLL in turn has a detailed contract with RasGas.[4] These types of contracts had to essentially be in detail as there was need to pass on the risks as arising out of ship voyages, price linkages, destination restriction clauses, etc. In the downstream, that is, between GAIL and RLNG buyers, there was emergence of provision that GAIL would pay liquidated damages for short supply under a particular scenario. Also contracts provided for greater magnitude of “take-or-pay” percentage (more than 90%). An outstanding feature of RLNG contracts has been the daily contractual quantity (DCQ) in terms of energy terms rather than volumetric references. This paved a way subsequently, as gas contracts by and large provide for gas price on energy terms.

In line with regulatory developments, provisions were also made in GSTA such that, in the event of unbundling of sale and transmission of gas, necessary changes may be incorporated in aspects related to gas transportation. Similarly on account of any directions from Govt. of India or Regulatory Authorities, etc., the seller can assign the GSTA partly/fully to any of its associates/subsidiary to perform its obligation under the contract.

The objective of this chapter is to provide a detailed understanding of the contractual developments and detailed operability on contracts in midstream and downstream of the gas value chain in India.

2 Present Practices of Gas Contracts in the Midstream/Downstream

Gas agreements/contracts are generally on back-to-back basis. This means a gas trader passes on commercial terms and risks it has in it upstream agreement, to the downstream customer. Over the years, import of gas in the form RLNG has brought some sophistication in the gas contracts, as the contracts for domestic gas would not

[4]Rasgas website, http://rasgas.com/Customers/PetronetLNGLimited.html.

involve elaborate gas nomination process. This is simply because the value chain in the case of RLNG is longer than that involved in domestic gas supplies. The process of imported gas has increased the number of sellers, and regulatory framework has brought into few types of contracts for gas sale and purchase.

Few gas agreements/contracts that are in vogue include the following:

Gas sales contract (GSC), gas sales agreement (GSA), gas sales and purchase agreement/contract all serve same purpose. The major difference, if any, that can occur can be pursuant to fact that transportation part be also included in the said agreements/contracts. If transportation of gas is included, then such a type of contract/agreement is referred to as bundled contract/agreement. Given the regulatory development in terms of unbundling of transmission and marketing activities (PNGRB 2012), a service contract (transportation) has to be separate contract—at arm's length. A gas seller, if a pipeline owner, has to provide for separate gas contract like GSA, gas sales and purchase agreement (GSPA),[5] and a separate gas transportation agreement (GTA).

In many cases, before GSA, GSPA, etc. is executed, parties enter into an memorandum of understanding (MoU), or heads of agreement (HoA), to agree on cordial basis about their intent to agree into further commercial contracts. These MoU or HoA are not legally enforceable, unless, specifically required to be so for executing parties. These types of contracts are to have sort of first movers' advantage and get into rapport building exercise and give both parties time to understand their deliverables and also provide confidentiality frame work for the information shared.

At times, parties though in agreement as seller and buyer relation are unable to execute a detailed contract for lack of convergence or due to paucity of time. In such cases, parties are found to have executed term sheets, which broadly emphasize on certain commercial factors such as price, tenure, take-or-pay, billing, and payment. These term sheets briefly mention legal framework and provide very little details about gas measurement and nomination patterns. These term sheets provided for a time frame within which a detailed GSA is to be executed. These types of contracts work, when the parties are either already related to each other in gas business or have some directives from a government department for gas supplies (in case of certain production sharing contracts—PSCs). These term sheets are also handy when tenure of supply is for a very limited period, say about one or two years.

Emergence of RLNG has also provided for gas supplies to be available on spot basis (about 1 month tenure or more), mid-term agreements (2- to 5-year duration), or long-term agreements. Mid-term or long-term contracts are essentially having same framework, except for difference in tenure.

---

[5]Gas Sales Purchase Agreement (GSPA) GeoEnpro Petroleum Limited; http://geoenpro.com/httpdocs/nio/docs/4%20%20GeoEnpro_Model%20Gas%20Sales%20and%20Purchase%20Agreement%20(GSPA).pdf.

Spot agreements, since being for shorter period and given the recurrence of supply based on convergence of the parties, are dealt under Spot Gas Sales Agreements. These types of agreements are detailed like mid-term agreement/long-term agreement; however, supply, which is for short period, is covered by means of Gas Sales and Purchase Notice (GSPN). Under this type of agreement, all details about payment term, dispute resolution, measurement and calibration, "take-or-pay," etc. are provided in the main contract, except gas price, duration of supply, quantity, and start and end dates are provided in the GSPN. As such, when supplies have to be maintained on spot basis, only GSPN, as per the agreed format, is executed as other terms are already agreed under spot agreement. Here, spot agreement is like a dormant document which only becomes active with the signing of GSPN for the period mentioned in GSPN.

3 Gas Sales Contract (GSC)

Gas sales contracts (GSC) are signed between the gas producers as sellers and entities involved in transporting/marketing of gas as buyers. Producers who are parties to production sharing contract (PSC)[6] signed with the Government of India also execute joint operating agreement (JOA)[7] among themselves for gas production from a particular gas-producing field.

If there is more than one producer in the field, they may agree to sell their respective participating interest share of sales gas to the buyer. The seller agrees to sell a daily contract quantity (DCQ) at the delivery point of the buyer at an agreed price. Gas sales quantity is generally measured in terms of MMBtu/day. Agreed specification of sales gas gets mentioned in the GSC. Gas specification may cover net calorific value, pressure and temperature at delivery point, hydrocarbon dew point, etc. Buyer makes its nominations based on anticipated ability to take delivery of gas from the seller at a uniform scheduled flow rate. Sometimes a range of 5% variation is allowed in the scheduled flow rate. There may also be some additional gas available from the existing field beyond the contractual obligation, which may be offered to the buyer on reasonable endeavor basis.

Majority of the gas sales contracts have "take-or-pay" clause. In some contracts, the buyer is supposed to take delivery of at least 90% of the contracted volume on a quarterly basis. Whether the buyer takes the delivery or unable due to some reasons, it has to pay the seller up to 90% of the value of sales gas contract volume in a quarter. Contract quantity gets reduced by aggregate shortfall quantity due to any force majeure or planned maintenance/shutdown, or unaccepted off-spec gas supply.

---

[6]Model Production Sharing Contract, DGH Website; http://dghindia.gov.in//assets/downloads/56ce987441809MODEL_PRODUCTION_SHARING_CONTRACT(MPSC).pdf.

[7]Joint Operating Agreements, Oil & Gas Financial Journal; http://www.ogfj.com/articles/print/volume-11/issue-10/features/joint-operating-agreements.html.

Force majeure generally includes natural calamities such as earthquakes, fires, or phenomena beyond control such as riots, war, and invasions, etc.

The contract may also detail out the extent of planned maintenance. Generally, 30-day advance notice is given by either buyer or seller, and efforts are made to reduce the effective downtime of the system.

Nomination Procedure by Buyer:

Buyer informs the seller that it wishes to nominate quantity of gas in accordance with the contract, for each day of the following week, no later than a stipulated time in the preceding week. The nominations are on energy basis (in MMBTUs). The conversion of energy to volumes is done on the weighted average net heating value of the previous fortnight at the delivery point for the buyer. If the buyer fails to inform the seller, then nominated quantity of gas becomes equal to the DCQ (in MMBTUs) in the relevant week.

In the event, due to unforeseen circumstances, any buyer requires a change in nomination for a particular day in the week, the buyer informs the seller of the changed nomination quantity (in MMBTUs) early in the previous day, which the sellers will schedule on a reasonable endeavor basis.

Scheduling Procedure:

After the nomination is received from the buyer, the seller within a stipulated time frame informs schedule of gas delivery to buyer for the following week. Seller's schedule of gas delivery may also include additional gas. However, inability to deliver scheduled additional gas will not constitute as any shortfall.

Allocation and Invoicing procedure:

Meter reading at delivery points forms the basis of allocation to the buyer and invoicing. The seller issues an allocation statement to the buyer in respect of gas delivered in each fortnight, specifying the allocated/measured quantities delivered at the delivery point.

Seller gives the buyer a fortnightly/monthly/quarterly payment statement. Buyer is supposed to make the payments within a stipulated time frame, say 7 days of receipt of the payment statements. In case the payments are not received within prescribed time limit, seller may suspend or stop the gas supply. The supply can be resumed after curing of the event. If the payment is not made within acceptable time limit (say 90 days), the contract can be terminated.

4 Gas Supply and Transmission Agreement (GSTA)

In the next part of the value chain, **Gas Supply and Transmission Agreement (GSTA)** is signed between the buyer and the seller, and this provides for the terms and conditions upon which the buyer purchases and seller sells gas. By this agreement, the seller agrees to procure, sell, and transport gas from its identified sources and deliver gas to the buyer at their agreed delivery point, and the buyer agrees to take and pay or pay for if not taken such gas, in accordance with the agreement. GSTA generally mentions delivery point, delivery pressure and temperature, deliverable volume and gas composition (minimum methane content, maximum non-combustible content in terms of CO_2 and N_2, minimum net heat value, maximum sulfur content, maximum moisture content, etc.).

As per the GSTA, buyer makes its nominations on its anticipated ability to take delivery of nominated daily quantities of sales gas and takes delivery of sales gas from the seller at the scheduled daily quantity at a uniform scheduled flow rate.

Seller installs and maintains at its own cost the piping control and regulation and metering equipment in the gas metering station and all other accessories inclusive of condensate collection and disposal system. The equipment installed by the seller remains the property of the seller, and it retains the rights to remove such equipment on expiry of the agreement. The buyer provides the utilities essentially required for installation, operation, and maintenance of metering station and allied equipment required for the supply of gas to buyer on payment of mutually agreed charges. The seller obtains all statutory approvals of the installation.[8]

Seller furnishes advance information about the projected annual plan profile of gas delivery as well as plan for maintenance. The same is revised on a quarterly basis.

During the contract period, if there is any additional gas available to the seller, the seller uses reasonable endeavors to sell and deliver such additional gas to the buyer and the buyer uses reasonable endeavors to purchase and receive the same, at the delivery point at the agreed sales gas price.

Seller and buyer's right for planned maintenance is also captured in the GSTA. Extent of fall in volume delivery or offtake can be partial or complete. As an example, seller may keep the right for planned maintenance, during which up to 10 days the DCQ may be reduced by 50%. Similarly, buyer will also keep the provision for reduction in DCQ for planned maintenance. In addition to these planned maintenance periods, the seller will also be entitled to shutdown in respect of producer's maintenance at producing facilities.

Any non-performance or delay in performance by either seller or buyer of any of its obligations may be excused if this is caused by force majeure events.

The buyer pays the seller for each MMBTU of gas on net heat value basis delivered and taken at the delivery point at an agreed sales gas delivery price.

[8]Statutory approvals may include the following:

1. Approval from PESO/CCE (Chief Controller of Explosives).
2. Approval from BIS (Bureau of Indian Standards) for all flameproof instruments manufactured in India.
3. Certificate from statutory body for hazardous area approval for all electrical items mounted on the skid etc.

The sales gas price is calculated excluding any statutory taxes and levies. The buyer reimburses seller any tax (inclusive of sales tax and/or VAT) payable by seller to the government, or local authority, on account of the sale and/or transfer of title of sales gas to the buyer at the delivery point. Royalty payments are borne by the producers. All costs downstream of the delivery point are borne by the buyer.[9]

In addition to the sales gas price, the buyer also pays gas transmission charges, marketing margin and service charges, as applicable, to the seller.

The seller gives the buyer a fortnightly payment statement for sales gas supplied. Following the end of a quarter, the seller also gives the buyer a quarterly payment statement for sales gas supplied during the quarter. These days most of the invoicing is done electronically.

The seller raises the invoices/supplementary debit note for each fortnight/quarter, as the case may be, and the buyer agrees to pay the invoices so raised in full within a stipulated time frame (say 3 days) of presentation of the said invoices.

At the signing time of the GSTA, the buyer also submits a Letter of Credit (LC) of value generally for 16 days' (or a fortnight) supply of gas at maximum contracted quantity (along with transmission charges, marketing margin, and service charges as the case may be) with coverage of value of gas for a period of three fortnights. The validity of LC is normally kept for a period of one year and needs to be renewed before expiry.

If for any reason the payment on any invoice raised by the seller is delayed or any disallowance is made from the invoice, the seller may present the invoice/supplementary debit note(s) for full amount or the amount not paid, as the case may be, to the bank against the LC and draw the amount. The LC should automatically be reinstated after invocation.

In case payment is made after the due date, interest is levied on the delayed payments.

Nomination procedure by buyer:

The seller informs the buyer about the deliverable volume and any additional gas (in MMBTUs) for each day of the following week by a specified timeline in the preceding week.

Based on the deliverable volume, the buyer informs the seller that it wishes to nominate the amount of desired offtake in the following week no later than an agreed timing in the preceding week. The nominations will be on energy basis (in

[9]To maintain safe, uninterrupted/continuous gas supply to the buyer, the seller is required to install and operate some equipment at the buyer's premises like Pressure Regulatory System (PRS), Metering Skid, Telemetry and Telecom system etc. For all such facilities the buyer is required to pay all charges mutually agreed between the buyer and the seller. In case the Buyer requests any advisory/consultancy services related to safety audit, energy audit, energy calibration, instrument calibration, maintenance of buyer's pipeline facility downstream of the delivery point, such value added services may also sometimes are provided at a cost.

MMBTUs), and the conversion of energy to volumes will be done at a notional heat value (e.g. 9000 kcal/SCM).

If the buyer fails to nominate, then buyer will be deemed to have nominated a quantity of gas equal to its deliverable volume (in MMBTUs) in respect of each day during the relevant week.

In the event of any unforeseen circumstances, if the buyer requires to change the nomination for a particular day in the week, it will inform the seller of the changed nomination quantity (in MMBTUs), generally one day in advance, which the seller will schedule on a reasonable endeavor basis.

Similarly, if on any day, if the buyer wishes to cease receiving delivery of gas either due to an emergency or force majeure event, it will immediately inform the seller with all relevant information and the seller will use reasonable endeavors to accommodate such requests from the buyer.

Scheduling procedure by seller:

Based on the nomination from the buyer, the seller informs the schedule of gas delivery to the buyer at the delivery point for each day of the following week by a specified timeline in the preceding week. Seller's schedule will include quantity, including any additional gas.

In the event of a change in daily nomination by the buyer for a particular day in the week, the seller will notify the buyer the revised schedule of delivery on reasonable endeavour basis on preceding day.

Allocation procedure:

Each day the sales gas delivered and sold to the buyer at the delivery point under the GSTA gets calculated on the basis of the gas quantity measured (in MMBTU) at the delivery point.

A process flow diagram (Fig. 1) of interaction between the seller and the buyer is included. Depending on whether it is domestic gas or regasified LNG (RLNG), the initial tie-up may be different but otherwise the balance part of the process more or less remains same.

5 Gas Transmission Agreement (GTA) and Capacity Tranche (CT)

Gas transmission agreement (GTA) is executed between the transporter and the shipper for the transmission of gas. The shipper may be having its existing facilities or may be creating facilities, which will require gas as fuel/feedstock or has its plans to sell gas to other consumers.

Generally, the transporter conducts demand estimation of customers alongside its pipeline network on routine basis and identifies potential customer/shipper and their gas requirement.

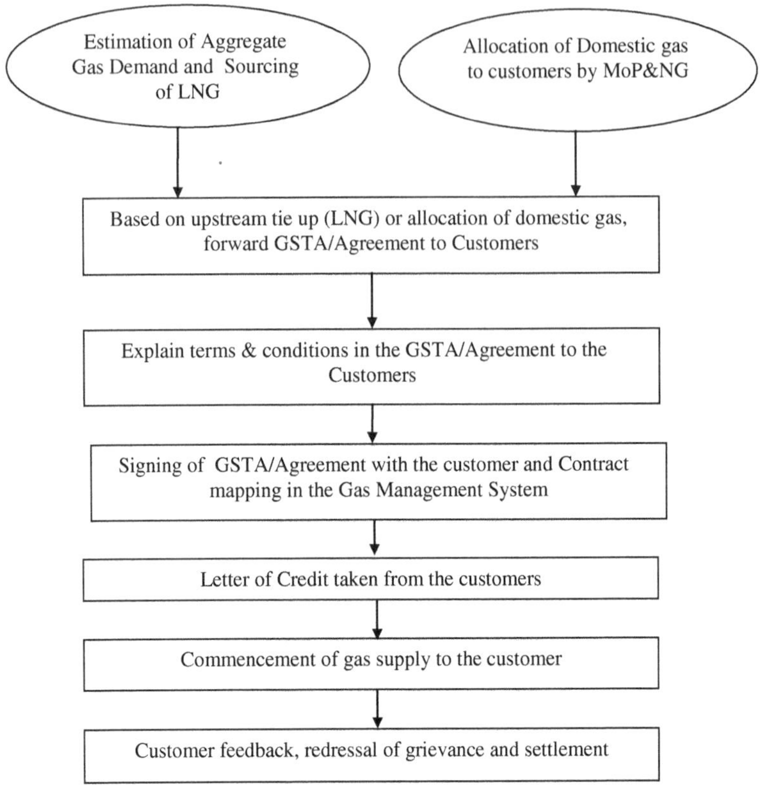

Fig. 1 Process flow diagram—RLNG or domestic gas supply. *Source* Prepared by the author

Shipper may also issue request to the transporter to provide necessary arrangements for receipt of gas at the delivery point(s), transmission of gas, and delivery of certain quantities of gas at the redelivery point(s). This comes as a request to book capacity in the transporter's pipeline. The shipper agrees to pay transmission charges and other charges to the transporter.

If transmission capacity is available, the shipper and the transporter enter into one or more agreements in relation to one or more redelivery points, by which the shipper will contract for specified capacity for a specified duration for the transportation of the gas through the transporter's facilities, and the transporter will provide gas transmission services for such duration. Each transaction is governed by the terms of the GTA and by other terms that may be agreed in writing between parties confirming such transaction, e.g., capacity tranche (CT). A "capacity tranche" is a quantity of transmission capacity reserved by transporter for a particular transmission purpose of shipper.

In the event, at any time, the shipper wishes to enter into a CT Agreement, the shipper submits a proposal to the transporter to provide gas transmission services to the shipper by entering into a ship-or-pay CT agreement or a reasonable endeavor (RE) CT agreement, as the case may be. The "Ship-or-Pay CT agreement" means an arrangement under which the transporter reserves transmission capacity to provide gas transmission services to shipper on a firm basis. The "RE CT agreement" means an arrangement in which the transporter uses reasonable endeavors to make transmission capacity available to provide gas transmission services for the shipper. After this, the transporter responds to the proposal (within given time frame, say 3 days) confirming that such spare capacity exists in the transmission pipeline to satisfy the shipper's request and specifies the transmission charges and other terms and conditions at which the services will be made available.

The transporter may also respond saying that spare capacity is not available to satisfy the request or ask for more time to respond as a detailed technical study may be required or accepting that it is not feasible to provide access. If the transporter confirms to the shipper that they accept the proposal, both the parties execute a ship-or-pay CT Agreement or RE CT agreement (as applicable) within a reasonable time frame. Each CT agreement is identified by a unique name, delivery point, and redelivery point.

The transporter agrees that from the relevant CT start date and in accordance with the terms and conditions of the GTA it will transmit gas delivered by shipper, up to the maximum delivery quantity (MDQ), from the delivery point(s) till the redelivery point(s).

In case of ship-or-pay CT agreement, the shipper pays for the transmission charges on the aggregate allocated quantity, whereas in RE CT agreement, the shipper is liable to pay transmission charges only for the gas offtaken by shipper at the redelivery point.

Further, the shipper provides land and other utilities (free of cost) to transporter for installation, operation, maintenance, and modification, if required, of transporter's facilities at redelivery points.

The title of the gas always remains with the shipper. All risks associated with the gas from the time it is delivered by the shipper to the transporter at the delivery point until redelivered by the transporter at the redelivery point, is borne by the transporter.

Execution of GSA by the shipper, for relevant volume of gas, is generally a condition precedent for effectiveness of shipper's obligations in GTA. Similarly, approval of any new pipeline by transporter, required for the proposed gas transmission, may also be a condition precedent for effectiveness of transporter's obligation in GTA.

At the time of signing of GTA or CT agreement, the shipper makes a security deposit to the transporter to ensure timely and appropriate performance of GTA. The shipper also arranges to submit a bank guarantee at the time of signing of GTA or CT agreement. The transporter pays interest charges to the shipper on the amount of security deposit up to the gas commencement date/start date, as per the provisions of relevant CT agreement. If the shipper does not commence drawl of gas by start date, the transporter has the rights to terminate the agreement and forfeit the security deposit as well as encash the amount of bank guarantee. At the time of starting gas drawl, if the shipper furnishes LC of requisite value, the transporter should refund the security deposit and also return the bank guarantee (within a specified period of gas commencement).

In general, transmission rates are subject to change as per the directives from governmental instrumentality. Transmission rates also include provisions for transmission losses (TL), line pack (LP), and system use gas (SUG) relevant to the pipeline as per the standard operating practices. Generally, the transmission charges are exclusive of taxes. Shipper reimburses the transporter the amount of any taxes leviable for providing the transmission services.

In respect of a month where there is a shortfall in the redelivery of gas, the transporter is obligated to pay "liquidated damages" or LD for the shortfall quantity. The LD may be equal to the product of the shortfall quantity in the relevant month and applicable transmission rates of the relevant CT Agreement. In case of any shortfall in redelivery, the transporter gives equivalent credit to the shipper for LD in the subsequent fortnightly invoice.

As usual practice, the transporter delivers fortnightly invoice to the shipper and within three/four days of receipt of such invoice shipper pays the due amounts to the transporter.

As support for the shipper's payment obligations under the GTA, the shipper submits an unconditional, revolving, standby irrevocable Letter of Credit (LC) from an acceptable bank to the transporter. The amount of the LC is generally equivalent to transmission charges of 16 days (or a fortnight) aggregate MDQ. Further the shipper's LC provides coverage of the value of three fortnight transmission charges in a contract year.

The process followed by the transporter is depicted in Fig. 2.

6 Gas Sales and Purchase Notice (GSPN)

Gas Sales and Purchase Notice (GSPN) is a brief agreement between the seller and the buyer and is issued pursuant to terms and conditions of gas sales agreement (GSA) earlier executed between them. The execution of GSPN by the parties forms basis by which the seller agrees to sell and deliver gas to the buyer and buyer agrees to purchase and take delivery of gas. It covers the total contract quantity, daily contract quantity, contract price with detailed break-up of the transmission charges, and marketing margin and supply period with start date, etc. The delivery of gas generally is on reasonable endeavour (RE) basis.

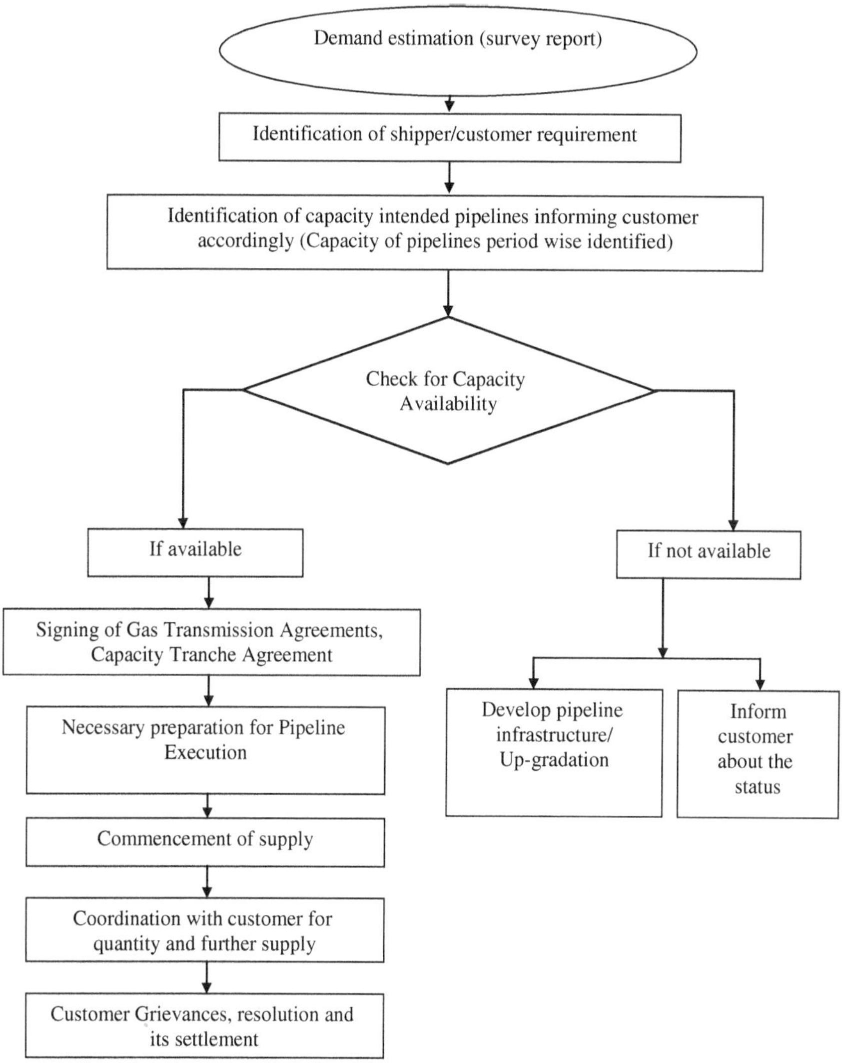

Fig. 2 Process flow diagram—gas transmission. *Source* Prepared by the author

As spot RLNG becomes available, communication is sent to the prospective customers to bid online. Online bidding happens for specified time duration (say one hour). The customers bid the gas quantity and price as per their requirement. Allocation is confirmed based on the bids received and availability of spot RLNG. A process flow diagram is presented in Fig. 3.

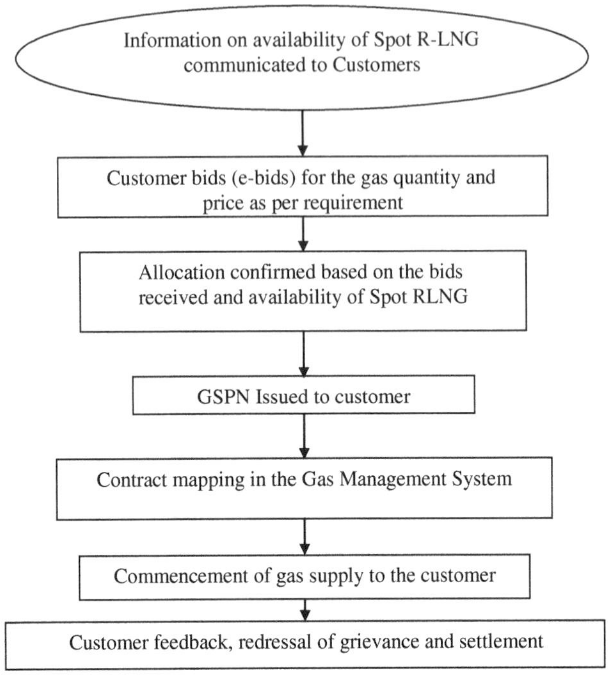

Fig. 3 Process flow diagram—GSPN on spot RLNG. *Source* Developed by the author

7 Issues and Challenges with Midstream/Downstream Gas Contracts

The issues and challenges of the gas sales contracts come from the very nature of the business. Gas industry by its nature is capital intensive and operations work in tandem in upstream, midstream and downstream. By and large, gas storage has been a constraint, and at least in the context of Indian industry, storage facilities are by means of regasification tanks and the line pack capacity of gas pipeline network. As is possible, in many business operations, a warehouse/storage facility can be used to leverage a distress situation; however, in present Indian gas industry, storage is an unworkable proposition.

Gas availability from the mother earth is in the form of associated natural gas (ANG) or non-associated natural gas (NANG). This association is in terms of oil coming along with the gas. When oil is produced, the gas continuity is inherent, and at the same time, when oil is not there, a stoppage of supply in downstream with consequent capping of the well may destroy the gas reservoir in certain conditions. As is obvious, one of the challenges in gas contracts is to maintain continuity of supply, and therefore operations and obligation under the contracts have shaped up accordingly.

A challenge, therefore, at the outset, is to have clarity of rights and duties, obligations and options, commitments and discretions. Seller on the one hand tries to include the risks and provisions of its upstream contract, and buyer on the other hand attempts to harmonize operational parameters and market adverse conditions. It would not be out of place here to mention that gas market so far has been that of seller's market, although, with passage of time, certain flexibilities have come in system. An example would be that no penalty on failure of buyer would be imposed, in case, the gas seller is not to pay any penalty to its upstream supplier. This flexibility has developed in a scenario that when government has made gas allocations for subsidized gas and when there is a dearth of availability of such gas, a failure of particular buyer, in fact is no commercial disadvantage to seller, as said gas is available in network and taken by other designated customer.

At the beginning of the discussions of gas sales contracts, issue is to precisely arrive at the daily gas consumption of the buyer (called daily contract quantity-DCQ). This quantity confirmation becomes the basis of all obligations and calculations under the contract. Few years back, DCQ used to be referred in volumetric references; however, in recent times, gas prices are on energy basis (in dollar terms) which has consequently resulted DCQ also in energy terms.

Gas contracts, whether for domestic gas or imported gas (RLNG), the basic framework remains same. There shall be a DCQ, magnitude of commitment from buyer (take-or-pay clause) for gas offtake, seller's shortfall (commitment of supply by seller), applicable gas price, payment mechanism, dispute redressal mechanism, operational clauses-measurement, gas specs, etc.

So far, there has been a tendency in gas sales contracts to provide terms and conditions on back-to-back basis. This means, a gas seller, who is buyer in upstream, passes the associated risks, terms, and conditions, to its buyer in downstream. This has been so, as supplies to gas buyers in downstream have not worked on pooling mechanism or gas majors have not acted as aggregators in strict senses.

A challenge that has emerged of late is the conversion of small liquid fuel customers, who work on 5–6 days a week basis. For these type of customers, there is a need to cater to some extent, on tailor-made conditions basis, as otherwise, a fit for all agreement, is bound to put them in distress when it comes to small buyers/customers annual off take commitment. Of late, many gas selling entities have designed contracts to suit small customers, wherein the commitment level for annual offtake has been far less, in comparison with large-volume-based customers who work on 24 × 7 basis. Also, there is a challenge in terms of financial securities that has to be in place for sellers' payments. Since the contracts provide for 'take or pay' commitments on annual basis and contracts provide for irrevocable letter of credit (LC), customers and banks have been unable to come on board in many cases. Banks read the default by customers (in extreme case situation) for the entire year and count liability accordingly, which makes the LC with huge outflows. This remains a challenge for many customers. Recently, some of gas marketing companies have provided for a combination of few days advance and a part security by means of LC for customer comfort.

An issue, before the parties as on date, is that of linkage of gas price under the long-term/mid-term contracts. By and large, RLNG gas contracts have been linked to Brent Crude, and due to fall of Crude, these contracts have not aligned to changes as formulas are designed to protect production costs. On other hand, gas availability, on spot basis, has been very low, as they are free from long-term gas price/formula and forced customers to sustain either on spot basis or run their units on alternate fuels. Solutions to this is not easy, as gas trader passes on formula to downstream customer and since contracts do not incorporate economic hardship clause; this has resulted in take-or-pay situation. While, seller may be made good for losses for a particular year by LC in place, however, further supplies and security is question mark. It would appear that given the present market scenario, wherein shale gas has dominated economic climate, for being very cheap source of energy in USA and making the country from net importer to net exporter, a mid-term contract or price reopener clause may be built in for avoiding economic distress. An alternate method is to have portfolio of gas supplies and arrive at a pooled price by incorporating shale gas supplies, as shale supplies to few upstream buyers have been on the basis of linkage to Henry Hub gas price (not linked to crude).

8 Discussion/Conclusion

Gas sales agreements or contracts, whatever the nomenclature, are to inherently have a balance in implications. The purpose of any contract is to spell out the rights and obligations of the parties in normal business course and also to have clarity of obligations, in case of defaults. It would be apt to say that a good gas sales contract starts with the aptitude of the parties negotiating and executing it. Attempt of a party to have a contract one-sided is akin to raise distress to other party and consequently results into discomforts of the other party. Therefore, a major challenge, before the parties executing the gas sales contract, is to create a win–win situation or to have optimum balance, as otherwise, a long-term relation would be difficult to be maintained. It may not be exactly possible to have reciprocity of commercial rights as buyer and seller are operating in different environments. A force majeure in upstream is not comparable to a force majeure in downstream. However, there need to be sincere efforts to recognize each other's environments as far as they are linked to gas operations. A challenge, is therefore, to make the agreement as explicatory as possible, to avoid confusions and reduce risks at either side. Also, challenge is there to create transparency in working and benchmarks and linkages need to be in place, such that contract is working at its own terms, rather on interpretations and workings on either sides.

The major issues that come from gas prices are not actually in the hands of traders. Gas trader/aggregator only passes on what it gets from upstream. Therefore as long as domestic gas supplies are in deficit, dependence on RLNG would essentially have its inherent risks in terms of price, at least in next couple of years, by the time, shale gas situation would be pegged in equilibrium.

References

Business Standard (2006) GAIL, ONGC sign gas sale agreement. 8 July 2006

PNGRB (2012) Concept paper on unbundling of activities on transportation and marketing of natural gas. 11 Sept 2012

Natural Gas Pricing

Pramod Paliwal

Abstract Natural gas commands an important position in the global primary energy basket. Natural gas prices are mainly a function of market supply and demand. As with any commodity, the prices of natural gas are also determined by the demand–supply equilibrium. However, it would be interesting to understand the uniqueness of natural gas and the subsequent context with its pricing. Given its physical and chemical properties, natural gas witnesses significant logistical constraints and challenges, and hence, although a hydrocarbon, it is to be treated differently as compared to crude oil in all economic and business contexts. Unlike crude oil, natural gas is still wanting of a structured international market. The overriding mechanism for the international gas trade nevertheless remains oil indexation. The phenomenon of international natural gas trade and pricing is under evolution also because technology in infrastructure and logistics has enabled natural gas intra-regional movements. Starting with Natural Gas International Demand and Supply Matrix, this chapter discusses distinctiveness of natural gas pricing followed by international context of natural gas pricing including the conventional oil indexation-based pricing. Natural gas pricing in India has later been discussed in detail with a brief mention of the concept of natural gas price pooling in India. This chapter concludes with a reality check on the prospects of evolution of global natural gas trading hub.

Keywords Natural gas · Pricing · Indexation · Gas trade

1 Natural Gas International Demand and Supply Matrix

Natural gas commands for an important position in the primary energy basket globally. Natural gas demand is rising in all markets, i.e. industrial, commercial and domestic use, and it is estimated that it is going to increase in future.

P. Paliwal (✉)
School of Petroleum Management, PDPU, Gandhinagar, India
e-mail: pramod.paliwal@spm.pdpu.ac.in

S.K. Kar and A. Gupta (eds.), *Natural Gas Markets in India*,
DOI 10.1007/978-981-10-3118-2_5

Natural gas-proved reserves are estimated to be around 187.1 tcm (trillion cubic metres)[1] worldwide at the end of 2014. Out of these proved reserves, non-OECD countries hold 89.6% and the rest are shared by OECD, European Union and Former Soviet Union. Non-OECD countries which consist of Middle East have a proven reserve of around 79.8 tcm, but its reserve-to-production ratio is the highest, which is a major challenge in meeting the rising demand.

The International Energy Agency in its recent yearly World Energy Outlook talks about the increasing position of natural gas in the primary energy basket of the world which having already crossed 20% plus is all set to command almost quarter of the pie by 2030. Within this, LNG is going to contribute even more significantly.[2]

Apart from its traditional usages, natural gas also complements the world's efforts towards unconventional energy by providing an appropriate support to power generation through renewable sources. However, it is to be seen as to how fast it becomes an important player in the global primary energy fuel matrix. Pricing has an important role in this.

As discussed, it is the market supply and demand forces which determine natural gas prices. Given the fact that natural gas may not have appropriate alternatives in the immediate run has a bearing on the pricing of this commodity, sometimes prices also have a bearing on the supply and demand of natural gas. On the demand side, the factors that play an important role are natural gas demand from different market segments, prices of close fuel alternatives such as crude oil, and temperature variance in the consuming regions. In addition, natural gas-based power generation also is one of the important factors on the demand side. Natural gas pricing—like any other commodities—is not immune to the forces of demand–supply.[3] However, it would be interesting to understand the uniqueness of natural gas and the subsequent context with its pricing.

2 Distinctiveness of Natural Gas Pricing

In the pan-global natural gas market currently different prices prevail in different regions and only that can loosely be termed as the pricing mechanisms of natural gas, if any. However, given the fact that there are regional pricing disparities, this makes the pricing of natural gas rather complicated phenomena. While indeed, even the regional prices are determined by certain hub-based transactions, still to compare natural gas pricing with that of crude oil pricing internationally remains a

[1]BP Statistical Review of World Energy 2015.

[2]http://www.ey.com/GL/en/Industries/Oil—Gas/Global-LNG–New-pricing-ahead—LNG-demand-growth.

[3]http://www.eia.gov/tools/faqs/faq.cfm?id=43&t=8.

challenging issue. Oil price-based indexation is indeed one of the conventions for pricing of natural gas.[4]

In underdeveloped gas markets such as India, natural gas pricing is based on referencing the domestically produced commodity's prices with the prevailing prices in other important producing and consuming regions. Experts however have questioned the conventional relationship between the oil and natural gas prices[5] (Melling 2010).

In the well-developed markets such as USA and UK, natural gas pricing is at an evolved state with both these markets using both spot and future pricing mechanisms. However, it shall also have to be seen in other regions of the world emulating US and UK experiences in pricing of this commodity.

The phenomenon of international natural gas trade and pricing is under evolution also because once isolated, regional gas markets are now practically interconnected through the increasing trade in liquefied natural gas (LNG). Next section deals with the international context of natural gas pricing that begins with the conventional approach of linking natural gas prices with that of crude oil.

3 International Context of Natural Gas Pricing

After having understood the distinctiveness of natural gas pricing in the previous section, the following section deals with the international context of natural gas pricing. This contextualization is important to get a feel of various factors prevailing across the world that have a bearing on natural gas prices. Starting with (crude) oil indexation, discussed below is natural gas pricing in major producing and consuming regions of the world, excluding India, as the Indian context is dealt with in a separate section.

3.1 Oil Indexation

Oil Indexation: Two main pricing mechanisms are common for natural gas pricing. The first was created in the USA and later on spread to the UK. It became known as "hub", or local market, pricing. Prices are thus changing from one local market to the other. The other system, conceived by the Netherlands in the early 1960s and broadly preferred by natural gas producers, is that of "oil indexing". This mechanism consists in linking the price of natural gas to its closest substitute (traditionally

[4]http://www.canadalngexport.com/press-release/10-challenges-facing-the-natural-gas-industry/.

[5]http://carnegieendowment.org/files/gas_pricing_europe.pdf.

gasoline of heavy oil). The practice of oil indexation was followed by Asia and is now one of the common pricing formulas for natural gas[6] (Grandi 2014).

Historically, natural gas prices have been region specific because of the low quantities exchanged. Since, unlike crude oil, gas was not a mutually substitutable commodity, there perhaps was no need to link one price to the other. Liquefied natural gas (LNG) technologies, allowing the liquefaction of gas in order to move it without pipes, were still too expensive (and less prevalent) in the 1960s. Realization costs were still too high for this technology to spread (with practically no economies of scale in place) at the pace it is witnessed today. Exchanges were primarily conducted through gas pipelines. Therefore, markets could operate according to their own independent mechanisms. The natural gas market started to evolve in the 1950s when the giant Groningen field was discovered in the Netherlands, followed by the new fields in the North Sea. The functioning of the new system, called Market Value Analysis (MVA), was interesting: a market price was set according to interfuel competition in every sector of the market. In those years, the whole European continent was in a process of "energy substitution" between coal and gasoline for domestic heating. Gasoline was thus chosen as the reference price for natural gas. Prices were renegotiated periodically and were affected by the distance of the end-users from the Dutch border, as well as by the storage capacity of each country.

Dutch producers were held responsible for the transmission from their wells to the national border in order to guarantee supplies for peak demand. Incidentally, this pricing mechanism will later become famous as the "take or pay" clause (*this clause obligates buyers to buy the entire amount of committed gas, or otherwise pay for the quantity that they have failed to draw*), necessary to cover the costs of longer distances and huge investments in pipeline transmission systems. (ibid) The LNG trade witnessed today has its antecedents in the contract which took place between Algeria, France and UK in 1964 with gas prices indexed with oil for long-term contracts. Simultaneously, apart from oil indexation, hub-based pricing mechanism was also evolving in the UK.[7] The gas trading mechanism in the UK deals with trading of gas from the offshore sources including LNG terminals, pipeline and other storage infrastructure. The UK hub (known as National Balancing Point) reflects a single price based on calorific value. Of late since late 2000, natural gas prices are gradually getting detached from the crude oil prices. It is believed that in the coming days, this detachment shall be full and complete. Natural gas thus is envisaged to be priced independently at global level. However, it remains to be seen when this happens. Thus, it would be also contextual to examine the issue of evolution of a global natural gas trading hub (Fig. 1).

---

[6]http://www.energybrains.org/docs/EA/EnergyBrains_EA_NatGasPricing_LG_2014.pdf.

[7]World LNG Report—2013, IEA, PFC Energy.

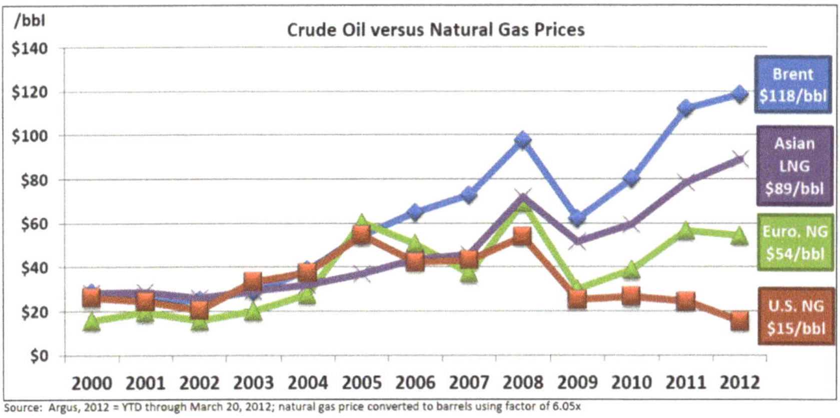

Fig. 1 Global natural gas and crude oil prices. *Source* http://www.econmatters.com/2012/04/oil-and-natural-gas-ratio-explodes-to.html

3.2 Natural Gas Pricing in the USA

Owing to the developments in the natural gas value chain technology, the USA has witnessed a significant increase in natural gas production. This increased production has created market disruption by displacing conventional sources which has also made customers the beneficiary of not only an increased supply base but also competitive prices of the commodity. The US demand for natural gas is said to increase significantly due to developments taking place on all fronts including the demand coming from gas-based power plants, transportation and industrial segment. This has also (unprecedentedly) created potential for exporting liquefied natural gas (LNG). As discussed, the increases in US natural gas production have resulted from unconventional shale gas resources, which have become more accessible and economic due to advancements in new technologies such as horizontal drilling and hydraulic fracturing (also known as fracking).

The shale gas revolution has led to US natural gas supply growth that has surpassed demand growth. As a result, (nearby) Canadian imports and other conventional supplies have been drastically reduced, prices have fallen substantially, and price volatility has moderated to some extent.

US natural gas market consists of NYMEX futures contract-based futures market and trading market for physical deliveries which (deliveries) are taken at well-defined locations in USA.[8]

[8]https://en.wikipedia.org/wiki/Natural_gas_prices.

*The Henry Hub is perhaps the best-known of all natural gas trading points in North America. Located near Erath, Louisiana, the pipeline interchange moves gas from across the US Gulf Coast through a header system operated by Sabine Pipe Line, a subsidiary of Chevron Corporation.

The interstate and intrastate pipelines that come together at the interconnect include: Gulf South Pipeline, Southern Natural Gas, Natural Gas Pipeline Co. of America, Texas Gas

3.2.1 Futures Market

The standardized NYMEX natural gas futures contract is for delivery of 10,000 mmbtu of energy at an interconnecting hub known as Henry Hub which is based in Louisiana. The delivery is usually spread over a month and consists of varying number of days. Usually, monthly contracts generally see expiry a few days in advance of the first day of the month of delivery after which the traders have the alternatives of settlement of future positions or opting for accepting physical delivery.

3.2.2 Physical Market

Physical prices at the beginning of any calendar month at any particular delivery location are based on the final settled forward financial price for a given delivery period, plus the settled "basis" value for that location (refer Sect. 3.2.3). Once a forward contract period has expired, gas is then traded daily in a "*day-ahead market*" wherein prices for any particular day are arrived at on the day before by traders on the basis of prevailing demand–supply dynamics. Mean price of individual daily trades is called index price for particular market (which may be different from the earlier settled future price). Traders determine the final transaction price by suitably adjusting the nearest physical market price (see Footnote 8).

3.2.3 Basis Market

In order to tackle the variability of market dynamics between Henry Hub and other physical trading points, traders concurrently trade in financial "basis" contracts as well. The terms and conditions are akin to the underlying futures contracts (see Footnote 8).

3.2.4 Derivatives and Market Instruments

Size and typicality (particularly its evolved nature) of US natural gas markets enable the players to enter into transactions using relatively complex instruments such as options, derivative and swaps (see Footnote 8).

(Footnote 8 continued)

Transmission, Sabine Pipe Line, Columbia Gulf Transmission, Transcontinental Gas Pipe Line, Trunkline Gas, Jefferson Island Pipeline and Acadian Gas. With the operational flexibility that comes from two compressor stations, Sabine currently possesses the ability to transport 1.8 Bcf/d across the hub.

In addition to being a very liquid spot trading point, Henry Hub is also the standard delivery point for the NYMEX natural gas futures contract in the US. The contract began trading in April 1990 and is traded 18 months into the future and is used as a primary financial hedging tool in the marketplace. (http://www.platts.com/price-assessments/natural-gas/henry-hub).

3.3 Natural Gas Pricing in Europe

As also discussed in Sect. 3.1, gas pricing has been gradually decoupling from oil prices, hubs have become a major basis of gas pricing in north-west Europe, and the trend has been witnessed in central Europe, as well. Even Russian gas prices have come down to hub levels due to the downward price spiral of long-term contracted prices happening due to continental European hub gas phenomena post-2008.

However, since gas flows in Europe are under long-term contracts, hub pricing dynamics in turn is also influenced to an extent by contractual pricing. At the same time, hubs are also increasingly influenced by international demand–supply forces and also by prices in other consuming regions.

Interestingly, European hub prices from time to time are also susceptible to price fundamentals prevailing in Russia.[9]

3.4 Natural Gas Pricing in the UK

As discussed, natural gas after arriving from various production sources in the UK is traded at National Balancing Point (NBP). The gas sellers and buyers at NBP have to comply with various regulations prevailing in the UK and EU which is governed by Uniform Network Code of the UK Amendments to the UNC, if any have to be allowed by Office of Gas and Electricity Markets (www.ofgem.gov.uk).

Incidentally, the National Balancing Point remains Europe's most prominent trading hub which has been in function for almost 25 years, now. NBP along with another important trading hub called TTF (Title Transfer Facility of the Netherlands) is a barometer of Europe's natural gas wholesale trade.

As mentioned, earlier the UK natural gas supply side comprises gas received from multiple sources including the UK's domestic production, gas coming through pipeline network from countries, such as Norway, and LNG supplies received from other destinations of the world. The demand side is represented by utilities, gas-based power plants, consumers from all segments and speculators, as well. The hub facilitates OTC trade of the commodities with the interaction of the players.

3.5 Natural Gas Pricing in Japan/Korea

Since the early Japanese LNG imports in the 1970s, Japan, South Korea and Taiwan (followed more recently by other Asian markets such as India, China and Thailand) cumulatively accounted for the majority of global imports of LNG. Reference price indicates for long-term LNG transactions in Japan, Taiwan and

---

[9]Stern and Rogers (2014).

South Korea. The prices are published on a monthly basis by Japanese government authorities, and they reflect the average crude oil import prices into Japan. For the past couple of decades, international LNG trade in Asia has been based on the JCC pricing mechanism. The Japan customs-cleared crude is the average price of customs-cleared crude oil imports into Japan as reported in Japanese customs statistics. The data to calculate JCC are published by the Japanese government on a monthly basis. The prices are for raw and crude oil imports in yen per kilolitre, at the dollar yen exchange rate and the total imports into Japan of all commodities for the particular month. JCC prices are available at the Petroleum Association of Japan (http://www.paj.gr.jp/english/).

3.5.1 Platts' JKM™ (Japan Korea Marker) Gas Price Assessment

Incidentally, it would not be out of context to mention the (professional) "Platts JKM™ (Japan Korea Marker) Gas Price Assessment" to understand the applied aspects of LNG pricing dynamics in Japan and Korea.[10]

Japan Korea Marker (JKM™): The LNG benchmark price assessment for spot physical cargoes imported by Japan and South Korea is known as Platts JKM™. With these two markets—particularly Japan—accounting for highest share of LNG imports globally, the Platts JKM™ is considered an important pricing benchmark.

LNG market place is witnessing that conventional trading patterns are undergoing rapid change; whereas earlier the consignments were transacted largely through obscure one-to-one transactions, the LNG market of current times engages in relatively transparent deals for all types of cargos. In addition, the traditional players are joined in by new-age speculators, which bring liquidity to spot market.

JKM™ gas market assessment: Based on the prevailing trades, bids and offers, Platts JKM™ price assessment reflects the value worth trading of LNG at the given delivery period. Further, some bids may undergo change, but only those bids and offers that are tweaked in rational range are considered. Platts LNG assessments are closing market prices prevailing at predefined market timings. Also, predefined is the standard ship and cargo size.

The Platts JKM™ price discovery process is undertaken by Platts wherein they survey the broadest possible entities and stakeholders of the market. Platts keeps in regular touch with all market players by all possible communication means.

Entire market intelligence, viz. shipping economics, supply and demand, transaction activity and even information related to weather, is considered for the Platts JKM™ price assessment process. Platts publishes all reported bids, offers and trades on a daily basis.

Platts JKM™ Price Assessment—The Evolution: Platts has been the pioneer to initiate LNG price assessments for cargoes trading in Asia on a daily basis, by the

[10]http://www.platts.com/price-assessments/natural-gas/jkm-japan-korea-marker.

means of Platts JKM™ in 2009. Later with the evolving natural gas market in Asia and increasing trading activity, the Platts JKM™ has been increasingly used as a price benchmark for trades not only in Asia but also outside.

As per Platts data, while long-term contracts remain the predominant mode of natural gas trade, spot trades have also been increasingly gradually and they represent approximately around 1/5th of the LNG volumes traded annually.

As more players enter into LNG market, they preferably look for shorter term and spot deals. Platts JKM™ assessment caters to the interests of such players by providing them a pricing mechanism that is based on empirical information and is transparent. Further with rising non-associated (standalone) natural gas production and technological strides in shale gas production, Platts feels that the need to decoupling natural gas pricing from oil indexation (as also discussed earlier in Sect. 3.1), together with delinking LNG prices from other commodities, gains importance.

3.6 Natural Gas Pricing in Middle East and North Africa (MENA)

Despite having share of world gas reserves, MENA region in recent past has contributed to less than about a quarter of international gas trade movement—both pipeline and LNG. Energy pricing and gas pricing in the region are complicated phenomena owing to local dynamics. Also, the pricing of gas further gets complicated by economic reasons typical to the region. In a nutshell, the natural gas pricing in MENA and sub-Saharan Africa is below not only international prices but also marginal cost of new domestic gas supply. The prices range between US$0.75 and US$3.0/mmbtu.

Of late some countries of the region (such as Egypt and Bahrain) are attempting to non-subsidizing their domestic gas prices.[11]

3.7 Natural Gas Pricing in Russia

Russian economy and regulators have been trying various approaches to natural gas pricing both in the pre- and post-Soviet era. Major Russian producer GAZPROM has been prominently factored into various approaches to Russian gas pricing. However, of late Russian policymakers have felt the need to make their major producer GAZPROM invest into new exploration and production opportunities and hence want to move away from earlier low prices to market determined prices resulting into gas prices being determined on netback parity* basis. At the same

[11]http://www.regulationbodyofknowledge.org.

time, complex Russian socio-economic and international dynamics make the authorities feel to also revisit the concept of inflation linked price rises.[12]

3.8 Natural Gas Pricing in Brazil

Brazil has a dual regulation mechanism guiding pricing of natural gas—one for producers, importers and transporters, who come under the command of National Petroleum Agency, and the other one for local distribution companies, who come under the jurisdiction of regulatory agencies of respective states. Individual Gas Supply and Purchase Agreements determine the imported prices, whereas domestic prices are open to negotiation between concerned players. Natural gas prices in Brazil are comparatively higher than its South American and European counterparts. Also, the price difference between domestic and commercial/consumers within Brazil is significant. Domestic gas prices are bundled, and thus, it is difficult to identify price breakup between the price of commodity (natural gas) and transportation/distribution. Similarly, it is a bit unusual that often domestic gas prices are more than those imported from nearby countries.[13]

Having an understanding of natural gas pricing issues in major global markets, the Indian context shall now be discussed in detail.

4 Natural Gas Pricing in India

Given the stage of market development and relatively low share of natural gas in India's energy basket, the questions have been asked, whether natural gas pricing in India were to be based on some international pricing mechanism or simply decided by government or other regulatory bodies.

For many years, Government of India decided natural gas prices, keeping the affordability of consuming sectors into consideration. This was known as administered pricing mechanism (APM). At the same time, some consumers were

[12]James Henderson, "Domestic Gas Prices in Russia—Towards Export Netback?", NG 57 OIES, UK, November 2011.

**Pricing assessment or pricing formula based on the effective price to the producer or seller at a specific location or defined point. For example, liquefied natural gas (LNG) netback prices may be determined by the market natural gas price at market destinations less the cost of pipeline transportation, regasification, waterborne shipping and liquefaction. Crude oil may be priced on the market value of its refined products, or natural gas priced based on the natural gas market price less the cost for delivering from the defined point to the market location.* (*Source:* http://www.risk.net/energy-risk/glossary/2040843/netback-price).

[13]Gomes (2014).

importing LNG at prevailing market prices through long-term contracts as well as spot mechanisms. Imported LNG prices reflected demand–supply dynamics.

However, while pricing mechanisms such as APM would have been good for some consumers, they also had the tendency to not only create market distortions but also would potentially discourage natural gas producers (as they would always be wanting of an appropriate market-based price that would give them an adequate return on investment).[14] Moreover, APM-like methods for natural gas pricing unwittingly invite government interventions.

Hence, another option of pricing of domestically produced natural gas in India is to benchmark Indian natural gas prices to some of the major trading hubs and also to those prevailing pricing mechanism in LNG production sources such as Qatar and Australia.

The Japanese benchmark, i.e. Japan customs cleared (JCC), incidentally is also benchmarked on international crude oil prices. India also has been determining its contract prices with major suppliers such as Qatar and Australia benchmark with JCC.

However, linking domestically produced natural gas prices in India with international prices is fraught with price volatility and foreign exchange currency risks.[15]

India has historically attempted to balance both pricing mechanisms. The RIL's (Reliance Industries Limited) KG-D6 (gas produced from D6 block in Krishna–Godavari offshore basin, east coast of India by RIL) gas pricing demonstrates this.

The KGD6 pricing was based on a formula that had international price of crude oil as a major component (with floor and cap prices), a constant and an exponential. This resulted in a price of $4.2/mmbtu for RIL and was eventually accepted by Empowered Group of Ministers, Government of India, in 2007.[16] However, even that was not without its own share of controversies. As it turned out to be, the formula that was fixed for 5 years eventually made everyone—government, RIL, critics, R-ADAG, NTPC etc.—unhappy. So much so that when it was up for revision, it witnessed intense politicking most of it without much basis, of course.

After the initial pricing formula, Indian authorities devised a pricing model that was based on market-linked pricing of natural gas prevailing it the evolved gas markets such as USA and UK. This mechanism came in place after high-level committee headed by a prominent economist made certain recommendations aimed at long-term growth of natural gas sector in India.[17] This new pricing was to be effective for all new contracts, starting April 2014.[18] It did indeed take effect. However, the government that took office in May 2015, though continued with the

[14]Not all pricing formulas used in these agreements are public but the pricing has to be done at 'arms-length'—that is without collusion or influence.

[15]Ravichandran (2013).

[16]Rangarajan (2012).

[17]Ministry of Petroleum & Natural Gas (2013).

[18]Aramane (2014).

international linking, expanded the benchmarks though, i.e. in addition to Henry Hub and NBP, it also considered markets such as Russia and Canada to arrive at the price.

4.1 Pricing Under Different Phases and Profit Sharing Mechanisms

4.1.1 Phase: Pre-1999, Pre-NELP (New Exploration and Licensing Policy)[19]

Administered Price set by government (e.g. PMT gas to power sector); **Pricing Market Linked**: linked to global crude oil prices; **Profit Sharing**: Royalty + Tax only.

4.1.2 Phase: 1999–2012, NELP[20]

Pricing* set through competitive "arm's length" bidding (a mechanism where bids were invited from prospective future buyers to "purchase" gas from prospective

---

[19]Please refer the explanatory note on NELP on the next page.

[20]***Pricing under the National Exploration and Licensing Policy (NELP) regime***:

With an objective to promote E&P activity and involve more players-particularly private (both domestic and foreign)-players, Government of India, in 1997–98 launched a New Exploration & Licensing Policy (NELP). One of the prominent features of this policy was remunerative pricing of domestically produced natural gas. Since natural gas market was at a nascent stage, no major precedents for pricing were available, a 'price discovery' mechanism on 'arms length basis' was evolved with the interaction of major consumers and producer of natural gas. Though not perfect, but still it was considered a way forward from nowhere. To provide legitimacy to this pricing mechanism, Government of India through its Empowered Group of Ministers (EGOM) agreed and approved a formula for pricing natural gas that was applicable to natural gas produced by Reliance Industries Limited (RIL) in Krishna Godavari (KG) offshore basin on the East coast of India (Incidentally RIL happened to be the only major private sector player to start natural gas production in substantial quantities under NELP). The formula was accepted in 2009 for duration of 5 years and was subject to revision in 2014. The formula {Selling Price (SP) = $2.5 + (CP − 25)^{0.15} + C} is explained as follows:

SP: Selling Price of KG basin RIL natural gas in USD/mmbtu; USD 2.5 was the base price of natural gas; **CP**: Lagged Price of crude oil with a base ($25) and ceiling ($60); **C**: Constant (apparently as a proxy for demand) representing the outcome of bids-in the process of interaction between major consumers and producer of natural gas towards the 'evolution' of natural gas price as discussed above. Incidentally it was not approved by government and hence was considered as zero; **0.15**: The exponential 0.15 resulted into an S-curve with comparative inelasticity at the upper and lower ends of the curve, which was also reflective of its significance for both buyers and sellers respectively at respective ends. This formula gave a price of $4.20/mmbtu.

producers). Also, administered prices continue; **Profit Sharing**: Royalty + Tax + portion of profits (after cost is recovered by producer).

4.1.3 Natural Gas Pricing Recommendations by Rangarajan Committee (2012)

Recommended Market-linked Pricing: This suggested mechanism was based on the average of arm's length prices in two contexts. The idea was that Indian producers should also receive a (remunerative) price similar to that is received by other global producers (of course, the spirit of arm's length pricing was kept in consideration). That would make E&P activities in India an attractive proposition. The two pricing contexts were as follows:

(i) Netback price for exporter's origin (well head) from where India exports majority of its natural gas (LNG), for example Qatar. The netback price was to be calculated by deducting transportation/insurance charges, ocean loss, liquefaction charges, etc.
(ii) Weighted average prices prevailing at Henry Hub, NBP and that of netback prices at JCC.

The price that Indian produces would receive would be the average of the above two prices and would be applicable to all types of consumers. The two respective prices had their own structured way of computation in terms of past 12 months' prices, etc. The suggested mechanism was open for review after 5 years. The basis of profit sharing between government and producers was revenue earned, i.e. on revenue sharing basis (and not on profit sharing). This was again to encourage the domestic producers and bring in transparency in costs and pricing.

4.1.4 Natural Gas Pricing Recommendations by Kelkar Committee (2013)

Government of India appointed Kelkar Committee submitted its report in 2013. The committee took into account a Production Sharing Contract between producers and government that was linked to investment multiple. The spirit of recommendations of this committee was that natural gas prices in India should gradually be market determined, government interference in pricing should be taken out, and producers should aim at transparent arm's length pricing. Contrary to the Rangarajan Committee, this mechanism suggested a cost recovery-based profit sharing between government and domestic producers.

4.1.5 Cabinet Committee on Economic Affairs (CCEA), Notification 2014

The Cabinet Committee of NDA government (that assumed office in May 2014) took a different approach for gas price determination, by modifying the Rangarajan formula (refer Sect. 4.1.3) as follows:

(i) New formula did not have JCC, and Indian imported LNG components. The idea was to remove price volatility from earlier suggested formula.
(ii) Similarly instead of Henry Hub prices for Canadian consumption, Alberta natural gas reference price was used.
(iii) Instead of NBP price for Russian consumption, actual Russian natural gas price was used for former Soviet Union (FSU) countries.
(iv) Suitable adjustments wherever applicable for transportation and treatment charges were considered.
(v) Quarterly price revision was replaced by biannual. This was done with an aim to bring about relative short-term stability in prices.

The revised natural gas price formula would replace all prevailing pricing mechanisms and shall be applicable uniformly to all entities—public as well as private.

(i) The approved formula is as follows:

$$P = (V_{HH} * P_{HH} + V_{AC} * P_{AC} + V_{NBP} * P_{NBP} + V_R * P_R)/(V_{HH} + V_{AC} + V_{NBP} + V_R)$$

where

(a) V_{HH} = Total yearly consumption of natural gas in USA and Mexico.
(b) V_{AC} = Total yearly consumption of natural gas in Canada.
(c) V_{NBP} = Total yearly consumption of natural gas in European Union and Former Soviet Union but excluding Russia.
(d) V_R = Total yearly consumption of natural gas in Russia.
(e) P_{HH} and P_{NBP}, respectively, are the annual average of daily prices prevailing at Henry Hub (HH) and National Balancing Point (NBP) after deducting the transportation and treatment charges.
(f) P_{AC} and P_R, respectively, are the annual average of monthly prices at Alberta Hub and Russia after deducting the transportation and treatment charges.

(ii) Prices were to be revised biannually. The price and volume data used for calculation of applicable price were determined as to be that of trailing four quarters' data with one the lag of one quarter. The initial price was determined on the basis of prices prevailing in the duration of 1 July 2013 and 30 June 2014 and so on. The announcement of upcoming prices would be made 15 days in advance of the applicable half year.

(iii) The notified price was applicable with effect from 1 November 2014 and would be on gross calorific value (GCV) basis as the input prices (in different regions) in the formula are also on GCV basis.
(iv) The revised gas price is applicable to all gas produced from nomination fields given to public sector companies, NELP blocks and all the pre-NELP gas blocks where Production Sharing Contracts provides for government approval of gas prices and coal bed methane blocks. Government however allowed for some exceptions as following:

   (a) Small and isolated fields in nomination blocks were excluded from the formula as these have peculiar operating conditions applicable to them.
   (b) In case of prevailing price contracts, the formula was to be applicable only after expiry of such contracts.
   (c) Exclusion for some time period was granted to certain prevailing prices which were based on certain specific PSCs.
   (d) Exception was also applicable to pre-NELP blocks where government approval was not given under the PSC.[21]

4.2 Pricing of Indian LNG Imports

It would be also contextual to discuss the pricing of LNG imports in India. LNG imports to India started around 2004. Qatar is the top destination exporting LNG to India. Petronet LNG and GAIL purchase and market LNG in India. Petronet LNG/GAIL imports LNG under both arrangements, i.e. spot and long-term contract bases.

The first long-term LNG supply contract between India and RasGas of Qatar signed in 2004 was for a volume of 5 mtpa, at a price of $2.53/mmbtu. The duration of this agreement was for a period of 5 years. The contract ended in 2009, after which it was agreed to index the LNG import price with crude oil. The contracted quantity was also increased to 7.5 mtpa in 2010. The pricing formula for imports by Petronet LNG Limited from RasGas on FOB (Free on Board) basis in USD/mmbtu was arrived as follows:

[21]http://pib.nic.in/newsite/PrintRelease.aspx?relid=110696

However with effect from January 2016, Petronet LNG—RasGas Qatar long term contract has undergone a revision. This revision is to ensure that imported LNG prices from RasGas are benchmarked with immediate crude oil prices i.e. past 90 days (instead of the average of past many months)-because that would be more realistic. Under the 25-year contract, RasGas of Qatar will supply 7.5 million tonnes a year of LNG it sells to Petronet LNG at (a revised) 12.66% of 3-month average price of Brent crude plus a fixed component of about $0.60/mmbtu and shipping cost per mmbtu from Qatar to Indian LNG terminal in Dahej, Gujarat. (http://articles.economictimes.indiatimes.com/2016-01-03/news/69486436_1_petronet-lng-rasgas-oil-price).

$$Po * JCC_t/\$15$$

where Po: \$1.90/mmbtu; JCC_t: 12-month mean of the JCC price and t referred to the month in which the price calculation applies. JCC_t in the formula was subject to a cap and a floor as follows:

$$\text{Cap: } [(60 - N) * 20 + (N * A60)]/60 + 4$$
$$\text{Floor: } [(60 - N) * 20 + (N * A60)]/60 - 4$$

where $N = 1$ for January 2009, which progressively increased by 1 every month subsequently up to December 2013, following which it stood at 60, and $A60 = 60$ months' average of the JCC price.

Further after adding shipping (\$0.30), insurance (\$0.0017), customs duty (\$0.50) and regasification charges (\$0.60), total imported price of LNG was \$10.44/mmbtu.

A deviation from JCC-linked pricing has recently occurred when (effective January 2017, for a period of 20 years) GAIL and Cheniere Energy (USA) {www.cheniere.com} entered into a contract for importing 3.5 mmtpa from Cheniere's Sabine Pass. The pricing formula consists of 115% of Henry Hub plus a fixed capacity charge of \$3/mmbtu.

4.3 *Natural Gas Price Pooling in India*

As Indian natural gas market looks for growth, the need for tackling concerns of differential pricing from various sources for multiple consuming sectors has become paramount. That is because multiple prices for the same commodity leads to inefficient consumption and seriously impairs competitiveness of the players in the same consuming industry. The concept of pooling of prices was thus conceived. Price pooling is a mechanism wherein the consuming sectors can look forward to relative stability and affordability in natural gas prices irrespective of the source of supply. This is turn would also enhance natural gas supply security by not only widening the supply base but also bringing about a balance between suppliers and consumers. At the same time with appropriate pricing leading to increased natural gas usage, natural gas infrastructure development also gets a boost. This also takes care of the relatively expensive LNG imports in India. With imported LNG prices pooled with domestically available relatively cheap gas prices, affordability for price sensitive consuming sectors such as fertilizer manufacturers, power plants and city gas distribution increases (Mercados 2010). In a nutshell, creation of such pooling mechanism is aimed at reducing the price volatility and providing a boost for natural gas infrastructure development.[22]

[22]http://petroleum.nic.in/docs/reports/gaspricepooling.pdf.

The natural gas price pooling mechanism has been conceptualized from the perspective of enhancing energy security. It is proposed that the volumes of domestic supplies and LNG be pooled for sectors such as power and fertilizer. The mingling of high-priced LNG and low-priced domestic gas will ensure that high-priced LNG volumes are absorbed in the Indian market. Thus, existing and planned regasification terminals in India would be utilized and more terminal capacities could be planned. The overall economy and people at large could benefit owing to increased power availability and less reliance on fertilizer imports.

The key distinguishing feature of the proposed pooling mechanism is that a nominated government agency will be responsible for allocating quantities and arriving at common prices for priority sectors (power and fertilizer) for gas produced from the NELP fields. Any policy move such as gas pooling should take into account the broad directional movement of India's energy policy; the impact of policy on the development of an associated value chain; and the risks associated with implementation.[23]

In a significant policy intervention, the Government of India has in 2015 decided to supply gas at a uniform delivered price to all fertilizer plants (for production of Urea) that are located on the gas transmission grid through a mechanism called "price pooling". This was done to tackle the price disparities in the prices of domestically produced natural gas and imported regasified LNG (RLNG). The policy intervention was also aimed further to address the variation in the conversion efficiency of plants (measured in Gcal/MT). Both the phenomenon resulted in cost disparities in urea production, and ultimately, the impact was not only on the producers but also on consumers and government (that had to provide subsidies resulting into market distortion). Government believes that a uniform natural gas price at the input stage shall be a more effective way to tackle the issues of urea manufacturers (PIB 2016). Similar price pooling mechanism for imported RLNG was approved by the Indian Government in 2015 to revive stranded gas-based power plants, so that they can start generating power (Zee News 2015).

5 Evolution of Global Natural Gas Trading Hub: Myth or Reality?

Crude oil market dynamics has evolved internationally over the years along with emergence of an association of prominent producers. Naturally, questions regarding similar dynamics and pan-global market for natural gas are also asked by the concerned. Answering these questions require the understanding of distinctive character of natural gas business including pricing dynamics. Evidence so far suggests that a whole host of factors may prevent the development of oil-like market for natural gas.

---

[23]India Infrastructure, May 2010.

It would be opportune to mention a study (Raymond et al. 2014) that attempts to find out medium-term integration in prices among natural gas markets in the USA, European and Asia. The study reveals some evidence of convergence among the Asian (Japanese, Korean, Taiwanese) and the UK prices. However the US price behaviour is distinct from this group (Asian/UK) of prices. The study does not arrive at any conclusions about a totally integrated international natural gas market. It seems that the possible integration between UK/European and Asian regions is basically due to the contractual mechanisms in both the regions (that are based on oil indexation) than anything else.

Natural gas, with close to quarter share, has an important place in world's primary energy mix. But given the production and consumption matrix, the producing and consuming regions are far away from each other and that creates its own unique logistical issues. As more and more LNG movement makes headway, gas transportation via pipelines is set to gradually come down. Moreover, high investment costs in fixed infrastructure and relatively inflexible contract terms and conditions also make pipelines less attractive globally and geopolitical issues complicate the matter even further. Juxtaposed to that liquefaction capacity is set to grow, and increased capacities likely be earmarked for exports, thus facilitating global LNG trade. LNG business is hugely capital intensive one and requires a long-term outlook with close cooperation between all major stakeholders.

Natural gas trade thus is not without its own challenges which increase given the cyclical nature of economic growth, complicated geopolitics and rising ecological concerns towards E&P companies.

Thus, it seems that an integrated global gas market in the near future is unlikely to happen. Unlike crude oil that has a global market, the absence of a world market for natural gas is likely to result into a situation, where exporters will find it difficult to coordinate production and trade, especially when it is important for them to secure long-term supply relationships in the light of additional capacities coming to the market. Moreover starting 2015, to envisage a gas cartel on the lines of OPEC is all the more difficult as the long-standing traditional buyers in Asia and Europe may be poached by new players (such as Iran and Australian LNG) who are set enter the market more assertively in near future.

Feygin and Revenkov (2007) believe that given the geopolitical peculiarities, it would be difficult to envisage a grouping of natural gas producers on the lines of OPEC, however that does not precludes the gas producing countries, evolving their own unique ways of interaction and coordination. In the long term, existing and emerging natural gas producing countries would indeed evolve a mechanism to come on a common platform but that would indeed also depend upon myriad of economic factors related to natural gas such as demand elasticity and market substitutability.

Acknowledgments The author would like to acknowledge Mr. Mridu Pavan Chakrabarty of PGP 14, School of Petroleum Management, Pandit Deendayal Petroleum University, for his research based inputs for this chapter.

References

Aramane G, Ministry of Petroleum and Natural Gas (2014) Domestic natural gas pricing guidelines (O-22011/3/2012 ONG-V). Retrieved from website: http://eac.gov.in/reports/rep_psc0201.pdf

Feygin V, Revenkov V (2007) The possibility of a gas cartel. RUSSIA in global affairs, vol 5(4)

Grandi L (2014) European gas markets: from oil indexation prices to spot prices? Version 1.0/June 2014

Gomes I (2014) Brazil: country of the future or has its time come for natural gas?, June 2014

Raymond R, Joyeux R, Ripple RD (2014) International natural gas market integration. Energy J 35 (4)

Melling AJ (2010) Natural gas pricing and its future Europe as the battleground. Carnegie endowment, 2010

Mercados Report on Price Pooling. http://petroleum.nic.in/docs/reports/gaspricepooling.pdf (2010)

Ministry of Petroleum & Natural Gas (2013) New gas pricing guidelines: "A ready reckoner" (97051). Retrieved from website: http://pib.nic.in/newsite/erelease.aspx?relid=97051

PIB (Press Information Bureau, 2016) as available on http://pib.nic.in/newsite/PrintRelease.aspx?relid=117888

Rangarajan C, Government of India, Committee on the PSC Mechanism in Petroleum Industry (2012) Report of the Committee on the production sharing contract mechanism in petroleum industry. Retrieved from website: http://eac.gov.in/reports/rep_psc0201.pdf

Ravichandran K (2013) Gas price hike: impact analysis. Gas price hike: impact analysis, 02. Retrieved from http://www.icra.in/Files/ticker/GasPriceHikeImpactAnalysis.pdf

Stern J, Rogers HV (2014) The dynamics of a liberalised European gas market: key determinants of hub prices, and roles and risks of major players, pp 11–27, Dec 2014

Zee News, 2015 as available on http://zeenews.india.com/business/news/economy/cabinet-clears-pooling-of-gas-prices-to-help-revive-stuck-power-plants-_121917.html

Feasibility of Price Pooling for Gas-Based Power Generation in India: A Reality Check!

Radha Krishna Tripathy

Abstract Gas-based power projects are on the verge of becoming bad assets for the country. While nearly 14 GW of gas-based projects are non-functional, 9 GW of projects operates at a suboptimal level of 25% or less. Domestic gas supply to power sector seems a distant dream with no major gas discovery in sight. In a bid to kick-start the plants and to enable the project developers to service their debts and cover up the operational expense, govt has started a scheme of reverse e-auctioning for supplying sport RLNG to power plants. This plan will help revive some of stressed gas projects in the country with the help of govt subsidy out of power system development fund. While this is a welcome step from the government and would function well in constantly depreciating spot RLNG price, the scheme provides only short-term visibility for operations and servicing of debt. The long-term operations and viability of these plants would remain dependent upon ramping up production of gas from domestic sources. At a time where long-term RLNG price hovers around USD 7 to 8 per million British thermal units (MMBTU), govt has to rethink its strategy and should move away from indirectly bailing out of gas-based power projects and open the sector for competition with necessary policy support. Gas price pooling is long debated in India, and finally, it was made effective for fertilizer sector from 1 July 2015. While the decision to introduce poling price concept would go a long way for fertilizer units' revival, the debate is still on for its implementation for the power sector. The feasibility of such an option for the power sector would be accessed in this paper along with government's preparedness in addressing the concerns of various stakeholders.

Keywords Gas price pooling · Imported LNG · Spot R-LNG · Policy support · Gas-based power projects

R.K. Tripathy (✉)
Athena Infonomics, Chennai, India
e-mail: radhakrishna.tripathy@gmail.com

S.K. Kar and A. Gupta (eds.), *Natural Gas Markets in India*,
DOI 10.1007/978-981-10-3118-2_6

1 Introduction

Natural gas contributes to almost 10% of the electricity generated in India. Power sector alone accounts for nearly 45% of the country's natural gas demand. Currently, out of the 21,000 MW installed capacity of gas-based power plants in India, more than 50% of the capacity is unutilized due to unavailability of gas, and rest of the plants are operating at a plant-load factor (PLF) of nearly 25% or less. Gas shortage is a major concern in the power sector. As on March 2014, against a requirement of 63 MMSCMD, only 21 MMSCMD[1]—out of which 17.55 MMSCMD was from domestic supply—was available to the gas-based generating stations. Imported liquefied natural gas (LNG) is an expensive option and hence unfeasible for the price-sensitive power sector. The deficit in gas supply has forced project developers to either shut down their gas-based power plants or convert the existing gas-based units into coal-based ones. These power projects are on the verge of becoming bad assets for the industry.

The situation is further aggravated by two directives issued earlier by the Empowered Group of Ministers (EGOM): first, to provide maximum domestic gas allocation to the fertilizer sector; second, to maintain 31.5 MMSCMD supply of gas till 2015–16. The directives also stipulate that a shortfall in requirement from the fertilizer sector be met on a priority basis by any incremental production from domestic sources. This has left the power sector high and dry in terms of gas supply. From March 2013 onwards, there has been no supply of gas to the power sector from the KG D6 basin. With the availability of domestic gas not expected to rise significantly and additional supply from existing fields minimal, there seems to be no respite for the power sector in the near future.

Liquified natural gas (LNG) is an alternative, but an expensive one. The price of LNG has fallen 58%, from $18.3 per million British thermal units (MMBTU) in April 2014 to $7.7 per MMBTU in September 2015—mirroring the drop in crude oil prices. As a result, the price of gas-based electricity has also plummeted, making it attractive to some consumers. Over the past five years, there has been an increase in RLNG imports. Currently, India imports around 41 MMSCMD of RLNG. With the existing infrastructure, it can only import an additional 20 MMSCMD of gas per year. Future RLNG imports will depend on the expansion programme of RLNG terminals in India and the international spot price for RLNG. In view of the prevailing situation of gas production and supply and the price-sensitive nature of the power sector, the government is promulgating gas-price pooling options to meet the power sector's gas shortage. The arrangement is to mix cheaper domestic gas with the expensive RLNG and supply it to all users at a common rate. This will invariably increase the cost of power supply and necessitate some degree of cost sharing by all the stakeholders and require financial support from the central government. On a pilot basis, govt has introduced a scheme for 2 years on reverse

[1]MMSCMD stands for Million Metric Standard Cubic Metre per Day.

bidding to kick-start some of the stranded power projects with subsidy assistance from power sector development fund.

This scenario is primarily the outcome of dwindling supply from the KG D6 gas basin and a fall in production from other domestic sources. It is expected that this arrangement will cater to the consumer's larger interest, help in reviving the gas-based units and enhance market development of the gas sector in India in some way. It will also help the government to reduce overall carbon footprint and thereby fulfil its larger commitment to climate change.

2 Natural Gas Is a Preferred Fuel Choice

Natural gas has emerged as the fuel of choice for many industries in India owing to its environmental benefits and higher economic efficiency. The power and fertilizer sectors are major consumers of natural gas and account for over 60% of the country's overall natural gas consumption. Also, considerable investments have been made by these sectors to establish new gas-based units over the last decade. Many naphtha and fuel oil-based fertilizer units have been converted into gas-based units because of the latter's benefits in comparison with alternatives such as fuel oil and naphtha as feedstock. Rapid growth of the City Gas Distribution (CGD) sector in the last decade also indicates a demand for natural gas among domestic and industrial consumers.

However, supply is not commensurate with demand. Unless a new source of gas is discovered in the interim, this deficit scenario is expected to continue for at least another two to three years. The KG D6 gas basin has failed to meet supply expectations and production has gone down over the years. The supply scenario (LS 2014) of domestic gas has compelled the country to import a higher volume of LNG, thus resulting in an increase in the economy's CAD (Current Account Deficit). With the government shielding the fertilizer sector by mandating a constant supply to it from all available sources, the power sector has borne the brunt of dwindling supplies.

Table 1 indicates that maximum demand for gas is expected from power and city gas distribution segments up to 2016–17. The demand will be stagnant for sectors such as fertilizer, petrochemicals, steel and other industrial sectors. Sectors such as petrochemicals and CGD are relatively insensitive to price changes, as both can afford expensive, imported LNG to compensate for the shortfall in domestic supplies.

According to British Petroleum's Statistical Review of World Energy (2014), India's gas production dropped by 16.3% in 2013 and since 2010, overall gas production has fallen by 33.8%. The current gas deficit has led to a situation where all consumers have been left competing for an increased share of gas supplies. The power sector has suffered the most due to a steep decline in production from KG D6. Furthermore, drastic cuts in allocation have left many of the existing and upcoming plants stranded in need of gas supplies. This has resulted in strong

Table 1 Sector-wise demand for gas up to 2016–17 in MMSCMD

Sectors	2012–13	2013–14	2014–15	2015–16	2016–17
Power	135	153	171	189	207
Fertilizer	58.22	63.52	109.3	109.3	109.3
City gas	15	19	24	39	46
Industrial	20	20	22	25	27
Petrochemicals/refineries/ internal consumption	54	61	67	72	72
Sponge iron/steel	7	8	8	8	8
Grand total demand	289.22	324.52	401.3	442.3	469.3

Source Ministry of Petroleum and Natural Gas (MoPNG)

lobbying from the power sector to review the order of priority and place it at par with the fertilizer sector for equitable gas allocation.

3 KG D6: Huge Blow to Power Sector

The power sector had high expectations from the KG D6 gas basin and its anticipated increase in production. As per the plan submitted by RIL in 2006, production from that gas source was to be ramped up to a level of over 80 MMSCMD. But failure on RIL's part to make good on its promises has left consumers in a sorry state (Fig. 1). A number of project developers, who had constructed their facilities

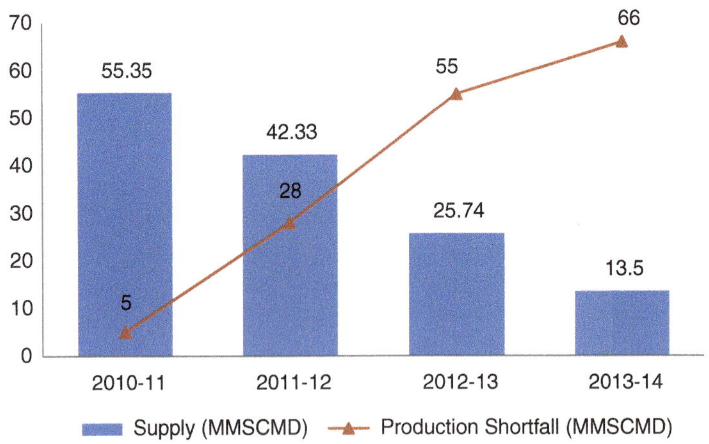

Fig. 1 Year-wise supply from the KG D6 block (MMSCMD). *Source* Directorate General of Hydrocarbons (DGH), MoPNG

anticipating assured future supplies from the KG D6 block, are currently scouting for alternate gas-supply arrangements.

From March 2013, supply to the power sector was halted after a drastic fall in production from the KG-D6 basin. At present, all gas produced from KG D6 is being supplied to the fertilizer sector. Between 2010 and 2014, the total production shortfall against the estimated amount was 154 MMSCMD (CEA 2014).

4 LNG Import Options

In this deficit situation, consumers are compelled to import LNG. Imported gas is expensive, prices of spot LNG being 3–4 times higher than the prevailing price of domestic gas in India in 2014. The price has come down to USD 7.7 per MMBTU in September 2015. Only the CGD entities and refineries are likely to have the capacity to pursue import options, while the core-consuming sectors such as fertilizer, LPG and power can only partially meet their requirements through the LNG import route. There are plans to set up several LNG terminals to increase the country's LNG import capacity. Currently, three LNG terminals are operational with less margin available for absorbing a higher LNG import. The shale-gas revolution in the USA, however, has opened up new avenues for India and it is expected that the international LNG price will remain low in short term for another couple of years with marginal fluctuations. That would give India some leverage in planning and designing its future strategy.

Indian companies led by GAIL have made early moves in the USA, acquiring assets and firming up future supplies, and many more acquisitions are in the offing. Other Indian companies, both public and private, are keenly scouting for assets in North America. In view of these fast-paced developments, India is gearing up to import a more substantial part of its gas sources.

5 Gas-Grid Development: Developer Dilemma

While the government is considering all possible options for gas sourcing, in order to increase supply and carry the gas to the consuming sectors, the associated infrastructure needs to be revamped. The government's ambition to extend the gas pipeline network by 15,000 km would entail almost doubling the existing capacity. However, the big question remains: What should be India's priority? Should it be gas production or gas distribution? In this deficit scenario, with existing pipelines operating at between 50 and 55% of their capacity, the plan of laying out additional pipelines seems unrealistic. The government, therefore, finds itself in a dilemma: Should it begin by addressing the issue of gas availability or should it first lay gas grids in the country?

With more gas bound to flow into the Indian Territory, the country is compelled to revamp all aspects of the gas infrastructure simultaneously. According to the Indian government's own projections, demand for natural gas will increase from 242.66 MMSCMD in 2012–13 to 378.06 MMSCMD in 2016–17. The projections concerning demand offer ample indications of the urgency of developing LNG terminals and laying natural gas pipelines in the country.

6 Gas-Based Power Generation in India: A Status Update

The gas-based power capacity of nearly 7500 MW (CEA 2014) is ready for commissioning, but due to the unavailability of fuel, these projects are stalled at different stages of implementation. Out of 7500 MW, one-third has already been completed and is awaiting the allocation of gas, while the capacity of 5400 MW is in an advanced stage of completion (Table 2).

At a time when India is in urgent need of power, gas projects are idling due to lack of fuel. Projects exclusively dependent on KG D6 gas (~3000 MW) are completely stalled.

A list of such plants (CEA 2014) is provided in Table 3.

As per a Central Electricity Authority (CEA) estimate, 86.6 MMSCMD of gas is required for existing power plants to operate at 90% PLF and an additional 50 MMSCMD for upcoming power plants to run at 90% PLF.

Gujarat has the highest installed gas-based capacity in India, totalling 4616 MW, followed by Andhra Pradesh at 3370 MW. The target for a new capacity of 8617 MW was set, based on the availability of gas from KG D6, with most of these projects being planned in the state of AP (4308 MW). A state-wise analysis of gas-based capacity dependent on KG D6 supply is presented in Fig. 2. Andhra Pradesh had pinned all its hopes on KG D6 gas and hence suffered the most due to non-production from the KG D6 fields.

Approximately INR 70,000 crores worth of investment is at stake in the power sector due to the unavailability of gas. While this is the case at one level, there are

Table 2 Gas supply to power projects (March 2013)

Category	Description	Capacity (MW)	Gas supply (MMSCMD)		
			APM	KG D6 (Allocation)	PLF (%)
1	Project commissioned before 2002	8044	17.55	0	32
2	Projects commissioned between 2002 and 2012	6996	9.86	0	24.5
3	Projects ready for commissioning	9109	0	0	0
	Total	24,149	27.41	0	21

Source MoPNG

Table 3 List of power plants which run only on KG D6 gas

Name of the project	State	Capacity (in MW)	Allocation of KG D6 (in MMSCMD)	Supply of KG D6 gas
Torrent Vatva (AEC Ahmedabad)	Gujarat	100	0.37	0
GEB, Dhuwaran (GSECL), Exp	Gujarat	218.62	0.44	0
GEB, Uttran (GSECL), Exp	Gujarat	518	1.45	0
Essar Power		300	1.17	0
Silk Road	Andhra Pradesh	35	0.1	0
Tanirbavi	Andhra Pradesh	220	0.88	0
Lanco Kondapalli Power Ltd (Exp)	Andhra Pradesh	366	1.46	0
Sriba Industries	Andhra Pradesh	30	0.12	0
RVK Energy	Andhra Pradesh	28	0.112	0
LVS Power	Andhra Pradesh	55	0.22	0
NDPL (Rithala)	Delhi	108	0.4	0
Pragati Power-III	Delhi	1000	0.93	0
Total		2978.62	7.562	0

Source Ministry of Power (MoP)

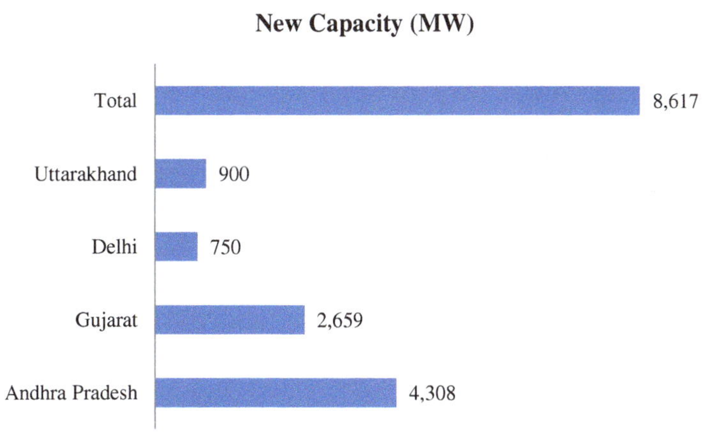

Fig. 2 New gas-based capacity coming up in different states. *Source* Central Electricity Authority (CEA)

concerns, at another, that an increase in the price of gas will result in power becoming costlier, leading to a scarcity of buyers in the market. The distribution companies (DISCOMs), for their part, are struggling with a staggering loss of more than INR 2 lakh crores and are likely to shy away from the high cost of gas-based power. This unfortunate combination of factors can lead to certain power projects becoming unviable.

In October 2014, the government raised the price of the gas to USD 5.6 per MMBTU. It was fully aware that it would have to cushion any increase in the power tariff ensuing from an increase in the price of gas by means of enhanced financial aid to the sector. The power sector has, therefore, suffered a dual setback: first, from the short supply of gas; second, from the recent price hike by the government. This has led to power producers coming together and lobbying for gas-sector reforms in the country.

The power sector demands a pro rata cut across industries, in view of the shortage of supply from the KG D6 gas basin, and a sharing by all consuming industries of any shortfall in gas production. It advocates price pooling for the gas sector, so as to keep it up and running, and expects incremental gas production, if any, to be diverted to meet its needs.

7 Gas Availability: Addressing Supply Side Concerns

On the supply side, with a major portion of the country's available gas being allocated to the fertilizer sector, the power sector will continue to suffer a gas deficit (BP 2014). As per the observations of the Ministry of Oil and Gas, there will be no gas available from the NELP blocks for the next two years. It is expected, however, that 4.5 MMSCMD of gas will be available in 2015–16. Assuming that all incremental gas will be supplied to the power sector from NELP, this sector is likely to be allocated 28.74 MMSCMD of gas by 2018–19 (Fig. 3).

As far as gas from sources other than NELP is concerned, it is already integrated with other sectors and unlikely to be diverted to the power sector, the reason being

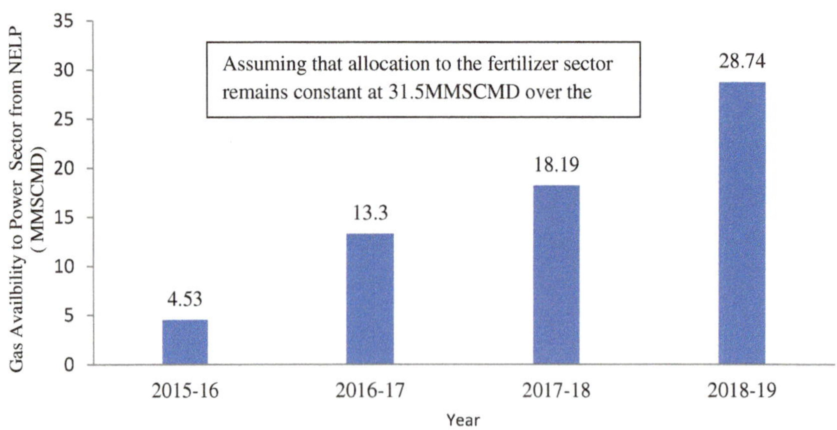

Fig. 3 Projected availability of gas for power sector from NELP up to 2018–19. *Source* MoPNG

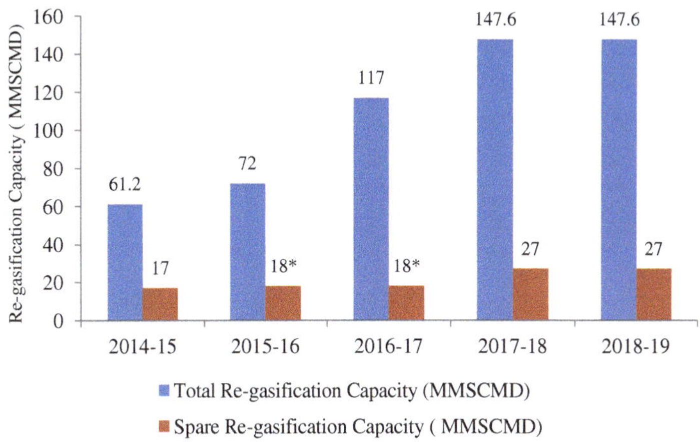

Fig. 4 Projected re-gasification capacity in India up to 2018–19. *Source* MoPNG (*Asterisk* During the monsoon, the re-gasification terminal at Ratnagiri is not in operation)

that many industries depend directly on gas for their operations and any decision to divert it elsewhere will have an adverse impact on them. The government's remaining option would be to focus on efforts aimed at discovering new gas sources and increasing the production of gas which can be supplied to the power sector. Alternatively, it could import RLNG and fuel the gas-based power plants. For the next three years, the expected capacity of gas available from the existing terminals at Dahej, Hazir and Ratnagiri is only 18 MMSCMD. The government is considering increasing this capacity to 27 MMSCMD by 2017–18 (DGH 2014) with the addition of re-gasification terminals along the country's east and west coasts, which would boost RLNG imports. These proposed capacities are at various stages of implementation and will take two to three years for completion.

With the proposed setting-up of re-gasification terminals, the capacity is projected to be 147.6 MMSCMD by 2018–19 (Fig. 4). It promises an increase in gas imports, whereas transport of the gas to various locations in the country remains an issue due to the insufficiency of gas pipelines required for the purpose.

To sum up, the government is running out of options for supplying gas to the power plants, because domestic gas production does not yield a surplus that could be diverted to other sectors. Thus, it is considering price pooling of gas and has had several discussions in this regard with concerned stakeholders and ministries.

8 Gas-Price Pooling for Power Sector: Suggested Options

A number of arrangements are possible for gas-price pooling (PMO 2014). While such pooling can be done for the entire supply of gas available and considered for all consuming industries, the implications would be too widespread for the government to handle effectively. Given below are two alternatives that the government is currently considering for the power sector after allowing gas pooling for the fertilizer sector:

Option 1:
Pooling of domestic gas exclusively with RLNG for stranded gas-based plants with a capacity of 16,107 MW and no gas supply from any other source. All power plants that have APM gas supply are excluded

Option 2:
Covers all gas-based plants (with both APM gas supply and non-APM gas supply) with a total capacity of 24,149 MW.

9 Proposed Operational Methodology

While Option 1 seems likely to meet the approval of all stakeholders with some funding from the government and subsidy to the end users, Option 2 risks facing stiff resistance from the power producers who already enjoy the benefits of the lower price under APM. The National Thermal Power Corporation (NTPC), one of the major beneficiaries of APM gas, is expected to oppose the move, though the increase in cost will be a pass-through. However, the government is considering going ahead with the price pooling for gas with some assumptions as stated below:

- Any incremental gas from NELP will have to be supplied to the power sector. This will help increase the volume of comparatively cheap domestic gas being channelled to the pool, thereby reducing the pooled price of gas to less than the anticipated amount. The price of domestic gas is capped at USD 6.5 per MMBTU for the power sector, without any further upward revision. This will effectively set the delivered price of gas at USD 7.96 per MMBTU, including all taxes and levies, and lend a certain direction to the calculation of the revenue gap and the means deployed to meet that gap with different tools designed by the government
- The price of imported gas (delivered cost) is to be capped at USD 17.38 per MMBTU (assumption on the basis of long-term sourcing of the gas at USD 13.5 per MMBTU). Although the international price is subject to fluctuation because of varying conditions, the government will try its best to keep the price around the assumption figure
- The PLF for gas-based power plants is considered to be at 65% with respect to gas availability from all sources
- The fixed cost for gas-based power plants is assumed to be at INR 0.85 per unit. The project developer has to absorb between INR 0.06 and INR 0.1 per unit of the fixed-cost requirement
- The sale of power from gas-based units will be capped at INR 5.5 per unit. This figure is assumed to be acceptable to all stakeholders of the power industry,

including DISCOMs backed by government-generated subsidies, and expected to be the higher end of the cost which can be absorbed from a gas-based power plant.

The revenue gap arising from this kind of arrangement is estimated to be around INR 30,000 crores over a three-year time period (PMO 2014). This gap will be addressed through a series of measures promulgated by the ministry which includes a customs-duty waiver on imported LNG, a waiver of the VAT (Value Added Tax) applicable to gas, a waiver of Central Sales Tax (CST) and funding support from the National Clean Energy Fund (NCEF), over and above the transportation-tariff and marketing-margin cuts offered by GAIL.

GAIL is acting as pool operator for gas pooling for the fertilizer sector and entrusted with getting imports after considering domestic availability and averaging out price of both, to deliver the fuel at uniform rate to all plants. For the power sector, GAIL has assured the government of its intention to take up part of the financial burden by reducing the transportation tariff by 20% for incremental gas flow in pipelines. Other gas transporters such as RTGIL and GSPC are expected to follow suit and agree to similar cuts in a bid to help the government implement this proposal. Approval from the regulatory board, PNGRB, is also necessary. The idea is to find the best possible way of utilising the existing pipeline infrastructure to the fullest extent. In its efforts to bring down the delivered price of gas for consumers and thereby help ailing gas-based power projects, GAIL may also be ready to accept an additional burden by reducing its marketing margin by another USD 0.1 per MMBTU from the existing USD 0.18 per MMBTU. As per the government's own estimate, the gap will be adjusted through the following measures (Table 4):

After taking into account all the aforesaid options, the balance gap in revenues could be met by the NCEF. The Ministry of Power anticipates bringing down the variable cost to approximately INR 4.65 per unit through the implementation of such measures.

A break-up of cost components for the delivered price of gas is provided in Table 5. It clearly indicates that there is ample scope for reducing the cost of delivered gas by waiving tax components and levied duties. By reducing the pipeline tariff as well, a major part of the revenue gap could be filled.

Table 4 Expected gap adjustment from different arrangements

Sl. no.	Components	Expected adjustment (INR crores)
1	Keeping the fixed cost at INR 0.85 per unit	1400
2	Waiver of VAT on sale of gas	13,100
3	Waiver of CST on sale of gas	2100
4	Reduction of pipeline tariff by 20% by pipeline operators	1800

Source MoPNG

Table 5 Cost component of delivered cost of gas (Domestic versus Imported LNG at 2014 price)

	Domestic gas (USD/MMBTU)	RLNG (USD/MMBTU)
Basic cost	6.50	13.50
Import duty	–	0.68
Re-gasification charges	–	0.65
Marketing margin	0.09	0.18
Pipeline tariff	0.34	0.34
Service tax	0.04	0.04
VAT	1.01	2.23
Total cost	7.98	17.61

Source MoPNG

As per 2014 gas price of USD 6.5 per MMBTU for domestic gas, the average cost of delivered gas at the consumption centre is estimated to be USD 7.98 per MMBTU, including all taxes and other charges. Similarly, for imported RLNG, the delivered price at the consumption centre would be USD 17.61 per MMBTU, taking into consideration the long-term average purchase price of RLNG at USD 13.5 per MMBTU. All calculations are made by taking this as the benchmark price for both domestic gas and imported gas. The situation has completely changed now with domestic gas price at USD 3.82 per MMBTU starting 1 October 2015 for a 6-month period and spot RLNG declining to a level of USD 7 per MMBTU. This works well for the new govt scheme of reverse bidding for subsidy support to stranded gas-based power projects on spot RLNG prices. Companies are now importing spot RLNG, and long-term RLNG supply was hit due to its higher prices. They are only maintaining with minimum takeoff from such arrangements. The lower prices of spot RLNG have given further comfort to the govt and they may go for a higher volume of gas import provided DISCOMs are willing to buy power at INR 5 per unit as envisaged. This is good time for the govt to go for price pooling concept as domestic and imported gas prices are at their lowest level.

While the government is engaged in the back calculations necessary for taking care of the revenue gap that might surface by pooling domestic gas and imported RLNG, several issues and concerns need to be addressed simultaneously. The calculations based on projected availability of gas from domestic sources (new discoveries as well as incremental supplies, if any) and projected import volume at a certain price may go haywire if the market throws up any surprises. It is clear that the kind of arrangement envisioned is fraught with a fair bit of risk and the government would, therefore, do well to have a Plan B ready to tackle such situations if they arise in the future. The surprise element can also come from the current Paris convention of COP21 on climate change. While the world debates on low-carbon growth, it may take its toll on the gas-based power generation in India and elsewhere given the investors drifting towards renewable energy generation at a huge scale.

10 Issues and Concerns

While the larger concern with the gas sector in India is to discover gas and attract private players to invest in and produce gas from the fields, there is a great impetus to formalise this price-pooling mechanism, so as to optimise utilisation of the gas and price it correctly for stakeholders.

A number of issues and concerns need to be addressed before getting into the nitty–gritty of price pooling for domestic gas. Consent from all stakeholders becomes a crucial aspect of implementing such a proposal. Apart from that, the import price of RLNG in the international market is a deciding factor for overall pricing. While the government can control and regulate the mechanism's internal structure, it is not in a position to exercise control over possible international price fluctuations.

Some of the critical issues and concerns for price pooling of gas would be:

1. **Availability of additional supplies of gas from domestic sources**
   (a) Additional supplies of gas are expected to be available from NELP sources (3.95 MMSCMD in 2014–15 and 5.78 MMSCMD in 2015–16), but these are merely tentative figures and there is no guarantee that any additional gas will be available from these sources.
   (b) The government cannot afford to ignore constant lobbying from other consuming sectors to supply them with additional gas and restore their supply.
2. **Price fluctuation of LNG in the international market**
   (a) Though the assumptions have been made at USD 13.5 per MMSCMD, this value may vary significantly because of international market factors and the demand and supply of gas.
   (b) Any upward swing in the RLNG price in the spot market has the potential of distorting the equations of price pooling in India.
   (c) Similarly, any downward movement of gas prices will favour such arrangements for a longer tenure and make the gas-based power projects viable.
3. **Re-gasification capacity and gas-transport infrastructure in India**
   (a) The surge in volume of RLNG imports requires additional re-gasification facilities in India and the connection of a dedicated gas-grid line to carry the gas to the consuming point. Any delay in and deviation from the addition to planned capacity will have a severe impact on the pooling mechanism.
4. **Treatment of the fixed cost of generation, irrespective of PLF**
   (a) The government's calculations take into consideration a fixed-cost recoverable by the plant operator at INR 0.85 per unit, irrespective of

the PLF. There is a possibility of project promoters refusing to accept this arrangement, as, by doing so, they would have to forego a considerable amount of their revenue.

5. **Probable support funding from NCEF may not be forthcoming**
   (a) The NCEF fund's basic premise is to support low-carbon footprint and the growth of clean energy in the country. The renewable sector that includes both wind and solar energy is firm in its stand that a considerable amount from this fund should be allocated to it to maintain its growth momentum. It is equally firm about preventing any diversion of this fund elsewhere.
   (b) Similarly, stakeholders from the coal sector argue that funds from NCEF should be reinvested in their sector alone for cleaner energy growth, such as coal gasification, CBM, CTL, and other similar options from coal.
   (c) In the midst of such conflicting demands, the government must decide on the best options available for utilising this fund, keeping in mind that any bid to seek financing from NCEF to meet a larger revenue gap may invite resistance from other sectors.
6. **Relaxation in terms of taxation and state government support**
   (a) Though the central government may cut its revenues through the waiver of central taxes and import duties, similar support from state governments may be difficult to obtain and convincing them to follow suit may require additional effort from the centre.
7. **The agreement on transport-tariff reduction from the gas transporters**
   (a) While GAIL, a government enterprise, may be receptive to any proposal put forward by the government, there is a possibility that other operators will resist the idea of reducing their transportation tariff and marketing margin.
8. **Regulatory help and support for implementation**
   (a) For this mechanism to be implemented in its true spirit, a deeper commitment will be required by the regulator in terms of adherence to guidelines and directives.
   (b) Any deviations from standard procedures have to be passed through the PNGRB—a difficult proposition. Convincing the PNGRB to approve such deviations would require greater co-ordination between all stakeholders.
9. **Willingness of DISCOMs to procure power at INR 5.50 per unit**
   (a) Though the government is considering capping the price of power from gas-based generation units at INR 5.50 per unit, it is also at a higher band and DISCOMs may be unwilling to procure power at this price (CERC 2014). Moreover, the recent bid for solar power sees solar power at INR 4.67 per unit and it may further reduce in times to come. With solar power

becoming very attractive, DISCOMs are very averse to buy power gas-based generation.
(b) Any subsidy arrangements or financial support from the central government would need to be formalised in accordance with the distribution utilities.
(c) Given the past experience of state-run utilities and the huge burdens of loss they have had to bear, persuading them to absorb any incremental cost of power would be an uphill task.

11 Conclusion

While the price-pooling mechanism seems to be a viable option for reviving India's ailing gas-based power generation facilities in line with fertilizer sector and can be implemented through collaborative efforts from all stakeholders, the larger issue regarding the implementation framework and the time period for which it will be sustainable remains to be resolved. Govt through its ongoing scheme has tweaked the gas pricing arrangement to a subsidy support assistance programme based on the operational performance with a minimum PLF guarantee. This may not last long with govt subsidy. The complete operationalisation and long-term operational success of the scheme would depend upon tying of other loose ends including contracting of gas supplies, DISCOM's appetite for this power, exchange rate movement and consent of all stakeholders including pipeline companies, re-gasification facilities and state governments. The need of the hour is a more proactive government that will make every effort to find ways and means of attracting more players to the sector and increase the country's gas production.

More assertive government interventions are required for market-based pricing in the gas sector, ensuring, moreover, that the pricing reflects demand–supply market dynamics. The current shielding of the fertilizer and power sectors for any upsurge in gas prices should be done away within a reasonable time frame. Such arrangements of price pooling as have been discussed can only offer a semblance of reassurance in the short run. What one needs to look at seriously is the larger picture in terms of energy security and power for all.

References

British Petroleum (June 2014) 63rd edition of the BP Statistical Review of World Energy report. BP Statistical Review of World Energy. http://www.bp.com/content/dam/bp/pdf/Energyeconomics/statistical-review-2014/BP-statistical-review-of-world-energy-2014-full-report.pdf

Central Electrical Authority (2014) Review of Progress of Thermal Power Projects under execution in the country. 148th CEA Quarterly Review (October 2014) http://www.cea.nic.in/reports/proj_mon/quarter_rev.pdf

CERC (2014) Summary of the comments and suggestions received on Approach Paper on Terms and Conditions of Tariff Regulations for the tariff period 1.4.2014 to 31.3.2019. CERC (Ref No.20/2013/CERC/Fin(Vol-I)/Tariff Reg/CERC) (June 25th 2014). http://www.cercind.gov.in/2013/regulation/Comments/Normative.pdf

Kelkar V, Parekh D, Ganguly A, Sharma RS, Mehta V, Bhattacharya A, Rao SV and Kapadia A (2014) Consultation paper by the Kelkar Committee: Towards a new E&P regime and roadmap for market determined pricing of natural gas. Committee on Roadmap for Reduction in Import dependency in Hydrocarbon Sector by 2030. Ministry of Petroleum and Natural Gas, Government of India (July 2014). http://www.dghindia.org/Admin/Document/Notices/44.pdf

Ministry of Power, Lok Sabha Starred Question No 252 on Gas based power plants, answered on July 24, 2014. http://164.100.47.132/LssNew/psearch/QResult16.aspx?qref=2096

Prime Minister's Office (2014) Summary record of discussion: gas pooling for power sector. Government of India, New Delhi, (1st September, 2014)

Part III
City Gas Distribution in India

A Comprehensive Review of City Gas Distribution in India

S.K. Kudaisya and Sanjay Kumar Kar

Abstract City gas distribution is going through transformation in India. A sincere attempt made to offer a comprehensive review of city gas distribution in country. This chapter provides critical insights into development of gas market and evolution of city gas distribution business in India. The authors discuss the causes of city gas distribution evolution. Important linkage between regulatory developments and city gas distribution has been discussed. The authors highlight the progress of city gas distribution system after constitution of Petroleum and Natural Gas Regulatory Board. The readers can find a comprehensive review of various bidding rounds as per the regulations developed by the Petroleum and Natural Gas Regulatory Board. This chapter provides meaningful insights on opportunities and challenges available in the city gas distribution business. It points out the critical factors responsible for the achievement or failure of commitments made by the entities to the regulator. The authors suggest that with the support of the Government, constant vigil and involvement of the Regulator, and professional management of entities, the city gas distribution business would gain greater penetration in India.

Keywords Natural gas · City gas distribution · Regulations · Compressed natural gas · Piped natural gas

1 Introduction

City gas distribution represents last—mile connectivity in the gas— value chain and it entails:

S.K. Kudaisya (✉)
Sabarmati Gas Limited, Gandhinagar, Gujarat, India
e-mail: kudaisyask@gmail.com

S.K. Kar
Department of Management Studies, Rajiv Gandhi Institute of Petroleum Technology, Jais, India
e-mail: skar@rgipt.ac.in

S.K. Kar and A. Gupta (eds.), *Natural Gas Markets in India*,
DOI 10.1007/978-981-10-3118-2_7

- supply of piped natural gas (PNG) at low/medium pressure to domestic households/commercial establishments and industries,
- supply of compressed natural gas (CNG) to transport sector.

1.1 Definition of city gas distribution (CGD)

CGD is defined in Section 2 (i) of PNGRB Act 2006,[1] as follows:

"City or local natural gas distribution network" means an interconnected network of gas pipelines and the associated equipment used for transporting natural gas from bulk supply high-pressure transmission main to the medium-pressure distribution grid and subsequently to the service pipes supplying natural gas to domestic, industrial or commercial premises and CNG stations situated in a specified geographical area.

1.1.1 PNGRB Act 2006

PNGRB Act 2006, vide 2(za) defined natural, as under:

"Natural gas" means gas obtained from boreholes and consisting primarily of hydrocarbons and includes:

i. gas in liquid state, namely liquefied natural gas and regasified liquefied natural gas,
ii. CNG,
iii. gas imported through transnational pipe lines, including CNG or liquefied natural gas,
iv. gas recovered from gas hydrates as natural gas,
v. methane obtained from coal seams, namely coal bed methane.

But natural gas does not include helium occurring in association with such hydrocarbons.

1.1.2 Definition of Local Distribution Entity

Vide 2 (y) PNGRB Act 2006, "local distribution entity" is defined as follows: "local distribution entity" means an entity authorized by the board under Section 20 to lay, build, operate or expand a city or local natural gas distribution network.

---

[1]PNGRB (www.pngrb.gov.in) (accessed on 5 April 2015).

1.1.3 Scope of CGD

A CGD network shall be designed to operate at a pressure as specified in the relevant regulations for technical standards and specifications, including safety standards for maintaining the volumes of supply of natural gas on a sustained basis to meet the following requirements, namely[2]:

(a) customers having requirement of natural gas up to 50,000 SCMD shall be supplied through the CGD network:
(b) customers having requirement of natural gas more than 50,000 SCMD and up to 100,000 SCMD shall be supplied:

   i. through the CGD network, or
   ii. through a pipeline not forming part of CGD network.

(c) customers having requirement of natural gas more than 100,000 SCMD shall be through a pipeline not forming part of CGD network.

2 Emergence of Natural Gas as Fuel

Historically, India has relied on coal to power its electricity sector, liquid fuels as feedstock and oil for its transport sector. Keeping in mind rapid changes in global energy scenario, including the acceptance of natural gas as cleaner and more efficient fuel, the Government of India firmed up long-term road map for securing energy needs of the country, with a focus on protecting environment as well.

With the discovery of natural gas fields (1960s) in Assam and Gujarat, the development of natural gas industry commenced, nearer the source of production.

After the discovery of south basin, Mumbai high fields by ONGC in 1970s, natural gas assumed importance in energy basket. Later discoveries were made in Gujarat, KG basin (Andhra Pradesh), Cauvery basin, Tripura, Assam, etc., which gave fillip to the usage of natural gas mainly for power, fertilizer sectors, and other industries. Natural gas was also allocated for domestic segment (Assam and Gujarat), which enabled the formation of CGDs in these states.

From 60s onward, gas was made available to various key sectors (power and fertilizers) and industries located near production sources.

In early 70s, Ankleshwar gas field was connected through transmission pipeline with Koyali, thus facilitating the usage of gas in IOCL refinery, other major industries and also domestic household sector in Baroda (Vadodara).

Towards late 70s, many gas fields and also marginal fields in Gujarat were connected through low/medium-pressure network which enabled gas usage in numerous industries at Kadi and Kalol. ONGC allocated natural gas at well heads to

---

[2]PNGRB (www.pngrb.gov.in) (accessed on 5 April 2015).

industries. Kadi Industrial cluster, took lead and transmission network from ONGC wells to Kadi Industrial area was established. Kadi pioneered gas application in ceramic, first time in country followed by glass, metal and other sectors—which in turn enabled rapid industrialization in the state of Gujarat.

2.1 Formation of Gas Authority India Limited

In 1984, Government of India set up a public sector undertaking, namely Gas Authority India Ltd (GAIL)[3] for transportation and marketing of natural gas. Pursuant to this, long-term action plan was firmed up to provide transportation infrastructure across the country. First cross-country gas transmission pipeline—HBJ (Hazira-Bijaipur-Jagdishpur)—1700 km long with a capacity of 18.2 MMSCMD was commissioned by GAIL in the year 1987–88. This facilitated the supply of gas from ONGC south basin fields near Hazira to fertilizers, power plants, gas fractionators, and industrial consumers in the state of Gujarat, Rajasthan, Madhya Pradesh, Uttar Pradesh and Delhi—a game changer. Subsequently, in 1997–98, capacity of the pipeline was enhanced to 33.4 MMSCMD flowing through a total length—2700 km.

GAIL has played a crucial role in creating gas transmission infrastructure at the pan-India level and currently operating numerous pipeline networks aggregating to around 10,000 km, which has enabled the adoption of natural gas in power/fertilizers/industries/CGDs alongside transmission routes.

2.2 Bhure Lal Committee (Statutory Committee)

In early 1990s, responding to the direction of Hon'ble Supreme Court (in matter of one PIL), GOI constituted a statutory committee, viz. the environment pollution (prevention and control), under Environment Protection Act 1986. Shri Bhure Lal, a senior bureaucrat, was appointed as the chairman of the committee. The committee was vested with the responsibility of the suggesting ways and other means of reducing pollution in Delhi and other large cities of India.

The Hon'ble Supreme Court of India also ordered that direction given by the committee shall be final and binding on all.

One of the recommendations of the committee[4] was using CNG as a transport fuel.

---

[3] www.gailonline.com (accessed on 6 April 2015).

[4] www.cpcb.nic.in/divisionsheadoffice/pci/3/02-clean%20fuel%20report (accessed on 6 April 2015).

Subsequently, considering the ground realities of clean slate, GAIL was entrusted responsibility to set up pilot projects for dispensation of CNG at Delhi, Mumbai and Vadodara. EPCA played very active role in developing this application.

On successful trial of CNG usage in dedicated vehicles, decision was taken for aggressive introduction of CNG as transport fuel, in a phased manner.

The above development later on proved to be a real game changer in history of CGDs.

2.2.1 Phasing Out Diesel Vehicles in Delhi

In mid-1990s, Smt. Sunita Narain, the director of Centre of Science and Environment (CSE), filed a law suit to force Delhi's buses, taxis and auto-rickshaws to convert to CNG, a cleaner fuel and also scrapping of all old diesel taxis and autos. The Hon' Supreme Court of India accepted the law suit and formed a committee, led by Shri Bhure Lal to enforce the judgement.

In December 2002, another landmark development resulted in driving out diesel driven buses by introducing CNG power buses. The green transport initiative resulted in introduction of 10,000 CNG buses, 12,000 CNG taxis, and 80,000 CNG cars.

2.2.2 Identification of Highly Polluted Cities and for Introduction of CNG as Transport Fuel, in a Phased Manner

Taking note of the recommendation of statutory committee, the Supreme Court in its 2002 and 2003 orders identified 17 cities, viz. Delhi, Mumbai, Agra, Lucknow, Kanpur, Varanasi, Pune, Faridabad, Patna, Ahmadabad, Sholapur, Hyderabad, Bangalore, Kolkata, Chennai, Jharia and Jodhpur.[5]

2.3 Long-Term Agreement for Import of LNG (Liquefied Natural Gas) and Setting Up LNG Import Terminal

In late 90s, Petronet LNG (PLL) was formed by GOI, to import liquefied natural gas (LNG) into country and set up LNG receiving/regasification terminals. The PLL entered into a long-term (25 years) agreement with Ras Laffen Liquefied Natural Gas Co. Ltd., Qatar, for sale and purchase of 5 MTPA LNG (which subsequently increased to 7.5 MMTPA). First terminal to receive LNG, regasify and supply regasified natural gas to customers through dedicated transmission pipeline (laid by GAIL) was commissioned at Dahej (Gujarat) in late 2004 by Petronet LNG Ltd.

---

[5]www.icra.in, January 2012 (accessed on 6 April 2015).

Additional availability of natural gas through the above-mentioned initiative facilitated aggressive marketing of natural gas in country.

PLL is a joint venture company, promoted by GAIL, ONGC, IOCL and BPCL, each having 12.5% equity holding.[6] GDG, France, with 10% equity holding was roped in as strategic partner for the project. Out of remaining equity, 5.2% is held by Asian Development (ADB), and balance 34.8% is held by the public.

First LNG terminal (2.5 TMTPA) in the private sector by Shell was commissioned in mid-2006 at Hazira, with a different business model. Hazira is global first for the LNG industry—the traditional model of gas imports calls for long-term contracts (over 20 years), but Hazira offers flexibility in contracts and matches customer demand profile by seasonality.

Capacity of the Dahej terminal has been expanded to 10 MMTPA, and further augmentation to 15 MMTPA is under progress (target: December 16). PLL has set up the second 5 MMTPA terminal at Kochi (Kerala) in 2013. The company has initiated actions to set up third terminal (5 MMTPA) at Gangavaram, Andhra Pradesh.

2.4 *Initiative of Gujarat State Govt.*

The Gujarat State Government in early 2000s, as a part of vision document, identified natural gas as a fuel of twenty-first century.

To achieve this objective:

i. An interim gas policy was announced by Gujarat Govt., and expression of interest was sought for distribution of natural gas in 5 zones.
ii. A state undertaking, Gujarat State Petroleum Corp. Ltd. (GSPCL), was established as a nodal agency to steer indigenous exploration and production of crude oil and natural gas and provide necessary infrastructure to supply natural gas to all segments of customers, across the state.

Award of no objection certificates for distribution of natural gas was finalized in the year 2003. (Ahmadabad was awarded to Adani Gas, a private entity, Gandhinagar, Mehsana and Sabarkantha districts to Bharat Petroleum Corp. Ltd. (BPCL), Bhavnagar to GAIL and rest of the districts to GSPC Gas). However, GAIL did not proceed to set up CGD at Bhavnagar. GSPC was, therefore, entrusted the task of providing necessary infrastructure for transportation of gas and ultimately provide the last-mile connectivity to end-users.

Gujarat State Govt. took exceptional lead in developing a vibrant gas distribution network and market.

The bold initiative of State Govt. along with the development of gas distribution network by Gujarat Gas Company Ltd. (GGCL established in 1988) propelled

[6]www.petronetlng.com (accessed on 7 April 2015).

Gujarat, in forefront of promoting gas culture, and eventually, Gujarat emerged as the largest consumer of natural gas in country.

GSPC group has since played a major role in promoting gas usage across Gujarat state and is engaged in gas value chain in totality, i.e.

- Production of natural gas,
- Operations of gas fields,
- Import of LNG through short-term and spot contracts at Dahej and Hazira terminals,
- Direct sales of regasified gas to fertilizer plants, power plants and other industries, besides making gas available to CGDs.
- Laying and building transmission network across Gujarat state and now through SPVs to other states. Through its subsidiary, GSPL is currently operating around 2500 km network).
- City gas distribution—GSPC Gas (as subsidiary), GGCL (taken over in 2013) and JV partner of Sabarmati Gas Ltd.

2.5 Identification and Road Map to Expand CGD Network to 330 Cities by PNGRB

With the backdrop of:

- Judicial intervention (early 90s) forcing the Central Govt. to aggressively introduce CNG as transport fuel in identified cities (to begin with Delhi and Mumbai), with a view to reduce ever-rising pollution.
- Judicial and regulatory monitoring enabled fast-track CGD projects in Delhi and Mumbai (late 90s).
- Support of Delhi State Govt.—including land allotment for putting a large number of CNG stations and also the lowest sales tax/VAT.
- Adoption of stringent emissions norm and action plan for next levels.
- Professional management of Delhi and Mumbai CGDs, leading to faster penetration, not only in CNG sector but also in domestic and commercial/industrial segments as well.
- Judicial intervention (late 90s/early 2000s) directing Govt. to ensure scraping all 15-year-old buses, taxis and autos latest by 2002–03 in Delhi and replacement through CNG as fuel only.
- Acceptance of CNG as transport fuel by private car owners (low-cost/lower emissions).
- Aggressive expansion of gas network in the state of Gujarat (Gujarat Gas Company Ltd., GSPC group).

In June 2006, immediately after formation, PNGRB embarked upon structured and long term planning for expansion of natural gas network across the country.

In the year 2008-09, feasibility of introducing Natural gas as CNG (transport) and PNG (domestic) as well as commercial/industrial segments, in 300 + geographical areas was established. The cities were classified into 4 categories, based on assessment of economic potential, which are summarized, as under:

- Type I: Metro cities with high population and a large number of vehicles (10+).
- Type II: Medium to large cities with limited industrial potential (10+).
- Type III: Highly Industrialized cities/towns/talukas (70+).
- Type III: Small cities/towns/talukas with limited industrial presence/potential (majority of cities falls in this category).

Simultaneously, a time bound action plan was drawn for awarding/authorizing for laying, building, and operating new gas transmission pipelines across the country (approx. 12000 km) on common carrier basis.

PNGRB, therefore, firmed up detailed action plan to award authorization to entities for the development of city gas infrastructure in more than 300 cities, within next 5 years, i.e. 2015.

However, for various reasons, some beyond direct control of PNGRB, between early 2008 and till July 15, only 23 new gas (of 4 bids of rounds, covering 35 GAs) could be awarded, while bids of 5th round of bidding (20 cities) concluded in June 2015 and bids for 10 GAs are under evaluation (no bid for 8 GAs, only 1 bid for 2 GAs).[7]

2.6 *Allocation of Domestic Gas to CGD Entities for CNG (Transport) and PNG (Domestic)*

In compliance with the directive of the Hon'ble Supreme Court of India/statutory committee, in the mid-90s the Central Govt. allocated domestic gas (APM) for catering requirement of CNG (transport) and PNG (domestic) segments in the cities of Delhi and Mumbai. Indraprastha Gas Ltd. (IGL) and Mahanagar Gas Ltd. (MGL), therefore, for a long period continued to receive APM gas which enabled rapid implementation/compliance with the directive, IGL and MGL, to cater increase in demand of these sectors had to utilize high-priced RLNG, which resulted in increase in retail selling price. However, both CGDs continue aggressive expansion and periodically passed on additional cost to end consumers.

GAIL, as a part of joint ventures, in many CGDs allocated RLNG (of long-term contracts); therefore, such CGDs continued to be benefited for a considerable period. However, to meet increased demand of these sectors, such CGDs also had to

[7]PNGRB and industry research papers (including Dolat Capital-January 2013, ATKerney-March, 2012).

procure RLNG, through short-term/spot contracts, at much higher price than APM gas as well as long-term contract RLNG.

MOP&NG, in mid-2009, in line with gas allocation policy 2008 and taking into consideration projection of huge production of natural gas (up to 80 MMSCMD) in stages, from new discoveries made by Reliance, allocated gas from KG basin to CGD entities.[8]

In line with the recommendations of Empowered Group of Ministers (EGoM) set up by the Central Govt., natural gas allocation (APM + new finds, etc.) to CGDs was restricted to 6 MMSCMD. In mid-2009, domestic gas from new discoveries in KG basin was allotted to CGDs authorized by Central Govt./ PNGRB to cater the needs of transport sector (CNG) and domestic households (PNG). This in turn propelled aggressive expansion of the network. However, due to sudden and steep drop in gas production at Reliance field in KG basin, supply of allocated gas was suspended from May 2011. Consequently, such CGDs had to procure RLNG through short-term/spot contracts entered by PLL, GSPC and GAIL, at international prices, which incidentally were 3–5 times of domestic gas price (4.2 $/MMBTU).

Thus, majority of CGDs had to depend on RLNG through long-term contracts and from 2011 onwards through short-term/spot contracts, at much higher price (3–6 times of domestic gas).

The above phenomenon resulted in the discrimination among CGDs for serving same segments' customers.

Aggrieved with this discrimination, one NGO in Ahmadabad filed a writ petition WP (PIL)—no. 47/11 and 54/11 before the Hon'ble High Court of Gujarat. The Hon'ble High Court of Gujarat, vide order dated 25.07.2012, directed it to the Govt. of India as below:

I. To allot natural gas for domestic and vehicular usage at the same rate to the city of Ahmadabad at which the same is supplied to Delhi and Mumbai, to enforce the right of equality.
II. To not discriminate between CGDs promoted by the central PSUs and other CGDs but also among Gujarat-based CGDs in the matter of allocation.

In order to comply with the above-mentioned order, MOP&NG, vide letter ref no L-16013/3/2012/GP II DATED 14 November 2013, issued guidelines for allocation of domestic gas[9] to cater demand of CNG (transport) and PNG (domestic) segments to all CGDs to GAIL India Ltd. Salient features are reproduced below:

- The proportion of supply of domestic gas vis-a-vis RLNG with respect to consumption in CNG (transport) and PNG (domestic) segments be kept uniform for all CGDs.

[8]Presentations in conferences.

[9]MOP&NG (www.petroleum.nic.in) (accessed on 15 April 2015).

To start with, actual consumption figures of CNG (transport) and PNG (domestic) segments of CGD entities for the year 2012–13 would be taken as a basis to determine the entity-wise supply of gas.

- GAIL would supply domestic gas to CGD entities at uniform base price. However, the delivered price of domestic gas to individual CGD may vary on account of transportation charges and local taxes and duties.
- Allocation of domestic gas for all CGD entities supplying CNG (transport) and PNG (domestic) stands cancelled/withdrawn except in the case of Tripura Natural Gas Company Ltd. and Assam Gas Company Ltd. (in view of operational issues).
- For supplying domestic gas to the CGD entities to cater the need of transport sector (CNG) and domestic household (PNG for cooking) about 6.4 MMSCMD of domestic gas (0.47 PMT + 5.93 APM) was allocated to GAIL. Thus on an average 80 % of actual demand/sale (8.2 MMSCM), in year 2012-13) was to be supplied uniformly to each CGD entities, without discrimination amongst CGD entities, subject to operational imperatives.
- In order to operationalize the above arrangement, GAIL would enter into necessary swapping arrangements with other entities wherever required.
- The requirement of any entity exceeding the allocation (out of total 6.4 MMSCMD) may be sourced by the CGD entity through RLNG or any alternate source, as may be available.

GAIL was directed by MOP&NG to finalize the operational modalities for these arrangements, which should come into effect at the earliest and not later than 30 November 2013.

The above guidelines were implemented by GAIL towards the end of Nov 2013.

MOP&NG, vide letter ref no L-16013/3/2012-GP II[10] dated 3 February 2014, issued revised guidelines for allocation/supply of domestic gas to CGD entities for CNG (transport) and PNG (domestic) sectors. Salient features/revisions are summarized below:

- Take into consideration the average consumption of natural gas in CNG (transport) and PNG (domestic) @ 8.32 MMSCMD. The allocation of domestic gas to GAIL, for supplying to CGD entities for CNG (transport) and PNG (domestic) segments, was increased to 8.32 MMSCMD with immediate effect. Thus, allocation would be supplied fully and was placed in no-cut category.
- The allocation of domestic gas to GAIL, for supplying to CGD entities for CNG (transport) and PNG (domestic) segments, shall be reviewed periodically.
- The allocation shall be revised based on the actual consumption in the previous period. The additional demand of natural gas arising between two reviews shall be met by the CGD entity through market price.

[10]MOP&NG (www.petroleum.nic.in) (accessed on 15 April 2015).

With the implementation of the above guidelines, all CGD entities are now having level—playing field and retail selling price of CNG (transport) and PNG (domestic) have been revised downwards by all CGD entities—a huge benefit to end consumer.

The guidelines coupled with announcement of Gas Pricing Policy 2014 by the Govt. of India are really historical decisions and shall enable rapid expansion of CNG (transport) and PNG (domestic) segments of existing CGD entities as well as favourable response during bidding rounds for the award of new geographical areas across the country.

2.7 *Revision of Domestic Gas Price*

The Government of India had constituted a committee under the chairmanship of Dr C. Rangrajan, the then chairman of the Economic Advisory Council to the Prime Minister in May 2012 to look into the production sharing contract (PSC) mechanism in petroleum industry. One of the terms of reference of the committee was to suggest structure and elements of the guidelines for determining the basis or formula for the price of domestically produced gas and monitoring actual price fixation.

In December 2012, the committee submitted its report on PSC mechanism in petroleum industry. The report also recommended a formula for pricing of domestically produced gas. The recommendations on gas pricing also mentioned:

- The guidelines shall applicable for all natural gas produced domestically, irrespective of the source (i.e. including CBM and shale gas).
- These guidelines shall not be applicable where prices have been fixed contractually, for a certain period of time, till the end of such period.
- The guidelines shall be applicable with prospective date only.
- Allocation of gas will be as per the prevailing gas utilization policy (GUP) of the Govt. of India.
- Domestic gas prices shall be notified in advance on a quarterly basis using the data for last four quarters, with a lag on one quarter.
- These policy guidelines shall be applicable for a period of five years with effect from 1 April 2014.

The recommendations of the committee[11] were considered by the Cabinet Committee on Economic Affairs (CCEA). Pursuant to this, on 10 January 2014, MOP&NG, VIDE REF NO 22011/3/2012-ongd.v) notified the Domestic Natural Gas Pricing Guidelines, 2014.

[11]MOP&NG-report on Gas Pricing Policy 2014 (accessed on 16 April 2015).

It is understood that based on the suggested formula, price of domestic gas w.e.f. 1 April 2014 would have been around 8.2 \$/MMBTU (almost double of price prevailing @ 4.2 \$/MMBTU).

The notification was to be implemented with effect from 1 April 2014. But the same could not be implemented as the Election Commission deferred the notification till the completion of 2014 Lok Sabha General Election process.

After the new Government assumed office, it is decided that the whole issue of gas pricing be compressively re-examined and directed that the Domestic Natural Gas Pricing Guidelines 2014 be kept in abeyance up to 30 September 2014, and till that time, the domestically produced gas may continue to be priced at the rate prevailing on 31 March 2014.

On 13 August 2014, the Ministry of Petroleum and Natural Gas (MOP&NG) constituted a committee comprising of the following members:

1. Shri P.K Sinha, secretary (power), GOI, member.
2. Shri Ratan Watal, secretary (expenditure), GOI, member.
3. Shri J.K. Mohpatra, secretary (fertilizers), GOI, member.
4. Shri Rajive Kumar, additional secretary (P&NG), GOI, member secretary.

The committee was to undertake a comprehensive re-examination of the issue of gas pricing. The committee was also mandated to consult major stakeholders before submitting its recommendations.

The new committee, after long deliberations and consultations with stakeholders, submitted its report to MOP&NG, which was finally approved by the Govt. of India.

Cabinet Committee on Economic Affairs (CCEA) chaired by Hon'ble Prime Minister approved the policy[12] on 18 October 2014.

Pursuant to the approval of CCEA, MOP&NG, on 25 October 2014, notified new guidelines for pricing of domestically produced gas. MOP&NG further issued:

- Sources of data for determination of domestic natural gas price as per the New Domestic Natural Gas Pricing Guidelines 2014.
- Pricing of gas produced from small/isolated fields (ref no L-12022/8/2014-GP II dated 19 December 2014).[13]

The recommendation also specified the following:

- There is no distinction between APM and non-APM pricing,
- Uniform price for national oil companies (NOCs) and NELP blocks,
- In the case of new supplies or where the duration of existing contracts has been completed by, the pricing would be determined by NOCs by calling bids an open competitive bidding process (shall be awarded too highest bidder). The

[12]www.pib.nic.in/newssite/pressrelease.aspx,dated (accessed on 16 April 2015).

[13]MOP&NG www.petroleum.nic.in (accessed on 16 April 2015).

reserve price would be as per notified on the basis of New Domestic Natural Gas Pricing Guidelines 2014,
- The pricing shall be determined for a period of six months.

Subsequently, Director General, Petroleum Planning and Analysis Cell (PPAC), Govt. of India, vide letter ref no. PPAC/Gas Pricing/Nov, 2014–March, 2015, dated 26th October, 2014, notified Domestic Gas Price @ 5.05 $/MMBTU (On Gross Calorific Value for a period from 1st Nov, 2014 to 31st March, 2015).

In line with the recommendations of the new committee and based on the input data, PPAC, vide letter ref no PPAC/Gas Pricing/April 2015 to September 2015, dated 31 March 2015, notified gas price @ 4.66 $/MMBTU (on gross calorific value basis).[14]

Subsequently, PPAC, vide letter ref no PPAC/Gas Pricing/October 2015 to March 2016, dated 30 September 2015, notified gas price @ 3.82 $/MMBTU (on gross calorific value basis).[15]

It is pertinent to record that with implantation of New Domestic Natural Gas Pricing Guidelines 2014 and revised allocation of domestically produced gas to all CGDs (for CNG (transport) and PNG (domestic)) segments, all CGDs are now having level—playing field as well receiving the gas at most competitive pricing—a huge positive impact.

3 Evolution of City Gas Distribution in India

GOI, taking into considerations limited availability of natural gas (huge gap in demand and supplies) and needing to provide fuel/feedstocks to power plants and fertilizers, motivated gas usage from production sources to nearby locations, in the initial years (60s, 70s and 80s). Domestic households were accorded lower priority, and therefore, very few CGDs could be established. With increased availability of indigenous gas and also import of LNG in mid-2000s, natural gas industry got the real thrust.

During the last 5 decades, numerous CGDs have been established, mainly in the state of Assam, Tripura, Gujarat, Maharashtra, Madhya Pradesh, Rajasthan, Delhi/National Capital Region, Uttar Pradesh and Andhra Pradesh, mainly due to concentration of gas transmission network in these states.

With this backdrop, CGDs are being discussed under various groups, in the following subsections:

[14]www.ppac.org.in (accessed on 16 April 2015).

[15]www.ppac.org.in (accessed on 12 January 2016).

3.1 CGDs in the Vicinity of Gas Production Sources

3.1.1 Assam Gas Company Ltd. (AGCL)

AGCL, a Govt. of Assam undertaking, was incorporated in March 1962 in Shillong with headquarters at Duliajan AGCL having distinction of being first gas distribution company in India. Gas transportation commenced in 1967 to Namrup (power plant) and nearby located tea estates. Subsequently, pipeline network has been established by AGCL and gas produced in nearby fields is being marketed in 6 districts of Assam—Dibrugarh, Tinsukia, Sivsagar, Jorhat, Golaghat and Cachar to various sectors (power/fertilizers/tea gardens and also to domestic households and commercial establishments).

Currently, AGCL, through network (10,000 km of steel pipelines, 30,000 km of MDPE pipelines with associated facilities) is providing natural gas to various power plants/fertilizers, around 2800 tea estates and 28,000 domestic households—a noteworthy achievement.

AGCL recently has taken decision to aggressively expand business to CNG segment as well as other areas of Assam (Guwahati, New Bongaigaon, etc.) and explore possibility to form joint venture with one of the public sector undertakings (PSU), e.g. Oil India Ltd. (OIL), Numaligarh Refinery Ltd. (NRL) and IOCL.

AGCL, through likely JV partner, also intends to establish gas-based power plants (of 5–60 MW capacities) in the state of Assam.

PNGRB, under regulation, vide letter ref no Infra//ii/1/2008-Vol-ii/Upper Assam, dated 6 February 2015, accepted the Central Govt., authorization to AGCL for CGD network in Upper Assam geographical area.

3.1.2 Vadodara Mahanagar Sewa Sadan (VMSS), Gujarat

In line with the vision of then State Govt. of Gujarat, VMSS has distinction to establish gas distribution in Baroda, way back in 1972, with allocation of domestic gas by GOI. In fact, Baroda (now Vadodara) emerged, as one of the first cities in the country to have piped gas supplies to domestic households/commercial establishments.

It may not be out of context that after bifurcation of Maharashtra/formation of Gujarat state in 1960, the State Govt. has drawn an ambitious plan for rapid industrialization of the state. Consequently, a large number of major industries were established in late 60s onwards—at Baroda, Koyali Refinery of IOCL (with equity participation of Gujarat Govt.), Indian Petrochemicals Ltd. (IPCL) and Gujarat State Fertilizer Ltd. (GSFC) as public sector undertakings.

Thoughtful and aggressive execution of Govt.'s vision, rapid industrialization followed in subsequent years (first industrial estate was set up in Vapi, in late 60s).

The infrastructure, in phases, was expanded to cater need of around 76,000 households and 2800 commercial establishments, through network of 13-km steel pipelines and 750-km MDPE pipelines with associated facilities.

Pursuant to the direction of Hon'ble Supreme Court of India/Bhure Lal Committee, GAIL set up pilot project for introduction of CNG to transport sector, towards 1992. CNG being cleaner and more efficient fuel was readily accepted by end-users. GAIL in stages set up around 9 CNG stations across the city.

Successful operation and management of CGD by VMSS for more than 4 decades is a remarkable achievement by any municipal corporation in country.

With a view to synergize strength of VMSS and GAIL and aggressively expand gas business, a joint venture (with 50:50 stakes), in name and style—Vadodara Gas Ltd. (VGL), has been formed and incorporated in September 2013.

VGL has drawn aggressive and ambitious plan to expand business in next 5 years (5 lakhs domestic households, 2000 commercial customers and 1000 industrial customers).

VGL shall be expanding to adjoining markets, and PNG and CNG distribution network shall be provided in Nandasari, Waghodia, Savali, Padra, Pore, etc., also.

Unfortunately, highly successful model of providing piped natural gas to residential areas/commercial establishment was not pursued further. If implemented it would have cost efficiency, high operating efficiency and lead to job creations at various levels to local people.

3.1.3 Gujarat Gas Company Ltd. (GGCL)

With a view to exploit opportunity provided by gas finds in south Gujarat, towards 70s and 80s, Govt. of India (through ONGC), in late 80s, allocated natural gas (0.17 MMSCMD) to Gujarat Industrial Investment Corporation (GIIC), an investment arm of State Govt.—Gujarat Amino Chem Ltd. (GACL) was incorporated in year 1987.

Then textile leader Mafatlal group took lead and a joint venture agreement with GACL was signed in June 1988 for purchase and distribution of natural gas to industries, commercial establishments and also residential areas in Ankleshwar and Bharuch markets of south Gujarat (hub of textile, chemicals, diamond business and numerous heavy industries).

The name GACL was changed to Gujarat Gas Company Ltd. (GGCL).

The JV implemented project very aggressively, and gas supplies to industries were commissioned in November 1989. With increased availability and market demand, GOI further allocated gas (0.3 MMSCMD) and for extension of network to Surat, diamond and textile hub of the country.

First CNG station at Surat was commissioned in February 1992.

JV approached financial market to raise resources towards late 1996. In the following year, i.e. 1997, British Gas (BG) Group acquired majority stakes from Mafatlal group, GIIC and others. Thus, BG group took full control of management and GGCL became a subsidiary of BG group.

BG group decided to exit from GGCL in the year 2011–12 and accordingly invited bids for transfer/sale of its equity stake. GSPC group, through GSPC distribution network, decided to bid for the same. Consequently, in October 2012, share purchase agreement (SPA) for 65.12% equity (83,518,750 shares) was signed between GSPC distribution network and erstwhile promoters (BGPAH) for acquisition. As per the SEBI guidelines, additional 7.58% equity was acquired by GSPC group from erstwhile promoters, thus total aggregating 73.70%.

In the year 2013, the process of transfer of equity with due approvals was completed, and since then, GSPC distribution network has taken management control of the company (GGCL).

GGCL and GSPC Gas (board of directors) had approved amalgamation of these entities to form Gujarat Distribution Network Ltd. (GDNL).

Recently, High Court of Gujarat sanctioned the scheme of arrangement of amalgamation of GGCL, along with its 2 subsidiaries (Gujarat State Gas Fin Service Ltd. and Gujarat Gas Trading Co Ltd.) and GSPC Gas to be known as GDNL.

GGCL, from mid-80s, has been the pioneer in city gas distribution and is in forefront and serves entire range of retail end of the gas value chain. Numerous applications of natural gas/commercial sectors were facilitated by GGCL. The area, being highly industrialized, picked up momentum from 2004 onwards (industrial segment being approximately 80% plus).

Gas was sourced through mixed portfolio—indigenous (PMT fields), long-term/short-term contracts and also spot purchase to cater ever-increasing demand. GGCL adopted very unique model for introduction of CNG—besides company-owned stations, a large number of CNG stations were set up through franchise model and few at Oil Marketing Companies.

GGCL adopted COST-PLUS model and was successful in periodically passing increase in procurement cost of gas as well as distribution and other operating cost to end-users and thus sustained profitability for a very long period. However, with drop in domestic gas allocation from 2009 onwards, GGCL adopted the following multipronged strategy to sustain top-line growth on year to year basis:

- Continued to assess and evaluate opportunities in high value and high volumes,
- Sectors (industrial/commercial)—additional connected load,
- Focus on gas supplies to high volumes with high margins only, not pursuing customers' segments which are not willing to pay higher prices, thus restricting gas supplies to low-margin sectors of industry,
- Extend network to new potential areas, unconnected zones,
- Intensifying utilization of existing network,
- CNG infrastructure in villages/towns and road junctions.

With the above strategy, industrial volumes, which declined in CY09, improved in CY10 and CY11. Due to a steep increase in RLNG prices on short-term/spot basis, volumes again dropped in CY12 and CY13. In view of drop in total allocation of domestic gas (from Nov 13), GGCL had also to source gas requirement for industrial/commercial segments through RLNG (short-term/spot contracts at very high price) which further eroded industrial volumes (many industries switching to alternate fuels). Consequently, share of industrial segment in total sales has declined significantly.

Table 1 Financial performance of GGCL (2011–2015)

	FY 14–15	FY 13–14	FY 12–13	FY 11–12
Net revenue (Rs. crore)	2512.24	3116.44	3157.56	2454.39
Profit after tax (Rs. crore)	312.09	359.33	286.649	286.649

Source Annual reports of GGCL

Table 2 Milestones agreed between GGCL and PNGRB for Surat-Bharuch-Ankleshwar

	Existing (November 12)	By December 2013	By December 2014	By December 2015
Steel pipelines (inch kms)	4300	4450	4500	4750
Compression capacity (kg/day)	451,000	475,000	499,000	522,000
PNG (domestic) lakhs	3.20	4.05	4.90	5.76

Source PNGRB

However, GGCL consistently sustained topline and PAT (Table 1).

The Hon'ble Supreme Court of India directed government to allocate cheaper domestic natural gas to CGDs to cater demand of domestic (PNG) and transport (CNG) segments. As a result of this development GGCL was allotted domestic gas to meet 80% (Dec. 2013) - 100 % (Feb. 2014 onwards) of its demand in the transport and domestic household segments. This in turn resulted in steep reduction of Retail Selling Prices of domestic PNG & CNG to end customers—leading to additional demand in these segments.

Gross revenue of Rs. 2512 crores in FY 15 was approximately 20% lower than FY 14 (Rs. 3112 crores. Profit after taxes, Rs. 312 crores in FY 15, was only 13% lower than PAT of Rs. 359 crores in FY 14, which indicates despite change of management control, business strategy of GGCL (COST-PLUS) continues. Average gas sales in FY 15 have further plunged from high of 3.5 MMSCMD (2011–12), 3.17 MMSCMD (2012–13) and 2.90 MMSCMD (2013–14) to 2.2 MMSCMD.

PNGRB, under Section 18 (i) of regulation, vide letter ref no Infra/ii/1/2008-Vol ii/Surat-Bharuch-Ankleshwar GA, dated 8 November 2012, issued the grant of authorization to GGCL for Surat-Bharuch-Ankleshwar geographical areas. PNGRB granted market exclusivity for three years (up to December 2015) and milestones to be achieved during the exclusivity period (Table 2).

GGCL has since firmed up plan for achieving agreed milestones (additional 2.46 lakhs PNG domestic connections and 1.21 lakhs additional CNG compressing capacity) in a time-bound manner.

However, till end 2013, progress in this regard had been slower than required to achieve the milestones at the end of each year, which is attributed to GGCL strategy to restrict gas volumes, in view of sourcing RLNG to meet incremental demand, through short-term/spot contracts (at much higher price, ranging 16–22 $/MMBTU).

Going forward, GGCL has signed long-term contract for procurement of 0.85 MMSCMD RLNG from GSPC. Procurement price for gas sourced for industrial/commercial segments is expected to be lower than spot contracts, thereby fuelling revival of industrial segment volumes as well as profits.

GGCL, currently serving around 5.00 lakhs PNG (domestic) customers, an increase of 1.70 lakhs customers in last 3 FYs (3.30 lakhs as on March 2012) is quite encouraging.

GGCL also serving 1700+ industries and 9200+ commercial establishments spread across various markets in south Gujarat, through a network—550-km steel pipelines, 4000 km MDPE pipelines along with associated facilities.

GGCL is operating 67 CNG stations to provide cleaner fuel to more than 2.5 lakhs vehicles. CNG sales of approximately 0.39 MMSCMD contribute 17.5% to total gas sales (approx. 2.2 MMSCMD).[16]

Role of GGCL in Gas Market Development in Gujarat

GGCL together with GSPC group has played a very key and vital role in developing a mature gas market of Gujarat. Many other states have learned lesson and intend to replicate the model.

Penetration to New Geographical Areas

Bhavnagar (Gujarat) Bid Round III (2010): On 5 March 2014, PNGRB, under Section 5 of regulation, granted authorization[17] to GGCL for the development of CGD network in Bhavnagar GA (8153 m^2 area).

The project milestones (Table 3) were submitted along with the Bid by GGCL, which were to be achieved by end of 5th year of authorization.

Going by the track record of most successful CGD penetration in domestic segment, milestone aggregation to 4.90 lakhs (around 100% of households in both districts) and similar milestones committed in other gas of 3rd round appear to be highly ambitious. This could be the result of GGCL (under earlier management) to desperately win the GA. GGCL has taken over CNG stations developed by group company GSPC Gas before the grant of the authorization (in favour of GGCL) and has commenced operations.

GGCL has also begun to lay/extend steel and MDPE network in the district of Bhavnagar and also in Botad (newly carved out district in August 2013, out of Bhavnagar and Ahmadabad districts), and connectivity to households is expected to commence towards the end of June 2015.

[16]www.gujaratgas.com (accessed on 5 May 2015).

[17]PNGRB, vide letter PNGRB/CGD/BID/R-3/2/GA Bhavnagar, dated 5 March 2014.

Table 3 Milestones agreed between GGCL and PNGRB for Bhavnagar

	2016	2017	2018	2019	2020
Steel pipelines (inch kms)	2695.73	2695.73	2695.73	898.58	NIL
Compression capacity (kg/day)					
PNG (domestic)	122,500	122,500	122,500	122,500	

Source PNGRB

Going by the release of maximum 50,000 domestic connections, each in CY13 and CY14 GGCL shall have to adopt a different strategy to bridge the huge gap created so far and achieve year-wise milestones in the end of 2nd/3rd year of authorization.

Thane (Maharashtra) Bid Round IV: After take over by GSPC group, GGCL revised its strategy of earlier management—focusing on its geographical areas in Gujarat. GGCL submitted bids in 4th round, for Thane district (excluding already authorized areas to MGL) in Maharashtra.

On 10 April 2015, PNGRB granted authorization[18] for the development of CGD network in Thane district (excluding already authorized areas to Mahanagar Gas Ltd.).

As per the revised criteria finalized for 4th round of bidding by PNGRB, considering a total area (8683 m^2) and population of 10.29 lakhs, during exclusivity period of 5 years (from the date of authorization), GGCL shall have to:

- Develop pipelines (steel + MDPE) network aggregating to 1800 in km (total area: 8683 m^2).
- 51,452 PNG (domestic) connections, @ 5% of total households (10.23 lakhs).[19]

GSPC group, now having majority stake in GGCL is expected to expand CGD business across various markets in India.

3.1.4 Tripura Natural Gas Company Ltd. (TNGCL)

With an objective to utilize natural gas from nearby gas fields, a joint venture of Tripura Industrial Corporation (TIDC) and AGCL, Assam, was formed in July 1990 for the development of city gas distribution in Agartala and extends beyond Agartala city. The joint venture was restructured in Feb 2005, and GAIL joined as major stakeholder.

[18]PNGRB, vide letter ref no PNGRB/CGD/BID/R-4/2013/08/GA-Thane, dated 10 April 2015.
[19]PNGRB CGD network-authorizations (accessed on 9 May 2015).

As on March 2015, TNGCL has provided PNG to 18,000 domestic households, 254 commercial establishments and 45 industries. Four CNG stations have also been set up catering need of approximately 6000 vehicles.[20]

PNGRB, under Section 17 of regulation, accepted the Central Govt. authorization[21] to TNGCL for the development of Agartala CGD network.

3.1.5 Charotar Gas Sahakari Mandali Limited (CGSML)

In the backdrop of phenomenal success of milk dairy cooperatives in the state of Gujarat and with motivation of then State Govt. of Gujarat, Charotar Gas Sahakari Mandali Ltd. was registered under Cooperative Act around 1998 for the development of city gas distribution in and adjoining area of Anand.

For natural gas supplies, CGSML tied up with a franchise of PNGC, viz. Intralink Pvt Ltd. (IPL), which has marketing rights of ONGC gas (this arrangement continued till April 2004, thereafter gas purchase agreement with GSPC).

Natural gas pipeline (68 km long) was laid to transport natural gas from ONGC fields at Bavda. The transmission pipeline from Bavda to Bakral was commissioned as fast-track project in June 2001. Simultaneously, laying of MDPE pipelines to provide last-mile connectivity to end-users (domestic households and educational institutes/commercial establishment) was undertaken by CGSM.

CGSML expanded MDPE network at a fast pace in Anand and 10 nearby cities/villages and is currently serving:

- Approximately 19,000 PNG (domestic) households,
- 122 educational and research centres,
- 75 plus trusts,
- 28 hospitals,
- 220 plus commercial establishment,
- 110 plus industries across the area.

One online CNG station also commissioned at Anand, which through 12 dispensers is marketing more than 18,000 kg/day CNG to approximately 6000 plus vehicles.

Due to non-availability of APM gas, CGSML purchases natural gas from GSPC at market-determined price; however, very high level of customer service is ensured to various sections of society.

CGSM has already crossed turnover of Rs. 100 crores and its march continues.

Analysis of the phenomenal progress made by CGSM clearly established a cooperative business model for providing natural gas, a cleaner fuel to various sections of the society.

[20]www.tngcl.com (accessed on 10 May 2015).

[21]PNGRB, vide letter ref no Infra/ii/1/2008-Vol-ii/Agartala GA, dated 24 February 2015.

Unfortunately, CGSM's successful model was not pursued further, which, if adopted, would have enabled low-cost, highly efficient business entities, benefiting all stakeholders.

PNGRB, vide letter ref. no S-Infra/ii/1/2008-Vol ii/ANAND KHEDA GA, dated 12 May 2015, granted authorization to CGSM for the development of network in Anand GA, along with Kanjari and Vadtal villages (in Kheda district).

3.1.6 Great Eastern Energy Corporation Ltd.

YK Modi group in the year 1993 signed MOU and licence agreement with Coal India Ltd. for exploration and production of natural gas from CBM (coal bed methane) in Raniganj south block (210 km^2 area). Exploration activities commenced in 2002, and gas reserves of 2.44 TCF (approx. 9 MMSCMD) are estimated. GEECL has developed as fully integrated energy CBM Company.

Between 2007 and 2009, GEECL:

- Commissioned gas gathering station and all related infrastructure,
- Steel pipeline (76 km) Asansol-Raniganj-Durgapur,
- Commenced commercial production/supplies to industries.

Currently, GEECL has signed contracts for supply of 42.50 MMSCFD gas to various industries and producing/marketing approximately 10.38 MMSCFD CBM gas. GEECL has entered into agreement with IOCL and BPCL for setting up CNG stations at OMC's retail outlets. GEECL has also entered into agreement with Tamil Nadu State Govt. for E&P activities in Mannargudi block (667 km^2 area), and proven reserves of gas @ 0.98 TCF have been estimated.[22]

PNGRB, as a part of 3rd round of bidding (January 2010), invited bids for the development of city gas distribution network in Asansol-Durgapur—issue/grant of authorization still awaited.[23]

4 Growth of CGD and Judicial Interventions

The Hon'ble Supreme Court of India directed the Central Govt. (early 1992) to introduce CNG, a cleaner and more efficient fuel to transport sector, to begin with in Delhi and Mumbai, in a time-bound manner, with sole objective of reducing pollution in major cities of country.

It was also ordered that Supreme Court shall be periodically monitoring progress with respect to the introduction of CNG as transport fuel and also have impact on pollution.

[22]www.geecl.com (accessed on 5 June 2015).

[23]PNGRB CGD Network-3rd round bids (Jan, 2010) (accessed on 5 June 2015).

GAIL was entrusted responsibility to set up pilot projects at Delhi, Mumbai and Baroda.

GAIL took immediate actions, and statutory committee (Bhure Lal) facilitated crossing various hurdles, and consequently, pilot projects were set up in the year 1992–93. Monitoring of air pollution/quality in Delhi and Mumbai established desired impact.

As a next and logical step, the Central Govt. authorized GAIL and BPCL to have a joint venture company for introduction of CNG in transport sector in Delhi. GAIL was also authorized to initiate similar action in Mumbai.

4.1 Indraprastha Gas Limited, Delhi

Indraprastha Gas Limited (IGL) was formally incorporated in the year 1999, as joint venture of GAIL, BPCL and Govt. of NCT. The gas transportation infrastructures along with dispensing facilities, 4 retail outlets of oil marketing companies, put up by GAIL, as part of the pilot project, were taken over by newly formed JV. IGL was subsequently authorized by MOP&NG to expand gas distribution network in adjoining Noida and Ghaziabad cities.

Currently, GAIL and BPCL hold 22.5% equity each, Govt. of NCT holds 5%, and balance 50% is held by financial institutions and public. IGL is the second listed gas distribution company listed in stock exchanges.

IGL has been pioneer in successful introduction of CNG in transport sector in NCT and has acquired leadership position in CNG segment in the country. It caters the largest fleet of CNG vehicles—around 7.70 lakhs. CNG segment sales contribute approximately 75% of total gas volumes. Other segments of city distribution business, i.e. PNG in domestic, commercial and industrial, were implemented as a robust business model by IGL, to cater need of fuel of various segments of society.

The Central Govt. for introduction of CNG as environmentally friendly fuel and other sectors allocated domestic gas to IGL. Due to the limited availability of domestic gas and increased demand, additional allocation of domestic gas could not materialize.

In order to meet the growing demand of CNG and PNG (domestic) and other segments, 2009 onwards, IGL had to resort to the procurement of RLNG (through long-term/short-term/spot contracts),facilitated by the promoters. IGL has been successful in passing incremental procurement cost of gas to end-users.

Pursuant to the implementation of revised policy of domestic gas allocation from the end of November 2013, IGL was allocated domestic gas at 5.66 $/MMBTU to meet the full requirement of CNG and PNG (domestic) segments—lower procurement cost was passed on the end—consumers by way of reduction in retail selling price of these two segments.

IGL has business model of playing the role of an enabler in the sector. Aggressive geographical expansion has established green culture in National Capital Region (NCR), which is a role model for other cities.

Delhi State Govt. has been very supportive—by way of allotment of land required for setting a large number of IGL-owned CNG stations and lower sales tax/VAT—nil on CNG and 5% on PNG (domestic)—lowest in any state.

Currently, IGL is operating around 300 CNG (in Delhi and Noida) including 50+ dedicated to Delhi Transport Corporation (DTC) and providing fuel to around 7.70 lakhs vehicles (NCR has the largest fleet of four-wheelers, in our country).

IGL is fully geared up to meet growing requirement of transport fuel in NCT and has drawn up time-bound expansion plan.

It is pertinent to record that due to regulatory compulsion, public transport (buses, taxis and autos) in NCT has to continue to use CNG as fuel.

In line with the Central Govt.'s vision, IGL has accorded high priority to domestic segment also. Approximately 73,000 additional domestic customers were commissioned in FY 14, and this trend continues and total domestic customer base has now increased to 5.25 lakhs, as against domestic customer population of 1.82 lakhs as on March 2010. Also, IGL is serving about 1500 commercial and 700+ industrial customers in NCR.

PNGRB, vide letter ref no Infra/2008/ii/1/Vol ii, dated 9 January 2009, accepted the Central Govt. authorization to IGL for the development of CGD network in NCT of Delhi. Marketing exclusivity period has expired. The milestones, agreed with PNGRB, have already been surpassed.

- As on January 2009, IGL had 1.32 lakhs PNG (domestic) customers—addition @ 35,000 customers per year in next 3 years,
- The current domestic connection is around 5.60 lakhs,
- As on January 2009, IGL had 23.76 lakhs compression capacity to increase to 36 lakhs kg/day, in the next 3 years,
- The current compression capacity is about 67 lakhs kg/day.

4.1.1 Ongoing Litigation with PNGRB

PNGRB, vide order dated 9 April 2012, determined the per unit network tariff and compression charges for city gas distribution network in Delhi, based on the submission of data by IGL in May 2009 and certain assumptions taken by PNGRB.

The tariffs were much lower than the submissions by IGL. Further, PNGRB made the determined tariffs applicable with retrospective effect from 1 April 2008. In its order, PNGRB stated that the modalities and time frame for refund of differential network tariff and compression charges would be decided subsequently. Aggrieved with the PNGRB's order, IGL filed a writ petition on 10 April 2012 against the order of PNGRB before the Hon'ble High Court of Delhi. The Hon'ble High Court of Delhi, on 1 June 2012, passed the judgement and quashed the PNGRB order dated 9 April 2012.

As was expected, against the above-mentioned order of Delhi High Court, PNGRB filed a special leave petition (SLP) before Hon'ble Supreme Court of India, on 27 July 2012. In the meantime, PNGRB continued to determine network tariff and compression charges applicable to other CGDs. However, in view of restraint by the court, orders were kept in abeyance.

Apprehending similar reduction in submitted data and applicability from 1 April 2008, many other city gas distribution entities (MGL, SGL, MNGL, CUGL, Avantika Gas and Bhagyanagar Gas Ltd.) approached the Hon'ble Court with the similar arguments, challenging the authority of PNGRB to fix network tariff and compression charges for own consumers of city gas distribution entities. It was submitted that PNGRB is authorized to fix network charges and compression charges for a third party selling gas in the authorized GA utilizing network on a common carrier basis.

The Hon'ble Supreme Court in judgement dated 1st July, 2015, upheld judgement of Hon'ble High Court of Delhi.

4.1.2 Performance of IGL

The final decision of the Hon'ble Supreme Court would have very significant impact on the performance of IGL and other existing CGDs. IGL, over the years, continues to sustain growth both in financial and in physical numbers, as is evident from the data in Table 4.

As a part of their strategy to consolidate CGD sector, in the year 2014, IGL acquired 50% equity stake in Central UP Gas Ltd. (a JV of GAIL and BPCL), a company engaged in city gas distribution business in the cities of Kanpur, Bareilly and Unnao in UP.

Counting the strategy, recently 50% equity stake in Maharashtra Natural Gas Ltd. (another JV of GAIL and BPCL) has been acquired. MNGL is authorized entity (by PNGRB in January 2009) for the development of city gas distribution network in Pune city including Pimpri and Chinchwad and with contiguous area of Hinjewadi, Chakan and Talegaon geographical area.

The company has also plans to participate in the bidding process of PNGRB, for setting up CGD business in new cities (participated for Jalandhar GA in 3rd round of bidding in the year 2010).

Table 4 Financial and physical performance of IGL (2011–2015)

	FY 14–15	FY 13–14	FY 12–13	FY 11–12
Net revenue (Rs. crore)	4607	3943	3380	2525
Profit after tax (Rs. crore)	437	360	354	306
CNG sales (MMSCMD)	306	2.82	2.75	2.56
Compression capacity (lakhs kg/day)		66.81	63.82	59.56
Total gas sales (MMSCMD)	3.84	3.79	3.67	3.33
PNG (domestic) lakhs	5.6	4.6	3.87	3.25

Source Annual reports of IGL and media reports

It is evident from the above that IGL is poised for the rapid development of city gas distribution business. Currently, having leadership in CNG segment on all India basis, IGL is the second largest CGD entity in country.

4.2 *Mahanagar Gas Ltd., Mumbai*

In the early 90s, the Government of India accorded its approval for GAIL to execute the Mumbai City Gas Distribution Project. Considering the requirement of technical know-how and foreign equity for setting up of India's first metropolis city gas distribution project and in accordance with the FIPB approval (February 94), GAIL partnered with British Gas of UK to form a joint venture company—Mahanagar Gas Limited (MGL). On 8 May 1995, MGL was incorporated with 49.75% equity stake of each, the remaining 0.05% equity to the Govt. of Maharashtra. It is understood that MGL has initiated actions to generate resources by diluting its equity through IPO route. MGL would be 3rd CGD entity after GGCL and IGL to go public.

In line with the Central Govt.'s directive, GAIL provided allocation of 1.5 MMSCMD of natural gas to MGL in a phased manner. The company got permission to commence business in July 1995. The infrastructure developed by GAIL, as part of the pilot project, was taken over by newly formed JV. Thane district, New Mumbai and other geographical areas were later on allocated to MGL by the Central Govt.

Under Regulation 17, PNGRB, on 21 January 2009, accepted authorization[24] by the Central Govt. in favour of MGL for the development of CGD network in Mumbai and Greater Mumbai (GA-1). Three years of market exclusivity (from the date of submission of performance bank guarantee of Rs. 10 crores) granted to MGL.

On 4 August 2009, PNGRB accepted the Central Govt. authorization[25] to MGL for expansion of CGD network in Thane district and adjoining contiguous geographical area (GA-2) with 5 years of market exclusivity.

Pursuant to the implementation of revised policy of domestic gas allocation from the end of Nov 2013, MGL was allocated domestic gas to meet the full requirement of CNG (transport) and PNG (domestic) segments only. Like other CGDs, MGL had to procure RLNG (through short-term contracts) at market-driven, much higher cost, to meet the demand of industrial and commercial segments. MGL, therefore, had to upward revise selling prices to these segments, which resulted in drop in sales volumes (many preferred to switch over to alternate fuels). Over the years, CNG volumes have registered a very moderate growth. CNG segment contributes approximately 73% of total gas sales.

---

[24]PNGRB, vide letter ref no S-Infra/ii/1/2008 Vol ii, dated 21 January 2009.

[25]PNGRB, vide letter ref no S-/Infra/ii/1/2008 Vol ii, dated 4 August 2009.

MGL, over the years, has pioneered gas distribution network in Mumbai, Greater Mumbai, Thane district and adjoining geographical areas and currently serving:

- Approximately 8 lakhs domestic households,
- Approximately 4.2 lakhs CNG vehicles through network of 180 CNG stations, spread across both gases,
- Achieved total gas sales of 2.47 MMSCMD.

MGL's authorized gas with almost 2 crores population offer natural gas potential of around 5 MMSCMD. MGL has drawn up firm and time-bound expansion plan to enhance its domestic customer population to 10 lakhs and CNG stations to 300, at an estimated capital expenditure of approximately Rs. 3000 crores. As a part of expansion strategy, MGL is:

- fast spreading CGD network beyond existing area of operation—Mira road-Bhayandar and Navi Mumbai,
- fast spreading its network to authorized areas—Kalyan, Dombivali, Amarnath, Badalapur, Ulhasnagar, Bhiwadi, Panvel, Taloja, Khargar, etc.

MGL also plans to develop CGD network in few more cities, through the bidding route.

4.2.1 Penetration to New Natural Gas Market

In 4th round of bidding by PNGRB, MGL participated in 3 GAs, namely Raigarh district (excluding already authorized area), Thane district (excluding already authorized areas) and Bengaluru rural and urban districts.

On 1 April 2015, PNGRB granted authorization[26] to MGL for the development of CGD network in Raigarh district (6848 km^2, excluding area already authorized).

As per the submitted bids, MGL committed to develop:

- 26,476 PNG domestic connections,
- 1800-km pipelines.

5 The Central Government-Authorized CGDs

The Hon'ble Supreme Court in its 2002 and 2003 orders identified 15 cities (Delhi, Kanpur, Lucknow, Varanasi, Pune, Faridabad, Patna, Ahmadabad, Sholapur, Hyderabad, Bangalore, Kolkata, Chennai, Jharia and Jodhpur) as the most polluted cities and the direction was issued for introduction of cleaner fuel (CNG) for

[26]PNGRB, vide letter ref no PNGRB/CGD/BID/4/2013/06/GA-Raigarh district, dated 1 April 2015.

transport sector. With a view to comply with the directive to the maximum extent possible, the Central Govt. authorized many CGD entities (PSUs) for the development of CGD network. Some of entities are discussed below.

5.1 Central UP Gas Ltd., Kanpur, Uttar Pradesh

Central UP Gas Ltd. (CUGL)—a joint venture of GAIL and BPCL—was formally incorporated in February 2005, under the Companies Act 1956. Later on, 50% equity stake was diluted by way of equity closure to the following FIs:

- Asian Development Bank—20%,
- Indian Development Fund—15%,
- IL&FS Trust Company Ltd—8.33%,
- Pan-Asia Infrastructure Asset Management Company Pvt Ltd.—6.67%.

All the above private investors exited in the year 2014 from CUGL, and Indraprastha Gas Ltd. (IGL), Delhi, has acquired 50% equity stake. GAIL and BPCL now hold 25% each.

CUGL was authorized to develop network in Kanpur and Bareilly districts. On 22 April 2009, PNGRB accepted the Central Govt. authorization[27] under Regulation 17. Recently, Unnao was also authorized.

However, development of the infrastructure has been significantly lower than the agreed milestones for Kanpur and Bareilly. The shortfalls are attributed to:

- Delay on part of local administration in according requisite approvals for laying of network,
- Lack of support and commitment of the State Govt.,
- Delay in the procurement of land for CNG stations,
- Higher taxes and duties, on natural gas, as compared to alternate fuels,
- Unwillingness of prospective customers for switching over to piped gas, due to high initial cost and easy availability of subsidized LPG,
- Not many high-rise buildings.

Financial and physical performance of CUGL is presented in Table 5.

5.1.1 Penetration to Other GAs: Round II Bidding

CUGL participated in the 2nd round of bidding process conducted by PNGRB and granted authorization[28] under Regulation 5 to develop CGD network at Jhansi (UP).

---

[27]PNGRB, vide letters-S-Infra/ii/1/2008-Vol ii-Kanpur GA dated 22 April 2009 and S-Infra/ii/1/2008/Vol ii/Bareilly GA, dated 22 April 2009.

[28]PNGRB, Letter Ref No. MI/CGD/JHANSI/Authorization, 26 February 2015.

Table 5 Financial and physical performance of CUGL

	FY 14–15	FY 13–14	FY 12–13	FY 11–12
Net revenue (Rs. crore)	190.00	195.20	146.74	113.04
Profit after tax (Rs. crore)	30.00	25.07	20.98	21.12
CNG sales (MMSCMD)	0.135	0.137	0.120	0.106
Compression capacity (lakhs kg/day)	NA	NA	NA	NA
Total gas sales (MMSCMD)	0.175	0.174	0.157	0.139
PNG (domestic)	4000	3000		

Source Annual reports (www.cugl.co.in/AnnualReports (accessed on 17 June 2015))
Despite steep reduction in selling price of CNG (consequent to allocation of domestic gas from Nov. 2013 onwards by the Central Govt.) and court mandate of CNG as fuel for public transports, CNG volumes continue to be more or less static. CNG sales with 75% contribution to total gas sales continue to be dominant segment. PNG (domestic) at both cities is far below the commitment of level of 97,500 by March 2014. Now with IGL's involvement in CUGL, faster expansion/penetration of market is envisaged

As per the winning bid document, CUGL committed to achieve 23,706 PNG (domestic) connections by March 2019. Physical infrastructure at Jhansi is yet to commence (land for city gas station has been acquired). It is understood that the delay is on account of availability of natural gas with desired specifications from GAIL.

5.2 *Green Gas Ltd., Lucknow, Uttar Pradesh*

Green Gas Limited, a joint venture of GAIL, IOCL and Government of UP, was formally incorporated in the year 2006 for the development of CGD network of Lucknow and Agra. The Central Govt. allocated domestic gas for promoting the usage of CNG and piped gas—commercial/domestic segments. PNGRB accepted the Central Govt. authorization for the development of CGD network in Agra,[29] whereas for Lucknow it is still awaited.

GGL and Adani Energy commenced laying the network and this resulted in total confusion. It is understood that steel pipeline network laid by Adani Energy would be taken over by GGL, on mutually agreed terms and conditions.

Even after almost 9 years, CGD network development has been far behind the targets.

In FY 2014–15, GGL has crossed CNG sales @ 0.175 MMSCMD, through 14 stations at Lucknow and Agra. It achieved PNG domestic connections of about 4000 in each city.

[29]PNGRB, vide letter ref no S-infra-/ii/1/ in 2009.

In FY 2013–14, GGL with net turnover of Rs. 160 crores (Rs. 124 crores in FY 2012–13) earned profit after tax of Rs. 22.68 crores (Rs. 20.82 crores in FY 2012–13).

In the light of huge gap in the commitments and actual physical progress, GGL needs to revisit its strategies.

5.3 Maharashtra Natural Gas Ltd. (MNGL), Pune, Maharashtra

MNGL as joint venture of GAIL, BPCL and Maharashtra State Govt. (5%) was incorporated in January 2006 for the development of CGD network in Pune city including Pimpri and Chinchwad along with the adjoining areas of Hinjewadi, Chakan and Talegaon GA. Later on, financial institutions acquired 50% equity stake in MNGL, which recently has been acquired by Indraprastha Gas Ltd. (IGL), Delhi. BPCL and GAIL hold 22.5% equity stake each.

On 1 June 2009, PNGRB accepted the Central Govt. authorization[30] of MNGL for Pune city. Table 6 presents the cumulative milestones committed by MNGL to PNGRB.

MNGL's progress w.r.t. domestic connections has been far behind. At the end of March 2014, MNGL achieved around 12,000 domestic PNG connections which are just 16.6% of the target. During FY 15, around 4000 additional households have been provided piped gas connections, taking the aggregate domestic PNG connections tallied to 16,000.

CNG continues to be dominant segment, around 0.3 MMSCMD—approximately 75% of total gas sales. Total gas sales have now crossed 0.4 MMSCMD. During the last 2 years, MNGL commissioned 12 CNG stations, and as on March 15, a total of 30 CNG stations were functioning.

Pursuant to the revised gas allocation policy by the Central Govt., MNGL has been allocated 100% requirement of CNG and PNG (domestic) segments from February 2014 (80% increase from Nov 2013) onwards which has resulted in steep downward reduction in selling price of these two segments. With this, demand of CNG and PNG (domestic) segments is expected to increase.

In FY 2013–14, MNGL achieved net revenue of Rs. 344 crores (Rs. 182 crores in FY 12–13) and earned profit (after tax) of Rs. 56.67 crores (Rs. 35.00 crores in FY 12–13). During FY 2015, net turnover is of Rs. 460 crores, and a profit after tax of around Rs. 70 crores is expected.

[30]PNGRB, vide letter ref no S-Infra/ii/1/2008-Vol ii Pune, dated 1 June 2009.

Table 6 Committed milestones (cumulative) by MNGL to PNGRB

	FY 13–14	FY 12–13	FY 11–12	FY 10–11	FY 09–10
PNG (domestic)	72,000	50,000	30,000	15,000	5000
Compression capacity (lakhs kg/day)	3.8532	3.625	3.398	2.32	1.254

In the light of huge gap in the commitments and actual physical progress, MNGL needs to revisit its strategies. With IGL's involvement in MNGL, now onwards faster expansion/penetration of market is envisaged.

5.4 *Avantika Gas Ltd. (AGL), Madhya Pradesh*

5.4.1 Operations

Avantika Gas Ltd. (AGL), a joint venture of GAIL and HPCL, was incorporated in January 2006, for the development of CGD network in Indore (including Ujjain) and Gwalior. Madhya Pradesh State Govt. was offered 5% equity stake. Later on, 3 financial institutes (IL&FS, IDBI Trustees and IDFC Ltd.) joined AGL. HPCL and GAIL now hold 22.5% each in equity.

5.4.2 PNGRB Authorization

PNGRB accepted the Central Govt. authorization to AGL for the development of CGD network in Indore (including Ujjain) GA[31] and Gwalior GA[32] in August 2009 and January 2010, respectively. Avantika Gas committed to achieve the following milestones by 2014:

- PNG Domestic connections: 107,000 (Indore GA) and 44,000 (Gwalior GA)
- CNG compression capacity (kg/day): 259,282 (Indore GA) and 74,782 (Gwalior GA).

5.4.3 Progress so Far

However, so far physical progress has been far behind as compared to the project milestones. As of March 2015, PNG domestic connections are around 3278—just about 2% of the target. The shortfalls could be attributed to:

[31]PNGRB, letter ref no S-Infra/ii/1/2008-Vol ii-Indore GA, dated 31 August 2009.

[32]PNGRB, letter ref no S-INFRA/II/1/2008-Vol ii-Gwalior GA, dated 4 January 2010.

- Delay on part of local administration according to requisite approvals for laying of network,
- Lack of support and commitment of the State Govt.,
- Delay in the procurement of land for CNG stations,
- Higher taxes and duties, on natural gas, as compared to alternate fuels,
- Unwillingness of prospective customers for switching over to piped gas, due to high initial cost and easy availability of subsidized LPG,
- Not many high-rise buildings.

AGL, during FY 2013–14, with average volumes of 59,900 SCMD, achieved net sales of Rs. 115.26 crores compared to Rs. 96.70 crores in previous FY. During the period under consideration, its PAT increased to Rs. 0.50 crores compared to Rs. 0.21 crores in the previous FY.

In the light of huge gap in the commitments and actual physical progress, AGL needs to revisit its strategies. With change of gas allocation policy, the transport and domestic segments may spur the growth.

5.5 *Bhagyanagar Gas Ltd. (BGL), Andhra Pradesh*

5.5.1 Operations

BGL was formally incorporated in the year 2002, as joint venture of GAIL and HPCL, and 5% equity was offered to Andhra Pradesh State Govt. for the development of CGD network in the state of Andhra Pradesh (to begin with in Vijayawada, Hyderabad and Tirupati).

Business activities in Vijayawada were commenced in 2005, whereas Hyderabad could be connected through EWPL put up by RGTIL, in 2009 (till such time marketing of CNG was to be done through Daughter Booster CNG Stations only).

5.5.2 PNGRB Authorization

On 9 January 2009, PNGRB accepted the Central Govt.'s authorization[33] to BGL, for the development of CGD network in Hyderabad. The agreed milestones: 2.66 lakhs PNG domestic connections and 8.49 lakhs kg/day CNG compression capacity was to be achieved at the end of 5 years marketing exclusivity i.e. March, 2014.

[33]PNGRB vide letter ref no S-Infra/ii/1/2008-Vol-ii-Hyderabad, dated 9 January 2009.

On 28 July 2009, PNGRB accepted the Central Govt. authorization to BGL,[34] for the development of CGD network Vijayawada GA. BGL committed to achieve 97,261 PNG domestic connections and 2.27 lakhs kg/day CNG compression capacity at the end of 5 years of marketing exclusivity.

5.5.3 Penetration to New GAs: Bidding Round I (October 2008)

BGL had aspiration to expand in other nearby gas. So, in the first round of bidding, BGL submitted bid for Kakinada GA. On 12 June 2009, PNGRB granted authorization[35] to BGL for the development of Kakinada CGD network, with 5 years of marketing exclusivity. BGL committed to provide 50,000 domestic PNG connections and set up 9 CNG stations by 2014.

Currently, BGL is operating in Vijayawada, Hyderabad, Rajahmundry and Kakinada Gas and also operating one auto LPG dispensing station at Tirupati.

5.5.4 Challenges Faced and Progress

BGL has encountered numerous challenges in implementing the projects. Few significant challenges are noted here:

- Delay in main line connectivity at Hyderabad,
- Allocation of domestic natural gas,
- Unusual delay in granting requisite approvals by local authorities,
- Lack of commitment of State Govt. in promoting natural gas business,
- Agitations/bandhs in connection with Samadhya, Andhra Pradesh (year 2012/2013).

BGL has achieved sales volumes of approximately 0.09 MMSCMD, predominantly transport segment through 29 CNG stations spread across GAs (Hyderabad 18, Vijayawada 8, Kakinada 1 and Rajahmundry: 1) including dedicated stations for APSRTC buses.

Penetration to domestic segment has been extremely low, only 3163 connections (at Vijayawada, Hyderabad and Kakinada) at the end of March 2015.

BGL achieved a turnover of Rs. 116.87 crores in FY 2013–14, which was about 24% higher than the previous financial year (Rs. 94.28 crores) and earned moderate PAT of Rs. 14.987 crores, which was significantly higher than Rs. 6.46 crores for previous FY.

In the light of huge gap in the commitments and actual physical progress, BGL needs to revisit its strategies.

[34]PNGRB, vide letter ref no S-Infra/ii/1/2008 Vol-ii-Vijayawada, dated 28 July 2009.

[35]PNGRB, vide letter ref no PNGRB/MI/CGD/I/Kakinada, 12 June 2009.

5.6 *Sabarmati Gas Ltd. (SGL), Gujarat*

5.6.1 Operations

Gujarat Govt. in the year 2003 granted no objection certificate to BPCL to develop CGD network in three districts of North Gujarat, namely Gandhinagar, Mehsana and Sabarkantha. In 2004, BPCL established a dedicated task force for execution of the CGD projects in allocated districts of north Gujarat. Incidentally, the districts had huge industrial clusters at Kadi, Kalol, Santez, Mehsana and Himatnagar. Many industries were already using natural gas from nearby production fields of ONGC.

BPCL after setting up city gas station at Santez commissioned first industrial customer, namely Specific Ceramics in May 2005. BPCL scaled up very fast, by mid-2006, around 11 industries with connected load of approximately 0.1 MMSCMD. Another CGS at Ambapur (from GSPL) was commissioned, and after laying about 10-km steel pipeline, the first CNG (online) station dedicated to GSRTC at Gandhinagar was commissioned in April 2006.

Gujarat Govt. being proactive, through GSPC/GSPL, facilitated fast-track execution of projects. With a view to synergize strength of BPCL and GSPC, a joint venture company was mooted. With due approval of the Central Govt., joint venture in the name and style of Sabarmati Gas Ltd. was incorporated in June 2006. In line with joint venture agreement, BPCL appointed managing director.

On the back of the groundwork by BPCL task force, SGL scaled up penetration to numerous industries/commercial establishments, set up CNG stations and later on provided piped gas to domestic segment as well. SGL declared the profit of approximately Rs. 0.56 crores in the very first year of inception (2006–07). Subsequently, outstanding performance (physical as well as financial) enabled entry of 3 financial institutions (IDFC/IFCI/UTI) and acquired 49.86% equity stakes, in the year 2009.

5.6.2 PNGRB Authorization

On 16 December 2009, PNGRB accepted the Central Govt. authorization[36] and SGL committed cumulative PNG connection of 99,000 by 2014 (Table 7).

5.6.3 Progress so Far

Prior to the incorporation of SGL, GSPC Gas, retail arm of the Gujarat State, provided gas network to Gandhinagar. During authorization process, it was therefore agreed that GSPC Gas shall not expand beyond identified areas.

[36]PNGRB vide letter ref no S-Infra/ii/1/2008-Vol ii, dated 16 December 2009.

Table 7 SGL committed milestones

	FY 13–14	FY 12–13	FY 11–12	FY 10–11	FY 09–10
Domestic PNG connections	99,000	75,000	50,000	35,000	20,000
Compression capacity (lakhs kg/day)	1.992	2.448	2.568	2.718	1.992

Source PNGRB

SGL, on the strength of the authorization, got allocation of approximately 0.78 lakhs SCMD domestic gas (from RILs to new finds in KG basin) for CNG and PNG segments. SGL took advantage of this allocation and very aggressively expanded network in nearby towns (Visnagar, Unjha, Mansa, Palej, etc.) besides district headquarters (Gandhinagar, Mehsana and Himatnagar) and towns Kadi, Kalol and Santez.

However, due to the steep drop in gas production at RIL KG basin fields, domestic gas supplies were suspended from May 2011 onwards. SGL, therefore, had no option, other than to source RLNG through spot contracts at very high price (14–20 $/MMBTU) in subsequent years, which has very significant adverse impact on industrial volumes (dropped from high of 0.68–0.45 MMSCMD, despite additional connections). With a view to retain industrial customers (205+), as a long-term strategy SGL decided to partly absorbed high procurement cost of gas, which along with substantial increase in CNG and domestic segments (30% from 15% 5 years back), resulted in net financial losses of Rs. 34.5 Crores in year 2012–13.

Pursuant to the allocation of domestic gas by the Central Govt. (from Nov. 2014 onwards) to meet the full requirement of CNG and PNG (domestic) segments, financial institutions have significantly improved.

SGL aggressively expanded PNG (domestic) markets, and approximately 20,000 new households were connected in the year 2011; however, subsequently due to higher price of natural gas as compared to highly subsided LPG (domestic), demand drastically dropped. Despite all efforts in FY 2014–15, additional 8400 households could only be connected. CNG sales, during FY 15, registered the growth of approximately 18% over previous year, which is attributed to penetration to new market (Idar) and drop in selling price of CNG.

By 2013–14, SGL achieved close to 79% of the domestic PNG connections' commitment given to PNGRB. The number of CNG stations increased 2.37 times, whereas the CNG sales grew by 2.16 times between 2010–11 and 2014–15 (Table 8).

Table 8 Financial performance and physical progress of SGL

	FY 14–15	FY 13–14	FY 12–13	FY 11–12	FY 10–11
Net revenue (Rs. crore)	908	937	860	690	453
Profit after tax (Rs. crore)	110.00	31.1	*(−)34.5*	7.50	27.56
Domestic PNG connections	86,000	78,000	71,400	61,700	46,100
CNG sales (lac SCMD)	2.145	1.822	1.863	1.593	0.987
Total gas sales (MMSCMD)	0.684	0.694	0.761	0.800	0.740
CNG stations	38	36	30	23	16

Source Annual reports

It is pertinent to record that under the Central Govt.-approved category, SGL is the most successful city distribution entity

Highly industrialized clusters/matured gas market, fast-track progress by BPCL task force, professional management and support of promoters have been key drivers

6 The State Government-Authorized CGDs

6.1 GSPC Gas, Gandhinagar, Gujarat

6.1.1 Operations

GSPC Gas Company was formed in the year 2006, on the directive of the State Govt. with the objective to cater natural gas requirement of all users in retail segment with the aim to provide the natural gas to every nook and corner of the Gujarat state and along with GSPC group to make state a truly gas-based economy.

GSPC Gas was directed to provide CGD network in all districts of Gujarat, except covered by GGCL/VMSS and awarded NOCs for Ahmadabad, Gandhinagar, Mehsana and Sabarkantha districts. Later on, owing to the slow progress of CGD projects, GSPC Gas was also instructed to provide CGD network in Gandhinagar city (part of SGL) and Valsad (part of GGCL Gas).

Over the years, GSPC Gas expanded the network very aggressively and is the fastest growing CGD Company in our country. Currently, GSPC is marketing more than 4.5 MMSCMD natural gas (highest by any CGD Company) to 5.4 lakhs plus PNG (domestic) customers, 1985 Industries (including largest cluster of around 730 in Morbi Ceramic area), 2000 plus commercial customers, approximately 0.70 MMSCMD CNG to 90,000 CNG vehicles through 164 CNG stations.

GSPC Gas is spread over 21 major cities and towns and more than 500 villages in various parts of the state and serves customers through steel network of 450 km+ and MDPE network of 8000 km+.

Major markets are as follows:

- South Gujarat: Sarigam, Vapi, Mora, Valsad, Navsari and Hazira,
- Central Gujarat: Karjan, Gandhinagar, Palej, Halol, Khambhat, Dahej and Nadiad,
- Saurashtra: Morbi, Rajkot, Thangarh and Surendranagar.

6.1.2 PNGRB Authorization

PNGRB, under Regulation 18(1), granted authorization to CGD network developed by GSPC Gas to CGD network developed in Nadiad, Navsari,[37] Surendranagar,[38] Rajkot[39] (Oct, 2013), Hazira[40] (July, 2014) and Valsad[41] (January 2015). Table 9 presents the existing domestic PNG connections and year-wise target to be achieved by GSPC Gas in various GAs.

Table 10 presents the existing CNG compression capacity and year-wise target to be achieved by GSPC Gas in various GAs.

It is pertinent to record that from modest beginning of:

- 8000 domestic connections in year 2006–07, GSPC Gas crossed 1.28 lakhs households in year 2009–10, 4.20 lakhs in year 2012-13, 4.92 lakhs in year 2013–14 and 5.41 lakhs as on March, 2015.
- 25 CNG Stations as on March, 2009, crossed 72 on March 2010, 152 on March 2013, 157 on March, 2014 and 164 on March, 2015.

Additionally, GSPC Gas is also operating 3 major online CNG stations in Gandhinagar (yet to be authorized).

Pursuant to the implementation of revised allocation policy of domestic gas by the Central Govt. (from Nov 2014 onwards), GSPC gas has been allocated 0.9 MMSCMD domestic gas to meet the full requirement of CNG and PNG (domestic) segments. This has resulted in steep drop in selling prices (30%) of CNG and PNG domestic gas and is expected to improve the financial performance in FY 2014–15 and beyond.

GSPC Gas, through special-purpose vehicle (SPV) GSPC Distribution Network Ltd. (GDNL), acquired 73.20% of equity of GGCL, from BG Asia Pacific Holdings Pte Ltd. and consequently took management control in June 2013.

After amalgamation, GDNL is the largest CGD Company in country serving more than 10.4 lakhs domestic customers, 233 CNG stations, 11,000 commercial customers and 2700+ industrial customers, with total sales volume of 6.50 MMSCMD, more than double of the 2nd largest CGD (IGL, Delhi).

GDNL aspires to acquire few more CGD companies, particularly operating in the state of Gujarat. It is evident from the above that GSPC Gas/GDNL shall play a very vital role in expanding CGD network/fulfilling vision/dream of Gujarat Govt.

---

[37]PNGRB, S-Infra/ii/1/2008-Vol ii-Navsari GA, dated 1 October 2013.

[38]PNGRB, S-Infra/ii/1/2008-Vol ii-Surendranagar GA, dated 1 October 2013.

[39]PNGRB, S-Infra/ii/1/2008-Vol ii-Rajkot GA, dated 1 October 2013.

[40]PNGRB, S-Infra/ii/1/2008-Vol ii-Hazira, GA, dated 17 July 2014.

[41]PNGRB, S-Infra/ii/1/2008-Vol ii-Valsad GA, dated 20 January 2015.

Table 9 Committed and existing domestic PNG connections

	Existing	Year 1	Year 2	Year 3	Year 4	Year 5
Nadiad	40,929	58,581	76,370	94,159	111,948	129,737
Navsari	55,208	67,185	79,162	91,139	103,116	115,093
Surendranagar	14,328	34,008	53,688	73,368	93,048	93,048
Rajkot	106,011	166,630	227,249	287,868	348,486	409,105
Hazira	26,928	Target of 5481 already achieved				
Valsad	73,503	81,821	90,140	98,458	106,757	115,095

Table 10 Commitment for CNG compressing capacity (kgs/day)

	Existing	Year 1	Year 2	Year 3	Year 4	Year 5
Nadiad	144,480	Target 72,300 already achieved				
Navsari	60,480	Ttarget of 64,140 already achieved				
Surendranagar	60,480	Target of 62,480 already achieved				
Rajkot	188,160	196,125	204,090	212,055	220,021	227,986
Hazira	227,986	Target of 5481 already achieved				
Valsad	166,320	Target of 86,760 already achieved				

6.1.3 Penetration to New GAs: Bid Round III (2011), IV (2013) and V (2015)

GSPC Gas participated (4 GAs—Jamnagar, Bhavnagar, Kutch west and Kutch east district) in 3rd round of bidding by PNGRB, in February 2011.

After a gap of more than 3 years, PNGRB granted authorization to GSPC Gas for the development of CGD network in Jamnagar[42] (January 2014) and Kutch west[43] (Dec 2014). The following milestones are agreed between GSPC Gas and PNGRB:

- Jamnagar: Domestic PNG connections of 92,000 per year for the first 4 years and nil in the 5th year of authorization (cumulative 3.68 lakhs),
- Kutch west: Domestic PNG connections 41,500 per year, first 4 years and nil in the 5th year of authorization (cumulative 1.66 lakhs).

Under the 4th round of bidding, GSPC Gas granted authorization[44] on 1 April 2015, for the development of CGD network in UT of Dadra and Nagar Haveli (area 493 km^2). GSPC Gas committed to achieve 3823 domestic PNG connections and 320.45-km pipelines in UT of Dadra and Nagar Haveli by 2019. GSPC Gas

[42]PNGRB, Infra/CGD/BID/R-3/4/GA JAMNAGAR, 17 January 2014.

[43]PNGRB, Infra/CGD/BID/R-3/4/GA KUTCH(West), 19 December 2014.

[44]PNGRB, vide letter ref no PNGRB/CGD/BID/4/2013/10/GA Dadra and Nagar Haveli, dated 1 April 2015.

submitted network tariff (from 1 to 25 years), Rs./MMBTU: 0.01 and compression charges (from 1 to 25 years) Rs./kg: 0.01.

GSPC Gas submitted bid for Banaskantha GA (Gujarat) only, out of 20 gas advertised by PNGRB, in 5th round of bidding. The bids are under evaluation.

6.2 *Adani Gas Ltd., Ahmadabad, Gujarat*

As mentioned earlier, as a part of interim policy of Gujarat Govt. to promote gas culture across the state, no objection certificate was granted to M/S Adani Gas Ltd. (a wholly owned subsidiary of Adani Enterprise Ltd.) for the development of CGD network in Ahmadabad and also in Vadodara (other than domestic segment).

Adani Gas was, therefore, set up towards 2005–06 with an objective to provide PNG to households, industries, commercial establishments and CNG to automobile sector in Ahmadabad and CNG at Vadodara.

Adani Gas was also granted NOC for the development of CGD network by other states, as given below:

- UP Govt. for Lucknow, Noida and Khurja,
- Rajasthan Govt. for Jaipur and Udaipur,
- Haryana Govt. for Faridabad.

6.2.1 Ahmadabad

Directive to use only CNG as fuel in public transport created immediate demand of CNG. GSPC supported Adani Gas through supplying RLNG of long-term contract for CNG and PNG (domestic) segments till allocation of domestic gas by the Central Govt. in November 2013. Currently, Adani Gas is serving automobiles through network of approximately 52 CNG stations (included dedicated to GSRTC) spread across the Ahmadabad and Vadodara cities and suppling about 0.47 MMSCMD natural gas (about 52% of total gas sales @ 0.8 MMSCMD). Adani Gas also connected approximately 1.90 lakhs domestic households, 1500+ commercial customers and 750 industrial customers.

Adani Gas has crossed total gas sales beyond 0.9 MMSCMD; however, due to switching of numerous industries to alternate fuel (in view of higher gas prices, in the year 2012/13), volumes declined. Even after 8/9 years, penetration to domestic segment is about 20% households.

Pursuant to the enactment of PNGRB Act 2006, Adani Gas has applied to PNGRB for the grant of authorization of its Ahmadabad network/operations is awaited.

6.2.2 Vadodara

In view of VMSS supplying natural gas to domestic sector and GAIL's CNG stations (under the directive of EPCA), Adani Gas's activities have been very limited—currently operating only 8 CNG stations in Vadodara.

6.2.3 Khurja (District Bulandshahr, UP)

On 4 December 2012, PNGRB granted authorization[45] to Adani Gas for the development of CGD network in Khurja GA (129 km^2 area). Adani Gas committed to provide 25,100 domestic PNG connections and develop 260-km of pipelines and 24,000 kg/day of CNG compression capacity by 2017.

Adani Gas commenced execution of CGD project and currently provides natural gas to many industrial customers.

6.2.4 Faridabad

Adani Gas commenced execution of steel pipeline network at Faridabad, and commissioning was delayed considerably due to connectivity with main/trunk pipeline and allocation of gas. Adani Gas currently:

- Supplying transport fuel, approximately 0.2 MMSCMD (50% of Total Gas sales) through 8 CNG stations spread across the city.
- Serving approximately 10,000 PNG (domestic) customers.
- Commercial establishments.
- Industries.

6.2.5 Lucknow

Pursuant to the grant of NOC by the State Govt., Adani Gas commenced execution of CGD project at Lucknow, and about 50-km steel pipeline was laid. Since the Central Govt. authorized Green Gas and accepted by PNGRB, Green Gas has already been operating. So, Adani Gas's marketing of natural gas could not materialize. It is understood that Green Gas and AGL are in dialogue for taking over the infrastructure laid by Adani Gas, on mutually agreed terms and conditions including equity participation of Adani Gas.

---

[45]PNGRB, vide letter ref no S-Infra/ii/1/2008-Vol ii-Khurja GA, 4 December 2012.

6.2.6 Noida

Pursuant to the grant of NOC by the State Govt., Adani Gas commenced execution of CGD project at Noida. Since the Central Govt. authorized IGL to operate in Noida and accepted by PNGRB, Adani Gas's marketing of natural gas could not materialize in Noida.

6.2.7 New Business Model

With an objective of being a major CGD player on a Pan-India basis, Adani Gas along with Indian Oil Corporation (IOCL) formed a joint venture—IOC-Adani. The consortium participated in 2nd, 3rd and 4th rounds of bidding conducted by PNGRB for the award of authorization for the development of CGD network by PNGRB.

The consortium granted authorization for Allahabad[46] (UP) and UT of Chandigarh[47] in Round II. Further, the consortium won Panipat and Daman GAs and granted authorization by PNGRB in the 4th round of bidding.

Milestones committed by IOC-Adani for Allahabad and Chandigarh (till 2017) and for Panipat and Daman (2020) are presented below:

- Allahabad (267 km^2)—2.25 lakhs PNG (domestic) connections and 480-km steel pipeline—as per the bid document,
- Chandigarh (727 km^2)—7.0 lakhs PNG (domestic) connections and 1750-km steel pipeline—as per the bid document,
- Panipat (1260 km^2 area)—11,782 PNG (domestic) connections and 650-km pipelines—as per the revised bidding parameters,
- Daman (51 km^2 area)—2519 PNG (domestic) connections and 33.15-km pipelines—as per the revised bidding parameter.

After 2 years from granting authorization by PNGRB, physical progress of development of CGD infrastructure at Allahabad and Chandigarh GAs has been far behind the agreed milestones.

6.3 Haryana City Gas

During the review of CNG implementation progress by EPCA and the Supreme Court (2003), the state governments were allowed to implement CNG through experienced parties. Many state governments took initiative and granted approvals for the development of CGD network in highly polluted cities.

[46]PNGRB, MI/CGD/ALLAHABAD Authorization, 8 May 2013.

[47]PNGRB, MI/CGD/ Chandigarh Authorization, 8 May 2013.

Haryana State Govt. granted approval to Haryana City Gas (HCG)—a private entity promoted by SKN Group for the development of CGD network for distribution of natural gas to transport sector and piped gas to domestic, commercial and industrial segments in Gurgaon, Rohtak, Rewari and Jhajjar districts.

HCG also got approval from the Rajasthan State Govt. for the development of city gas distribution network in Bhiwadi and Neemrana.

HCG commenced execution of CGD project towards mid-2000, and city gas station (from cross-country steel pipeline of GAIL) was commissioned near IFFCO Chowk, Gurgaon.

HCG has laid network in Gurgaon and also in Bhiwadi[48] and currently, it is:

- Operating 5 CNG stations at Gurgaon and actions initiated to augment to 12.
- Operating 6000 Domestic connections and actions initiated to serve 15,000 Domestic connections.
- Supplying approximately 0.03 MMSCMD natural Gas to Commercial and Industrial segments.

Consequence to State Govt.'s approval and direction of Supreme Court, HCG was receiving domestic gas (earlier allocated to IGL, Delhi, account Gurgaon). Pursuant to the revised policy of allocation of domestic gas to CGDs, from November 2013, HCG has been allocated domestic gas to meet the full requirement of CNG and PNG (domestic) segments. Gurgaon and Rohtak districts have high potential of introduction of natural with firm allocation of domestic gas, and aggressive expansion of infrastructure by HCG is expected in Gurgaon and neighbouring district (Rohtak and Jhajjar). It is to be noted that HCG is yet to receive authorization from PNGRB for its CGD network.

6.4 SITI Energy Ltd., Moradabad (UP)

UP State Govt. granted approval to SITI Energy Ltd. (SEL) for the development of CGD network in Moradabad (UP). SEL was incorporated under Companies Act 1956, in mid-2006. SEL is promoted by Essel Group, which has varied national and global interests in media, newspaper publishing, content broadcasting, film exhibition, theme parks, infrastructure development, etc.

On 30 November 2012, PNGRB under Regulation 18(1) granted authorization[49] to SEL for the development of CGD network in Moradabad. SEL committed to provide 127,100 domestic PNG connections and build 653-km steel pipelines, with CNG compression capacity of 70,800 kg/day by 2017 (Table 11).

---

[48] www.harayanacitygas.com (accessed on 5 July 2015).

[49] PNGRB, vide letter ref no S-Infra/ii/1/2008-Vol ii-Moradabad GA, dated 30 Nov 2012.

Table 11 Milestones agreed between SEL and PNGRB

	Existing (November 2012)	2013	2014	2015	2016	2017
Domestic PNG connections	Nil	500	20,000	50,000	95,000	127,100
Steel pipeline (inch km)	100	200	300	400	500	653
CNG compression capacity (kg/day)	3000	16,000	29,000	42,000	55,000	70,800

SEL faced several challenges in execution of CGD project. In the absence of commitment and support of State Govt./local administration/authorities, long delay in approvals is a routine.

After more than 7 years of incorporation, SEL is currently operating one online/mother CNG station and one daughter/booster CNG station at IOCL retail outlet and dispensing approximately 0.02 MMSCMD natural gas to automobiles. Moradabad has got very limited public transport/autos. Currently, CNG is not available in nearby areas, due to which conversion of private cars/vehicles has been very limited. State transport department has also not switched over to CNG as fuel. Like other cities, price difference in natural gas versus alternate fuel is key driver for adoption by the customers (environmental issue appears least concern).

SEL has commenced supplies of PNG to 1000 domestic customers,[50] commercial and industrial segments.

With allocation of domestic gas (as per the revised allocation policy) by the Central Govt. from November 2013. Retail selling price of CNG/PNG (domestic) has reduced considerably, and now onwards, higher growth is expected. SEL has firmed up plan for expansion of infrastructure.

7 Grant of Authorization by PNGRB Through Bidding Route

7.1 Bidding Round I (23 October 2008)

Six geographical areas (GAs) aggregated around 100 km^2 area attracted 18 bids from 10 entities. Bidding process concluded smoothly, and GAIL Gas emerged as the most successful bidder by bagging four GAs (Table 12). All successful bidders granted authorization in June 2009 and have executed CGD projects in all GAs.

[50] www.sitienergy.com (accessed on 15 May 2015).

Table 12 Authorization of successful bidders under Round I

No. of GAs	GAs	State	No. of bids received	Successful bidder	Grant of authorization
1	Kota	Rajasthan	3	GAIL Gas	06 June 2009
2	Sonipat	Haryana	5	GAIL Gas	12 June 2009
3	Mathura	UP	2	DSM Infra and Saumya Mining	18 June 2009
4	Kakinada	Andhra Pradesh	2	Bhagyanagar Gas (BGL)	12 June 2009
5	Meerut	UP	3	GAIL Gas	12 June 2009
6	Dewas	Madhya Pradesh	3	GAIL Gas	18 June 2009

Table 13 Authorization of successful bidders under Round II

No. of GAs	GAs	State/UT	Bids received	Successful bidder	Grant of authorization
1	Ghaziabad	UP	5	IGL	–
2	Allahabad	UP	2	IOC-Adani	8 May 2013
3	Jhansi	UP	2	CUGL, Kanpur	4 March 2014
4	Rajahmundry	Andhra Pradesh	3	Reliance Gas Corp. Ltd.	
5	Yanam	UT of Pondicherry	1	Reliance Gas Corp. Ltd.	
6	Shahdol	MP	1	Reliance Gas Corp. Ltd.	
7	Chandigarh	UT	4	IOC-Adani	8 May 2013

Source Compiled by the authors from PNGRB website

7.2 *Bidding Round II (13 February 2009)*

On the due date of submission of bids, only single bids were received in respect of 3 GAs, viz. Jhansi, Shahdol and Yanam. In line with Regulation 5 (9), PNGRB extended due date till 17 July 2009. In response, one bid for Jhansi GA was received. Unreasonable bids/litigations coupled with change resulted in long delay in evaluation/grant of authorization to successful bidder. Subsequently, PNGRB granted authorization to Allahabad, Jhansi and Chandigarh (Table 13).

In 2011, IGL challenged the authority of PNGRB for the grant of authorization. MOP&NG granted authorization to IGL for the development of CGD network, in Ghaziabad GA, in the year 2011.

Due to the winding up of Reliance Gas Corporation (RGPCL), 3 GAs (covering 184 km^2 area)—Shahdol, Rajahmundry and Yanam—were again included in the Round IV of bidding, by PNGRB.

Table 14 A brief details of GAs and successful bidders in the Round III

	Geographical area (GA)	State	No. of bids received	Successful bidder	Grant of authorization
1	Asansol-Durgapur	West Bengal	7	IGL	…
2	Ludhiana	Punjab	16	Jay Madhok	25 June 2015
3	Jalandhar	Punjab	12	Jay Madhok	6 June 2013
4	Panipat	Punjab	0	Cancelled	
5	Jamnagar	Gujarat	2	GSPC Gas	17 January 2014
6	Bhavnagar	Gujarat	2	GGCL	5 March 2014
7	Kutch west	Gujarat	8	Jay Madhok	12 March 2015
8	Kutch east	Gujarat	4	GSPC Gas	16 December 2014

Source Compiled by authors from PNGRB website

7.3 *Bidding Round III*

In the third round of bidding, PNGRB received 51 bids from 26 entities for 7 geographical areas and there were no bidders for Panipat. Ludhiana attracted maximum number of bidders (16) followed by Jalandhar (12). Jay Madhok emerged as the successful bidder in the Ludhiana and Jalandhar (Table 14).

7.4 *Bidding Round IV*

On 22 September 2010, PNGRB invited bids[51] for the following 8 GAs:

- Ernakulum district (Kerala),
- Rangareddy and Medak districts (AP),
- Nalgonda district (AP),
- Khammam district (AP),
- Alibag/pen (Maharashtra),
- Lonavala/Khopoli (Maharashtra),
- Guna (MP),
- Shahjahanpur district (UP).

During the open house session conducted by PNGRB, numerous clarifications were sought by entities. After a series of clarifications and extension of due date of submission,[52] PNGRB informed cancellation of Round IV of bidding for 8 GAs.

[51]PNGRB, vide EOI/CGD/BID/4, 22 September 2010.

[52]PNGRB, vide public notice, dated 4 Nov 2011.

7.5 Bidding Round IV

PNGRB in June 2013 amended the bidding criteria. Under the revised bidding criteria, the weightage of the network tariff was increased to 70% from 40%, and accordingly, weightage of compression charges was reduced from 30% to earlier 10%.

PNGRB also specified the minimum work programme for two parameters-inch km of steel pipeline and number of domestic connections in the first 5 years of operation for an authorized area.

Pursuant to the above, PNGRB reinitiated the bidding process, and for the Round IV (September 2013), bids were invited for 14 GAs, spread across nine states.

The last date of submission was fixed as 11 February 2014. However, due to various reasons including lack of bids for some GAs, the last date of bid submission extends 11 August 2014. In the meantime, PNGRB issued many clarifications, in response to queries raised by entities during open house. The minimum work programme was revisited and revised, through public consultation process. On due date of bid submission (11 August 2014), no bid was received for 5 GAs (Rangareddy, Nalgonda, Khammam, Guna and Shahjahanpur) and in line with Regulation 5(9), PNGRB extended due date by one month for these 5 GAs, although this no bid was received. At least 2 bidders in each of 7 GAs submitted:

- Network tariff (for 1–25 years), Rs./MMBTU: 0.01,
- Compression charges (1–25 years), Rs./kg: 0.01.

PNGRB advised all such bidders to submit additional performance bond amount. Finally based on the highest performance bond offered, grant of authorization was issued by PNGRB (Table 15).

7.6 Bidding Round V

On 19 January 2015, PNGRB invited bids for 20 GAs (Table 16) and on the due date of submission of bids (22 May 2015), no bid was received for 8 GAs (Budaun district, Aligarh district, Bulandshahr district excluding already authorized area, Latur, Dhar, Shivpuri, Bidar and Osmanabad districts) and single bid was received for 4 GAs (Muzaffarnagar, Banaskantha, Ahmednagar and Dahod). Therefore, PNGRB, in line with Regulation 5(9), extended last date of submission of bids till 22 June 2015.

Despite extension of submission dates:

- No bid was received for 6 GAs (Budaun, Aligarh, Bulandshahr, Shivpuri, Bidar and Osmanabad),
- Only single bid for 2 GAs (Dhar and Dahod),
- Bid for Muzaffarnagar, submitted by Gala Gas withdrawn,
- 23 bids received for remaining 10 GAs.

Table 15 Successful bidders in Round IV

No. of GAs	GAs	State/UT	No. of bids received	Successful bidder	Grant of authorization
1	Ernakulum district	Kerala	6	IOC-Adani	15 October, 2015
2	Bengaluru rural and urban	Karnataka	7	GAIL Gas	18 February, 2015
3	Raigarh district[a]	Maharashtra	3	MGL	01 April, 2015
4	Pune district[a]	Maharashtra	2	Mahesh Res. Pvt Ltd.	18 May, 2015
5	Thane district[a]	Maharashtra	3	GGCL	01 April, 2015
6	UT of Daman	UT	4	IOC-Adani	01 April, 2015
7	Dadra Nagar Haveli	UT	4	GSPC Gas	01 April, 2015
8	Panipat district	Haryana	5	IOC-Adani	01 April, 2015
9	Amritsar district	Punjab	4	GSPL	05 May, 2015

[a]Excluding areas already authorized
Source Compiled from PNGRB website

Table 16 PNGRB invited bids for GAs under Round V

Sl. no.	GAs	State/UT	Sl. no.	GAs	State/UT
1	East Godavari	AP	11	Tumkur	Karnataka
2	Belgaum	Karnataka	12	Latur	Maharashtra
3	Ahmednagar	Maharashtra	13	Dahod	Gujarat
4	Krishna Nagar	AP	14	Dhar	MP
5	Muzaffarnagar	UP	15	Haridwar	Uttarakhand
6	Godavari	Karnataka	16	Dharwad	Karnataka
7	Budaun	UP	17	Shivpuri	MP
8	Aligarh	UP	18	Bidar	Karnataka
9	Bulandshahr	UP	19	Osmanabad	Maharashtra
10	Banaskantha	Gujarat	20	Udham Singh Nagar	Uttarakhand

After evaluation of the bids submitted by total 8 entities, PNGRB granted authorization for the development of CGD network in 8 GAs (Table 17). Ahmednagar and Banaskantha are yet to be authorized by PNGRB.

Table 17 PNGRB granted authorization under bidding Round V

Sl. no.	Geographical area	State/UT	No. of bids received	Successful bidder	Grant of authorization
1	East Godavari district	Andhra Pradesh	2	APGDCL and HPCL	14 August, 2015
2	West Godavari district	Andhra Pradesh	2	APGDCL and HPCL	14 August, 2015
3	Udham Singh Nagar	Uttarakhand	2	IOC-Adani	30 July, 2015
4	Haridwar district	Uttarakhand	2	GAIL Gas and BPCL	03 August, 2015
5	Tumkur district	Karnataka	3	MEIL	20 August, 2015
6	Dharwad district	Karnataka	3	IOC-Adani	17 September, 2015
7	Krishna district	Andhra Pradesh	2	MEIL	26 September, 2015
8	Belgaum district	Karnataka	3	MEIL	15 October, 2015

Source Compiled from the website of PNGRB

7.7 *Bidding Round VI*

On 13 October 2015, PNGRB invited bids for 34 GAs under Round VI with submission of bids between 15 and 25 January 2016. Table 18 presents the GAs and the pipelines available for tap-off point.

As of mid-January 2016, PNGRB has granted authorization for 32 GAs during 5 rounds of bidding. As mentioned earlier, some of the reasons for delay/slow process of bidding/awards include lack of direct control of PNGRB over irrational bids and litigations.

Considering the current status and plan for next few months, PNGRB needs to revisit bidding parameters in terms of economic viability of some of the GAs. Although PNGRB accorded high priority for the award of additional cross-country pipelines, it is understood that most of the project execution is at slow pace, which is attributed to likely lower materialization of projected demand of natural gas in country.

Huge drop in production at RIL's KG basin field had very strong negative impact on overall production/availability of natural gas leading to the slow development of CGDs. Similarly, very high price of RLNG through short-term/spot contracts (16–20 \$/MMBTU) prevailing in the year 2011/2012 till early 2014 also adversely affected planning of gas linkage to power and fertilizer sectors.

With softening of RLNG (6.5–8.0 \$MMBTU) through short-term contracts and the Central Govt.'s decision of pooling of natural gas (indigenous and RLNG) for power and fertilizer sectors, demand of natural gas shall increase substantially, which in turn would facilitate faster completion of ongoing pipeline projects. In turn, many more cities along new pipelines would be covered for bidding by PNGRB.

Table 18 CGD networks available for bidding under Round VI

GA	State	Name of the pipeline	GA	State	Name of the pipeline
Dhar	Madhya Pradesh	HVJ, DVPL	Yamunanagar	Haryana	DBNPL
Dahod	Gujarat	HVJ, DVPL-I&II, HPGGG	Shahdol	Madhya Pradesh	SPHPL
Ahmednagar	Maharashtra	EWP	Gadag	Karnataka	DBPL
Banaskantha	Gujarat	HPGGG	Jhabua	Madhya Pradesh	DVPL
Saharanpur	UP	DBNPL	Nainital	Uttarakhand	KMKRPL
Raebareli	UP	HVJ Thulendi Phulpur P/L	Rewari	Haryana	CJHPL
Mainpuri	UP	HVJ	North Goa	Goa	DBPL
Ramabai Nagar	UP	HVJ Auraiya-Jagdishpur	Datia	Madhya Pradesh	HVJ
Chitradurga	Karnataka	DBPL	Rupnagar	Punjab	DBNPL
Bhiwani	Haryana	CJHPL	Fatehgarh Sahib	Punjab	DBNPL
Ratnagiri	Maharashtra	DBPL	Dahej-Vagra Taluka	Gujarat	EWPL, DVPL, Bhadbhut-Gana
Amreli	Gujarat	Tana-Amreli	Amethi	UP	HVJ Thulendi Phulpur P/L
Bathinda	Punjab	DBNPL	Ahmadabad[a]	Gujarat	Darod-Jafrabad
Auraiya	UP	HVJ	Anand[a]	Gujarat	Anand-Rajkot
Patan	Gujarat	Mehsana-Palanpur	Panchmahal	Gujarat	DVPL, Halol-Dahod
Baghpat	UP	DBNPL	Rewa	Madhya Pradesh	SPHPL
Etawah	UP	HVJ Vijaipur-Auraiya	Rohtak	Haryana	CJHPL

[a]Excluding the area already authorized
Source PNGRB, http://www.pngrb.gov.in/pdf/cgd/bid6/Tap13102015.pdf

8 Recent Developments

8.1 Draft Guidelines, Dated 5 March 2015 on Setting up CNG Stations/Granting Marketing Rights for CNG as Transportation Fuel by MOP&NG

MOP&NG issued the guidelines to secretary of PNGRB, C&MD pf IOC/BPC/HPC/GAIL and MDs of all CGDs and comments were invited within 15 days—so far no further communication.

Guidelines, if notified, shall have adverse impact on CGDs.

In the meantime, PNGRB has clarified as under:

The guidelines are still in draft stage and have not been finalized yet. As of now, the stated position of MOP&NG, as per their letter no l-16021/9/2013-GP dated 5 February 2014, that no entity may set up new CNG station (till further order) except these:

- Who are authorized by PNGRB under PNGRB Act 2006 to establish CGD network in their respective GA? or
- Who are deemed to have authorization under PNGRB Act 2006, for establishing city or local gas distribution network in their respective GA?

PNGRB further clarified that the issue of other Entities setting up CNG stations other than authorized Entity within authorized GA is at present subjudice in the Hon'ble Punjab and Haryana High Court.

8.2 National Green Tribunal

With a view to curb pollution in NCR, in April 2015, National Green Tribunal (NGT) issued an order to restrict life of diesel vehicles to a maximum of 10 years. This means all the diesel vehicles older than 10 years shall not be allowed to operate in NCR. Such a move may help in the growth of CNG vehicle in the NCR zone.

8.3 Proposed Action Plan by the Centre for Science and Environment

The Centre for Science and Environment (CSE) has proposed detailed action plan[53] to curb high pollution level in NCR. Salient points that are relevant for CGD are reproduced here below:

- Increase Cess on Diesel in Delhi to equate price in nearby states.
- Implement ban on 10 years old Diesel Vehicles (in line with directive of National Green Tribunal).
- Plan for upgraded introduction of Battery operated Vehicles-Buses, Two and three wheelers for transition to cleaner fuels.
- Purchase of 5000 Buses (CNG) in next 5 years, through Global Tender.
- Waiver off Taxes on CNG Buses.

Implementation of the action plan may have positive impact on sale of CNG in NCR.

[53] Times of India, Delhi edition, dated 20 April 2015.

8.4 *Exit of BG and FIs*

Exit of BG from GGCL may erode the possibility of MNCs entering in CGD business in India. FIs, after being part of CUGL and MNGL, for more than 5 years, exited. Keeping in mind the status of theses CGDs, the promoters facilitated entry of IGL (another JV of BPCL and GAIL) instead of new FIs.

8.5 *Prime Minister's Vision of Domestic PNG Connections*

Hon'ble Prime Minister, during Urja Sangam organized on 27 March 2015, reiterated his vision to connect 1 crore household with piped gas by 2019. Accordingly, MOP&NG, in a meeting held with CGDs on 9 May 2015, firmed up targets for each of 21 CGDs.

Minimum new connections: 944,600.

Aspirational new connections: 1,392,914.

CGDs were also advised to provide monthly status and also timely escalate constraints, if any. Progress shall be reviewed periodically. Consequently, a significant improvement in domestic PNG connections is expected.

8.6 *Diminishing Response of Private Players in Bids*

With the backdrop of litigations/long delay in processing bids/issue of authorizations under the Round IV of bidding, relatively lower participation by private players (without clutches of PSUs) has been observed. Out of award of 8 GAs, only 2 GAs (Amritsar and Pune rural districts) could be granted to private entities. Diminishing interest of private players in the CGD network developments may seriously impact future bidding and growth of CGDs in the country.

8.7 *Kelkar Committee Recommendations Relevant to CGD*

Kelkar Committee, appointed by GOI, submitted its recommendation[54] in September 2014 for energy security in 2030. Some of the recommendations shall be game changer for CGD sector:

(a) Strengthen and empower PNGRB: PNGRB must be provided with an independent category of staff with competitive remunerations and access to

[54]www.petroleum.nic.in (accessed on 20 May 2015).

professional experts. This will help in building the required capabilities for effective sector regulator.

(b) Unbundling of gas transportation, marketing or production (financial as well as legal). This will bring greater transparency to the system and enable complete open access to infrastructure.

(c) Leveraging PPP model of infrastructure and govt. sourcing of funds.

8.8 Fruitful Negotiation with Ras-Gas

Recent revision (January 2016) of agreement with Ras-Gas, Qatar, for supply of LNG would reduce landed price of LNG at PLL Terminal by around 50% of earlier price determination formula, which shall be close to prevailing spot LNG prices. This is a very significant development and is expected to bring relief to Indian gas industry.

9 Key Take-away

- All CGDs in Gujarat state have performed exceedingly well.
- Gujarat consumes more than 50% of total NG on all India basis, which is attributed to the early introduction (70s, 80s) of natural gas for industrial usage and high commitment level of Gujarat State Govt. enabled mature market.
- Even now, industrial segment contributes 75–80% of total NG consumption in Gujarat, Delhi and Maharashtra.
- Higher industrial volumes have facilitated faster penetration in other segments also.
- It is pertinent to note that industrial and large commercial segments have advantage of large volumes, pricing flexibility, low operation and service cost.
- CGDs at Delhi and Mumbai have been very successful on the back of judicial intervention. CNG segment contributes around 75% of total NG sales.
- Low priority to CGD sector under gas utilization policy (till Nov 13) hampered the development of CGD sector.
- Most of the CGDs in other cities are still struggling, and performance has been far below expectation/commitment, particularly in domestic segment. It is pertinent to note that domestic has disadvantage of very low volumes per customer, high cost of last-mile connectivity, high customer service cost and completion with highly subsidized LPG domestic fuel.
- Lack of support and commitment of most of the state governments, leading to delay in executions.
- Delay in the deregulation of HSD pricing and reduction in subsidy on domestic LPG adversely affected penetration of NG.

- Higher taxes/duties by numerous states leading to disadvantage in competitive pricing versus alternate fuels.
- Govt. support/judicial intervention is key factor in the development of CNG segment, more so in tier II and III cities.
- Establishment of PNGRB in the year 2006 facilitated uniform approach in all aspects of CGD infrastructure and business.
- Lack of effective monitoring of CGD's progress by PNGRB. Most of the CGDs are getting away without imposing penalties by PNGRB (in line with the terms and conditions of authorization).
- Non-materialization of PNGRB's vision with respect to extension to 300+ cities, due to numerous legal interventions, cartelization by few bidders, clarity and frequent changes in bidding parameters.
- Concentration of CGDs in few states due to uneven distribution of cross-country pipelines in country.
- Dominance of public sector in CGD business.
- Lack of customer focus and professional management of most of the CGDs.
- Consolidation of CGDs through acquisition/merger—a necessity—so as to have efficient, cost-effective and customer-centric approach.
- Need to have robust business model (product/segment mix and area) to attract investors.
- Need to have innovation in the development of CGD infrastructure so as to optimize the capital and revenue cost.
- To enhance the customer awareness on safety aspect of PNG versus LPG.
- GSPC group and GAIL shall be major player in CGD business followed by consortium of IOC-Adani. OMCs (IOC/BPC/HPC) through JVs shall continue to play a significant role in the growth and expansion of CGD business.

10 Conclusion

With recent major changes in policy for allocation of domestic gas to all CGDs to meet the full requirement of CNG and domestic segment demand, major constraint has been addressed by MOP&NG. With faster implementation of revised guidelines for pricing of domestic natural gas, uncertainty on this account has been addressed by MOP&NG. Target setting for domestic connections for each CGD and close monitoring by the parent ministry shall motivate the CGDs to comply with the targets. As against 3.43 lakhs new connections released in FY 2013–14, MOP&NG fixed minimum aggregate target of 9.44 lakhs for the FY: 2014–15. MOP&NG in line with the vision of Hon'ble Prime Minister aspires to increase domestic connection population from current 28.66 lakhs (as on March 15) to 1 crores in the next 4 years. Such a step will certainly bring rapid transformation even in the rural India.

MOP&NG planning dedicated CNG corridors along highways—will boost CNG segment. The recent judgement of NGT for restricting life of diesel vehicles to max 10 years followed by Administrative decision (supported by Judiciary) of no sale of Diesel vehicles exceeding 2000 cc Engines in NCR from Late December 15 to 31 March 16 and conversion of all Diesel Taxis to CNG by 31st March, 16—will strongly push CNG growth.

With revision of long-term contract for supplies of LNG, with Ras-Gas, Qatar, CGDs shall be able to retrieve lost volumes in industrial and commercial segments, which in turn shall improve assets utilization and profitability.

Commissioning of new cross-country NG pipelines in the next few years shall deepen reach of NG—new gas. PNGRB shall have to expedite bidding rounds and expeditious grant of authorization for new GAs. Sustained and concerted efforts are required by the transmission companies/CGD industry/PNGRB/MOP&NG to achieve the vision of Hon'ble Prime Minister.

Professional management of CGDs now becomes prerequisite which may lead to merger/acquisition/consolidation of existing CGDs. It is therefore evident that CGD business in country shall expand at a very fast pace from 2017 to 18 onwards, and during the next 2 years, many new CGD players/entities are likely to emerge.

Building and Sustaining Natural Gas Business in India

Sanjay Kumar Kar, Piyush Kumar Sinha and Bhashit Dholakia

Abstract Natural gas is a green fuel and the Government of India is highly committed to promote green fuel like natural gas to replace more polluting fuels. The chapter provides critical insights on demand–supply scenarios, enabling factors, challenges, and opportunities to build sustainable gas business in the country. It highlights the role of global factors such as demand, supply, pricing, and geopolitics in natural gas market developments in India. Author presents insightful analysis on natural gas business environment, industry growth, role of competition, new entrants, opportunities available, challenges experienced, and suggestions for sustainable natural gas market growth in the country.

Keywords Natural gas · Market development · Competition · Demand–supply

1 Introduction

In the global context, natural gas business is loaded with opportunities and challenges. The scenario is no different in the Indian context. Because the product is an international commodity and the global factors such as demand, supply, and price of natural gas as well as competing fuel drive global trade of natural gas. On top of that the geopolitical factors do influence the trading pattern of natural gas leading to change is business scenario in any local country.

S.K. Kar (✉)
Department of Management Studies, Rajiv Gandhi Institute of Petroleum Technology, Jais, India
e-mail: skar@rgipt.ac.in

P.K. Sinha
Marketing and Retailing, Indian Institute of Management, Ahmedabad, India
e-mail: pksinha@iimahd.ernet.in

B. Dholakia
Business Development, Adani Gas, Ahmedabad, India
e-mail: bhashit.dholakia@adani.com

S.K. Kar and A. Gupta (eds.), *Natural Gas Markets in India*,
DOI 10.1007/978-981-10-3118-2_8

In addition to the above global factors, the domestic factors such as energy policy, oil and gas exploration and production policy, pricing policy, and monetization of energy, subsidy to oil, natural gas, and competing fuel affect the natural gas market developments in any specific country, especially in India. Often, large and frequent price fluctuations lead to unpredictable market outcomes. Therefore, raising concerns over economic viability and sustainability of natural gas business. It is important for business managers to assess global, regional, and local developments for developing socio-economically sustainable and adaptable business models for long-term gas market growth and development.

This chapter presents critical insights on business environment analysis, industry analysis, global developments, and natural gas market opportunity assessment relevant to India. An attempt made to highlight existing or potential challenges pertinent to natural gas business in India. The chapter presents some suggestions to convert some of the challenges into opportunities to ensure growth and development of natural gas business in the country.

2 Business Environmental Analysis

2.1 *Political*

Political environment is one of the key components of business environment. A stable political environment is helpful to build political consensus on various enabling factors of business. In India, the political environment has been relatively stable since 2010 and the general election-2014 added further stability by giving thumping majority to the ruling party. Although there is coalition government in the form of National Democratic Alliance (NDA-II), Bharatiya Janata Party (BJP) has the absolute majority in the lower house. This political stability enables the Government to take many reform-oriented initiatives. The current Government has taken many important decisions, and the prominent among them are diesel price de-regulation, domestic gas re-allocation, and domestic gas price revision. Political stability proves very important to take such bold decisions.

2.2 *Economic*

Economic environment is relatively stable compared to last 3–4 years. The economic growth started to pick up and the economic survey report projects growth rate higher than 7.5% for the year 2015–2016.

Economic Survey (2014–2015) suggests that India has moved from a carbon subsidization regime to one of significant carbon taxation regime—from a negative price to a positive price on carbon emissions. The government levies excise duty on petrol and diesel, which considered one form of carbon tax intended to reduce consumption of polluting fuel. The Government of India increased effective rate of Clean Energy Cess from Rs. 100 to 200 per ton (Union Budget 2015–2016) to fund clean energy initiatives.

2.3 Social

Ever increasing concerns and awareness for green energy builds greater social inclination to adopt green forms of energy including natural gas. Some of the domestic, industrial, and commercial/institutional customers are self-motivated to adopt green energy. This puts peer pressure on the other customers to consider greener forms of energy. In states like Gujarat, the communities and the industrial customers demand for natural gas connections from the entities (Kar and Sahu 2012).

2.4 Technological

Technology positively affects the entire natural gas value chain. New technologies across the oil and gas value chain have been constantly improving exploration, production, transportation, and efficient distribution of natural gas. For instance, evolution of fracking technology enhanced recoverability of shale gas leading to shale gas revolution the USA.

Exploration and production of methane from sources like gas hydrates require advanced and commercially viable technologies. Gas hydrate resources in India are estimated at 1894 TCM, and these deposits occur in Western, Eastern, and Andaman offshore areas (Business Standard 2014). Commercially viable technologies available at affordable cost would be helpful to increase domestic gas production from non-conventional sources such as gas hydrates and shale gas in India. In the highly capital intensive and risky business like oil and gas exploration and production, role of commercially viable technology accessible to Indian companies at a lower cost could significantly reduce the energy poverty in the country.

In the midstream, especially to ensure safety and security of the pipelines technologies are used. Supervisory control and data acquisition (SCADA)-based automation systems deployed for safety shutdown systems in key locations of a pipeline. In the downstream, marketing entities are now offering support services through technology-enabled platforms. There is evidence that the customer services improving through technology adoptions. For example, billing and payment process has been moving towards paperless and cashless.

3 Natural Gas Industry Analysis

Proper analysis of the current natural gas industry in India holds key to devise suitable strategies for sustaining current growth, expanding customer base, and exploring future growth. This section uses Porter's five forces model to discuss about industry rivalry, buyer power, supplier power, threats from new entrants, and threat of substitutes to natural gas.

3.1 *Industry Rivalry*

Currently, rivalry among existing firms in single geographical area is low to moderate in most of the states. However, rivalry is very high (from price front) in markets such as Mehsana, Kadi, and Kalol. In the above-mentioned markets, GAIL Gas and Sabarmati Gas Limited are supplying natural gas. Rivalry becomes stronger when multiple suppliers vying for the same customer. This may be the reality after expiry of marketing exclusivity in many geographical areas. Over a period, CGD business may move towards consolidation, one can expect more merger and acquisitions in the CGD business. In 2012, Consolidation happened in Gujarat, where Gujarat State Petroleum Corporation (GSPC) group bought 65.12% stake in Gujarat Gas from British Gas (BG). GSPC Gas is one of the largest CGD companies in India. Table 1 presents performance in terms of CNG stations, sales volume, and domestic customer base of GSPC Gas in Gujarat. Acquisition of Gujarat Gas from BG group strengthens position of GSPC in Gujarat.

3.2 *Buyer Power*

Bargaining power of the buyers found to be low in case of CNG and moderate in case of PNG. Despite availability of alternative fuel at a cheaper cost for PNG buyers in the industrial and domestic segments, the buyers are not able to exercise any level of bargaining power. It is found that the bargaining power of the buyers is

Table 1 CNG stations, CNG sales, and domestic customer base of GSPC Gas

Year	No. of CNG stations	Sales volume (kg)	No. of domestic customers
2009–10	72	56,909,533	127,830
2010–11	110	87,533,136	240,976
2011–12	135	123,198,680	328,348
2012–13	152	140,296,160	420,023
2013–14	157	140,967,277	492,381

Source 15th Annual report, GSPC Gas

clearly influenced by other factors such as switching cost and environmental concerns.

On the other hand, in the recent times, the Indian LNG buyers enjoyed their bargain power in the international market. Gostelow (2016) highlights that in the recent past, Indian LNG buyers enjoyed opening the discount of 17.5 cents/MMBtu compared to the buyers from the North Asia.

3.3 Supplier Power

Bargaining power of suppliers varies according to the market conditions. The Government determines domestic gas price and quantity allocation for priority sectors. In the domestic context, the suppliers have limited power to exercise.

On the other hand, the R-LNG throws varied challenges and offers some scope for bargaining and negotiation. In the past, the price of LNG imported on long-term contract basis was found to be lower than Spot LNG. Therefore, bargaining power of a supplier having access to long-term contracted LNG used to be relatively high. However, in the recent past, the LNG dynamics changed due to availability of shale gas in the USA, and the USA becoming an exporter of LNG. Spot LNG market saw a transformation resulting in currently spot LNG selling at much lower price than long-term LNG. Those who had booked long-term LNG found no favour from the final consumers, so the suppliers had limited scope for manoeuvring.

On 4 August 2015, the ICIS East Asia Index (EAX) for spot LNG on a DES (delivered ex-ship) basis was assessed at $8.20/MMBtu for September and $8.25/MMBtu for October (Kazmin 2015). As of February 2016, the spot LNG prices have come down further. Gostelow (2016) from Platts reported that offers from sellers with unsold cargoes from the Middle East to India were becoming especially cheap, with February offers to India coming down to $5.80/MMBtu.

Supply glut coupled with low demand for LNG resulted in lower supplier power. Recently, Petronet LNG Limited (PLL) re-negotiated its long-term contract with RasGas, Qatar. As per media reports, RasGas has in-principle agreed to changing the current pricing formula based on a 60-month average of a basket of Japanese crude oil prices to a 3-month average of Brent crude, a move that will lower cost of LNG to $7–8/MMBtu compared to $12–13/MMBtu as per the earlier contract. Further, PLL successfully negotiated penalty waiving off amounting to $1 billion for low gas off take. These negotiations are beneficial for both the parties.

3.4 Threat of New Entrants

Currently, threats of potential new entrants are low in many geographic areas where city gas entities are operating. However, the threats of new entrants in terms of competition are likely to increase in the near future because most of the entities won bids for GAs by bidding negligible network tariff. Practically speaking, the network tariff is almost zero, which means pipelines can be used by a third party as a common carrier/contract carrier with almost no fee.

The primary driver for growth of competition should have been the expiry of marketing exclusivity in many geographical areas. The Petroleum and Natural Gas Regulatory Board (PNGRB) grants the marketing exclusivity to a city gas entity for 3–5 years from the date of authorization. As of now, very few cities would attract new entrants because multiple entities may not find the market very attractive to operate. Experts from the CGD industry suggest that even in the future, threats from new entrants likely to be reasonably low. PNGRB grants pipeline infrastructure exclusivity to the first entity for 25 years, which means one entity controls the infrastructure.

3.5 Threats of Substitutes

Threats of substitutes like coal found to be very high and threats from other substitutes such as light diesel oil (LDO) and naphtha varies from low to moderate. Depending on the prices and comparative economics of substitutes, the level of threats changes time to time. In other words, price of substitutes may determine level of threats to natural gas.

4 Global Factors

4.1 Demand–Supply Dynamics

Demand and supply gap is the principal driver of global pricing. In turn, pricing positively or negatively affects demand and supply of natural gas. Weak European economic growth and a warm 2013–2014 winter reduced demand in Russia's primary market for gas exports (Doman 2015). The USA was one of the importers of natural gas a few years back. However, over the last few years, the USA became one of the largest producers of natural gas. Commercial availability of shale gas drastically changed the gas trading dynamics of the USA. The USA became net exporters of natural gas. In the second week of February, 2016, the tanker Asia Vision set off for Brazil with liquefied natural gas from Cheniere Energy's Sabine Pass terminal.

Weak demand in the USA considered as one of the factors behind low natural gas prices in the USA. The El Niño weather phenomenon has kept temperatures warmer than normal across much of the USA this year, reducing natural gas demand. High production coupled with low demand leads towards lower price of natural gas.

Natural gas futures for April delivery recently fell 6.5 cents, or 3.6%, to $1.726 a million British thermal units on the New York Mercantile Exchange. Prices traded as low as $1.69/MMBtu earlier in the session, on track for the lowest settlement since March 1999 (TWSJ 2016).

Japan, China, and India remain as large natural gas importing markets expected to continue to be in the centre of attraction in the future too. ExxonMobil (2015) suggests that each region will continue to rely on a unique mix of domestic and imported supply to meet its natural gas needs through 2040. Further, it anticipated several major trends: Asia Pacific will likely to overtake Europe as the world's largest gas importer; North America is expected to emerge as an exporter of natural gas, and Russia/Caspian is forecast to remain a significant gas supplier. Our understanding is that natural gas demand in India is going to increase in the short and long term.

4.2 Pricing

Demand and supply of natural gas does shape price of the commodity. However, other factors such as price, demand, and supply crude of oil affect the price of natural gas. As crude oil has a linkage with natural gas price determination, change in crude price does affect the price of natural gas, especially in case of long-term contracts. On the other hand, the global demand–supply scenario of natural gas primarily influences the sport price of natural gas. Couple of years back price of long-term contracted natural gas was cheaper, but recent past shale gas developments changed the scenario. Currently, spot LNG purchases are much cheaper than the long-term contracts, which disturb economics of pricing based on long-term contracts.

The domestic gas pricing holds key to meet government's environmental and social objectives. While delivering the keynote address at World Energy Congress 2015, New Delhi, the Minister of Petroleum & Natural Gas, reiterated that Government's pricing policy decision should not be influenced by the commercial decisions of the profit making enterprises. Rather domestic gas pricing decisions should be driven by social and environmental objectives of the government. Regarding the domestic gas pricing decision, the actions and initiatives taken by the Government seem to be in harmony with its social objectives.

The domestic natural gas marketing companies are quite happy with the pricing mechanism and gas allocation policy of the current government. Diverting natural gas from non-priority sector to CGD entities CNG (transport) and PNG (domestic) segments as per their actual consumptions is a very important step.

The E&P companies were building an argument to increase the domestic gas price but the new pricing formula resulted in bringing down the price (Fig. 1). The E&P companies were disappointed with the revised domestic gas pricing mechanism, whereas the marketing companies received the much-needed impetus. The

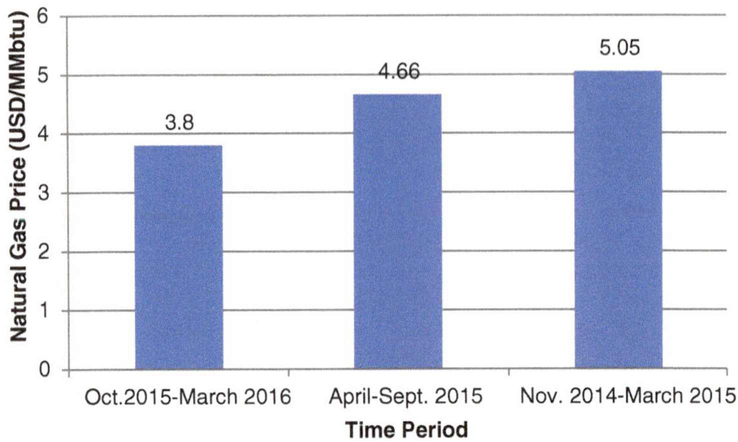

Fig. 1 Natural gas price in India

Budget 2016–2017 proposals raises the hopes of market-determined domestic gas price in the country. If this happens, the domestic producers would benefit the most.

Our discussions with some of the CGD companies suggest that the existing policies on gas pricing and allocation are quite encouraging for the growth and development of city gas distribution in the country. The allocation of cheaper domestic gas to CGD companies for CNG (transport) segment can spur the growth of CNG infrastructure and consumption in the 57 geographical areas authorized by the PNGRB Further, this would also excite bidders for new geographical areas across the country.

On the contrary, researchers like Sen (2015) opine that India lacks a well-defined vision on gas-based economy compared to countries such as Russia and China. Further, it is argued that the city gas distribution sector can possible accept higher gas prices due to lack of price controls on gas distribution companies. However, the argument does not factor in the price of alternative competing fuel available to the consumer. Research by Kar and Sahu (2012), Kar et al. (2016) suggest that industrial consumers in the city gas distribution sector look for price benefits along with other benefits such as supply security and service. It is to be noted that the price of LNG is determined by global factors beyond the control of CGD companies in India. Table 2 presents historical price of LNG and Natural gas in the global market.

In the year 2000, landed LNG price in Japan was just 50 cents higher than prevailing natural gas price in Henry Hub. By 2003, Henry Hub price became 86 cents costlier than landed LNG price in Japan—the trend continued until 2005. Since 2006, landed LNG price in Japan has been expensive than the Henry Hub price in the USA (Fig. 2). The gap became as big as \$11.98 in 2014 compared to 37 cents in 2006.

Table 2 Global LNG and natural gas price ($/million Btu) trend

	LNG		Natural gas		
	Japan	Average German import	UK	USA	Canada
	cif	price cif	(Heren NBP index)	Henry Hub	(Alberta)
1996	3.66	2.46	1.87	2.76	1.12
1997	3.91	2.64	1.96	2.53	1.36
1998	3.05	2.32	1.86	2.08	1.42
1999	3.14	1.88	1.58	2.27	2.00
2000	4.72	2.89	2.71	4.23	3.75
2001	4.64	3.66	3.17	4.07	3.61
2002	4.27	3.23	2.37	3.33	2.57
2003	4.77	4.06	3.33	5.63	4.83
2004	5.18	4.32	4.46	5.85	5.03
2005	6.05	5.88	7.38	8.79	7.25
2006	7.14	7.85	7.87	6.76	5.83
2007	7.73	8.03	6.01	6.95	6.17
2008	12.55	11.56	10.79	8.85	7.99
2009	9.06	8.52	4.85	3.89	3.38
2010	10.91	8.01	6.56	4.39	3.69
2011	14.73	10.48	9.04	4.01	3.47
2012	16.75	11.03	9.46	2.76	2.27
2013	16.17	10.72	10.63	3.71	2.93
2014	16.33	9.11	8.21	4.35	3.87

Note Btu = British thermal units; cif = cost + insurance + freight (average prices)
Source BP Statistical Review 2015

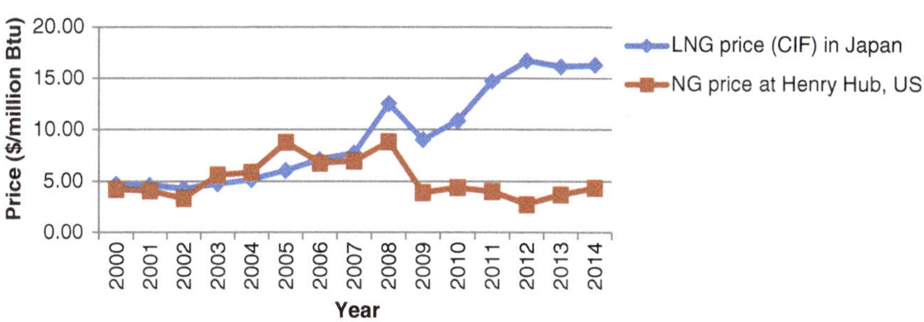

Fig. 2 Price movement of LNG (Japan) and natural gas (US)

A key reason for spike in LNG prices was the Fukushima nuclear disaster, which occurred in March 2011, forcing Japan to shut 47.5 GW of nuclear generation capacity.

As new LNG facilities came online, LNG prices moved downward. In 2015, the average Japan–Korea Marker (JKM) price was $7.45/MMBtu, and as of 3

February, the average price in 2016 came down to $5.73/MMBtu (Platts 2016). As of 3 February, 2016, the average price on Henry Hub has been $2.26/MMBtu, which is $3.47 lower than JKM price (EIA 2016).

4.3 Geopolitical Scenario

Of course, geopolitical scenario plays one of the important roles in shaping trade environment and influencing important decisions in the context of oil and gas business. For instance economic, social, and political situation in the Middle East or Russia seriously influence the global oil and gas trade dynamics. For example, sanctions imposed on Iran restricted sale of Iranian oil and gas in the International market. On 16 January 2016, the sanction was lifted which resulted in availability of Iranian oil and gas in the international market. Such a development would help India in a big way. Iran has 34 trillion cubic metres (TCM) of natural gas, which is about 18% global natural reserve (BP Statistical Review 2016). In the coming years, Iran is expected producing more than the current annual production of about 173 billion cubic metres and monetise the surplus natural gas. This development could be in favour of India.

India could actively engage with Iran to import natural gas through pipeline or LNG. Indian companies could strategically invest in acquiring and developing oil and gas assets in Iran. In the end, such a strategic move could help ensure access to Iranian gas.

Political stability and security situation in Afghanistan and Pakistan could play important role in completion of Turkmenistan–Afghanistan–Pakistan–India (TAPI) pipeline by 2019. TAPI pipeline would certainly improve natural gas supply scenario in India.

Availability of shale gas in the USA could be an important source of LNG for India. However, India does not figure in the list of either Free Trade Agreement (FTA) or Trans-Pacific Partnership Agreement (TPPA) of the USA. Therefore, India faces challenges of importing long-term LNG from the USA. Whereas, Korea and Japan, respectively, figure in the FTA and TPPA list of the USA, which makes long-term LNG import from the USA easier for Korea and Japan.

5 Assessment of Opportunities and Expectations

This section covers opportunities and expectations of various segments of natural gas customers in the country. Primarily, the natural gas customers fall under following segments:

- Industrial (PNG),
- Commercial (PNG),

- Domestic (PNG), and
- Transport (CNG).

The industrial segment is volume-driven and relative less price elastic. The transport segment fairly price elastic but offers volume. After implementation of new domestic price allocation policy, transport segment gets domestic gas, so for the time being price is not a concern for this segment. The commercial segment is reasonably price inelastic and offers large number with moderate volume. Each segment has its own priorities, needs, and expectation. Table 3 presents opportunities and expectations of customers in various segments.

5.1 Transport Segment

The transport segment has huge potential for growth in India. Vehicle penetration per 1000 people in India is the lowest among the developed and developing countries (MOSPI, n.d). However, growth of automobile is rapidly increasing in India. It is observed that improvement in road infrastructure and economic prosperity bound to increase vehicle penetration in India in the near future. Availability of domestic gas for transport sector is going to boast development in CNG infrastructure in the country. Currently, share of CNG vehicles in India is about 10–15% of the total passenger population. Table 4 presents status of CNG stations and CNG vehicles in India. It is evident that Gujarat, Delhi/NCR, and Maharashtra contribute 88% of the CNG vehicles in India. Therefore, 87% of the CNG stations are in Gujarat, Delhi/NCR, and Maharashtra.

Future CNG growth would come from states such as Uttar Pradesh, Madhya Pradesh, Rajasthan, Andhra Pradesh, and Telangana. Except NCR and some parts of Gujarat, CNG primarily consumed for intra-city travel due to lack of CNG infrastructure. In the next 5–10 years, more than 100 cities will have required CNG infrastructure leading to consumption of CNG for intercity travel purposes. Already green corridors developed in Gujarat, Delhi-NCR, and some parts of Maharashtra. Some more 'green corridor' projects are underway to facilitate use of clean fuel for long journey. Some of the green corridors are as follows:

- Delhi–Agra–Kanpur–Lucknow on NH2 with stations at Mathura, Firozabad, and Auraiya
- Vishakhapatnam–Vijaywada on NH5 with station at Kovvur.

Once the green corridors are put in place with adequate CNG filling stations, the light commercial and heavy-duty vehicle may run on CNG. In the USA, heavy-duty vehicles are successfully running. If demand for natural gas-driven heavy vehicles pick up, the manufacturers may produce more natural gas-driven heavy vehicles in India. This would serve the natural gas industry, auto industry, and environment

Table 3 Opportunities and expectations of consumers in various segments

Industrial segment	Commercialsegment	Domestic segment	Transport segment
Gas use: Industrial customers use natural gas for varied purposes such as heating, drying, and feedstock Size of customer base: Size of the industrial customer base is comparatively small with reference to domestic segment. However, this segment offers significant volume even in a single location Quantity of consumption: An industrial customer consumes 5–60 thousand SCM/day Price response: The industrial customers are relatively less price sensitive but seriously concerned about net financial savings from fuel heading Switching opportunity: The segment is seriously looking out for economical alternatives such as naphtha, diesel, coal, and electricity Expectations: Industrial customers demands for adequate pressure and higher calorific value of gas Customers in this segment demand for non-stop supply	Gas use: commercial or institutional customers use natural gas for heating, cooling, and cooking purpose Size of customer base: Similar to the industrial segments at the initial period but consume less volume compared to industrial customers Quantity of consumption: A typical commercial customer may consume 100–2000 SCM/day Price response: Like the industrial customers, the commercial customers are less price sensitive but seriously concerned about net financial savings from fuel heading May be price sensitive but may not have the negotiating power due to lack of volume Switching opportunity: The commercial customers look for cheaper alternatives and may demonstrate their behaviour when the need arises Commercial consumers look for financial savings from fuel consumption	Gas use: Indian domestic customers primarily use natural gas for cooking. However, natural gas can be used other purposes such as heating and cooling Size of customer base: Size of the domestic customer base is reasonably high. It may vary in different geographic areas. For example, IGL supplies gas PNG connections to 0.46 million customers in Delhi and NCR Quantity of consumption: A domestic customer normally consumes about 0.45–1 SCMD Price response: Domestic segment is highly price sensitive but do not have any negotiating power Switching opportunity: These customers show no intention to switch as long as natural gas is cheaper. The domestic consumer looks for financial savings from energy required for cooking Expectations: Consumers look for continuous supply	Gas use: Vehicles use natural gas as a transport fuel Size of customer base: The CNG-driven vehicle population may be large. For example, about 0.7 million CNG-driven vehicles are served by IGL Quantity of consumption: A small to medium size passenger car fills about 7–8 kg of CNG per filling Price response: The transport segment is price sensitive Switching opportunity: Given an opportunity, some of the customers of this may switch to alternative fuel Expectations: Customers are looking for greener, safer, and cheaper fuel; hence, CNG Safety is a concern. Customers do not want to wait for filling CNG in CNG stations Customers look for availability of gas with adequate pressure Customers look for financial savings over the competing fuel

Table 4 Status of CNG infrastructure and CNG vehicles in India (as of 30 September 2015)

State	Company Name	No. of CNG stations	No. of CNG vehicles	CNG vehicle (%)
Gujarat	Adani Energy Ltd., Gujarat Gas Ltd., (An amalgamated entity of Gujarat State Petroleum Corporation Gas Company Ltd. and Gujarat Gas Company Ltd., Sabarmati Gas Ltd., Hindustan Petroleum Corporation Ltd., Vadodara Gas Limited (JV of GAIL Gas Ltd. and Vadodara Mahanagar Seva Sadan)	361	811,880	33.4
Delhi/NCR	Indraprastha Gas Ltd. (IGL), New Delhi	324	804,289	33.0
Maharashtra	Mahanagar Gas Ltd. (MGL), Mumbai, Maharashtra Natural Gas Ltd., (MNGL), Pune, GAIL Gas Ltd.	212	530,961	21.8
Andhra Pradesh/Telangana	Bhagyanagar Gas Ltd. (BGL), Hyderabad	32	32,774	1.3
Rajasthan	GAIL Gas Ltd.	3	4010	0.2
Uttar Pradesh	Green Gas Ltd. (Lucknow), Central UP Gas Ltd. (Kanpur), Siti Energy Ltd., Adani Energy Ltd., GAIL Gas Ltd., and Sanwaria Gas Ltd.	41	97,545	4.0
Tripura	Tripura Natural Gas Co. Ltd., Agartala	5	7560	0.3
Madhya Pradesh	Avantika Gas Ltd. (Indore), and GAIL Gas Ltd.	21	20,508	0.8
Haryana	Haryana City Gas Ltd., GAIL Gas Ltd., and Adani Gas Ltd.	20	122,209	5.0
West Bengal	Great Eastern Energy Corporation Ltd.	7	2275	0.1
All India		1026	2,434,011	100.0

Note 1. Vehicle figures of IGL—Noida, Greater Noida, and Gaziabad is not available
2. After merger of GSPC & GGCL in May 2015, the average number of CNG Vehicles in GGL is based on average number of vehicles filled at Companies CNG Station per day
Source Author's analysis based on data collected from PPAC

well. Consumption of natural gas by a heavy-duty vehicle would be higher than passenger vehicles, which could be a favourable factor for growth CNG infrastructure.

5.2 *Industrial Segment*

The industrial segment is very attractive in terms of volume with higher amount of profitability. In some cases, consumption capacity of a large industrial customer may very much exceed actual consumption of all the domestic consumers of that geographical area.

From environment point of view, conversion of industrial customers from conventional fuel to natural gas helps reducing carbon and other toxic gas emissions. Natural gas has become the preferred fuel for various industrial applications as it is environment friendly, easy to handle, reliable, and requires no storage space. Some of the industries that use natural gas are as follows:

- Power,
- Ceramic,
- Automobiles,
- Textiles,
- Forging,
- Chemical,
- Bakery,
- Food Processing,
- Pharmaceutical, and
- Dairy.

Our market study findings suggest that most of the ceramic industry customers in Morbi and Thangadh (both towns are in Gujarat) benefited from use of natural gas supplied by GSPC Gas. Sabarmati Gas Limited serves the Dairy industry in North Gujarat by supplying Natural Gas to large customers such as Mother Dairy, Sabar Dairy, and Dudhsagar Dairy.

Due to early development of CGD network in Gujarat, natural gas penetration among the industrial customers is considerably higher than other States/Union Territories. As of 30 September 2015, 6044 industrial customers were using natural gas in the country, and Gujarat contributed 64% industrial customers (Table 5). The next best contribution comes from Delhi/NCR with 12.8% followed by Uttar Pradesh with 7.7% share. The states having low penetration and high potential for growth are likely to drive future growth in this segment. States such as Uttar Pradesh, Punjab, and Telangana, Maharashtra, Andhra Pradesh, and Madhya Pradesh would contribute in a big way in the future.

5.3 *Commercial Segment*

Most of the Geographical areas authorized by the PNGRB/State Government/ Central Government have reasonable potential commercial establishments. However, at the initial stage of operations the commercial segment is not in the priority list for many of the entities. Therefore, penetration of natural gas in the commercial segment is low. These customers are not very high volume customers, so they are always tertiary on the priority list.

In this segment, Gujarat has a commanding share of 72% (Table 6) followed by Maharashtra (12.6%) and Delhi/NCR (7.5%). Future growth would come from Punjab, Haryana, Rajasthan, Madhya Pradesh, Andhra Pradesh, and Uttar Pradesh.

Table 5 Status of industrial PNG connections in India (as of 30 September 2015)

State	City covered	CGD companies	Industrial connection	Share (%)
Delhi/NCR	National capital territory of Delhi (Including Noida and Ghaziabad)	Indraprastha Gas Ltd.	776	12.8
Maharashtra	Mumbai, Thane, Mira-Bhayender, Navi Mumbai, Pune, Kalyan, Ambernath, Panvel, and Bhiwandi	Mahanagar Gas Ltd., and Maharashtra Natural Gas Ltd.	160	2.6
Gujarat	Gandhinagar, Mehsana, Sabakantha, Nadiad, Halol, Hazira, Rajkot, Khambhat, Palej, Valsad, Navsari, Surendra Nagar, Ahmedabad, Vadodara, Surat, Ankeleswar, Bhavnara, and Anand	Sabarmati Gas Ltd., Gujarat Gas Ltd., Hindustan Petroelum Corporation Ltd., Vadodara Gas Ltd., Adani Gas Ltd., and Charotar Gas Sahkari Mandal Ltd.	3913	64.7
Uttar Pradesh	Meerut, Mathura, Agra, Firozabad, Kanpur, Bareilly, Lucknow, Moradabad, Khurja	Green Gas Ltd. (Lucknow), Central UP Gas Limited (Kanpur), GAIL Gas Ltd., Sanwariya Gas Ltd., Siti Energy Ltd., Adani Gas Ltd.	466	7.7
Tripura	Agartala	Tripura Natural Gas Company Ltd.	47	0.8
Madhya Pradesh	Dewas, Indore, Ujjain, and Gwalior	GAIL Gas Ltd., Avantika Gas Ltd.	74	1.2
Rajasthan	Kota	GAIL Gas Ltd.	9	0.1
Assam	Tinsukia, Dibrugarh, Sibsagar, and Jorhat	Assam Gas Co. Ltd.	392	6.5
Andhra Pradhesh/Telangana	Kakinada, Hyderabad, and Vijayawada	Bhagyanagar Gas Ltd.	3	0.0
Haryana	Sonepat, Gurgaon, and Faridabad	GAIL Gas Ltd., Adani Gas Ltd., and Haryan City Gas Distribution Ltd.	204	3.4
Total			6044	100.0

Source Author's analysis based on data collected from PPAC

Table 6 Status of commercial PNG connections in India (as of 30 September 2015)

State	City covered	CGD companies	Commercial connection	Share (%)
Delhi/NCR	National Capital Territory of Delhi (Including Noida and Ghaziabad)	Indraprastha Gas Ltd.	1652	7.5
Maharashtra	Mumbai, Thane, Mira-Bhayender, Navi Mumbai, Pune, Kalyan, Ambernath, Panvel, and Bhiwandi	Mahanagar Gas Ltd. and Maharashtra Natural Gas Ltd.	2761	12.6
Gujarat	Gandhinagar, Mehsana, Sabakantha,Nadiad, Halol, Hazira, Rajkot, Khambhat, Palej, Valsad, Navsari, Surendra Nagar, Ahmedabad, Vadodara, Surat, Ankeleswar, Bhavnara, and Anand.	Sabarmati Gas Ltd., Gujarat Gas Ltd., Hindustan Petroelum Corporation Ltd., Vadodara Gas Ltd., Adam Gas Ltd., and Charotar Gas Sahkari Mandal Ltd.	15808	72.1
Uttar Pradesh	Meerut, Mathura, Agra, Firozabad, Kanpur, Bareilly, Lucknow, Moradabad, and Khurja	Green Gas Ltd. (Lucknow), Central UP Gas Limited (Kanpur), GAIL Gas Ltd., Sanwariya Gas Ltd., and Siti Energy Ltd., Adani Gas Ltd.	198	0.9
Tripura	Agartala	Tripura Natural Gas Company Ltd.	300	1.4
Madhya Pradesh	Dewas, Indore, Ujjain, and Gwalior	GAIL Gas Ltd. and Avantika Gas Ltd.	31	0.1
Rajasthan	Kota	GAIL Gas Ltd.	1	0.0
Assam	Tinsukia, Dibrugarh, Sibsagar, and Jorhat	Assam Gas Co. Ltd.	1028	4.7
Andhra Pradesh/Telangana	Kakinada, Hyderabad, and Vijayawada	Bhagyanagar Gas Ltd.	46	0.2
Haryana	Sonepat, Gurgaon, and Faridabad	GAIL Gas Ltd., Adam Gas Ltd., and Haryan City Gas Distribution Ltd.	100	0.5
Total			21,925	100

Source Author's analysis based on data collected from PPAC

5.4 Domestic Segment

As per the PNGRB Regulations, the domestic customer should be given priority in terms of supply. While bidding, the bidders have to declare their domestic connection target during the period of marketing exclusivity. In addition, the Government puts significant thrust on increasing domestic PNG connections in the country. The existing domestic gas allocation policy is a reflection of the Government's vision and strategy. As of 30 September 2015, about 3 million domestic households were getting piped natural gas across the country. Out of those connections, Gujarat had a share of 48% (Table 7) followed by Maharashtra (29%) and Delhi/NCR (20%). As per the Petroleum & Natural Gas Regulatory Board (PNGRB), about 1.78 million domestic PNG connections are targeted in the next two years by the PNGRB authorized entities (PIB 2015). It is expected that by 2017, about 4.8 million households will have access to piped natural gas.

The Union Government set an ambitious target of achieving 10 million connections by 2019, which is more than threefold jump from the existing level. The target may look unrealistic from the outset but has certain merits. The Government is highly committed to move towards green economy and reduced dependence on LPG import. The honourable Prime Minister, Sh. Narendra Modi likes to achieve ambitious targets, and hopefully, the Ministry of Petroleum and Natural Gas under the leadership of Sh. Dharmendra Pradhan achieves the ambitious target.

The entities, which are lagging behind their domestic gas connection targets, are launching innovative schemes to achieve the targets. For instance, Avantika Gas Ltd. started a referral scheme and is giving cash incentive of Rs. 300/for every new connection (Ahuja 2015). Many of the entities are running campaigns to create awareness for piped natural gas connections.

Despite mandatory requirements for providing agreed number of domestic connections, in many geographical areas the entities are not able to achieve the target. This is a real concern for the CGD industry, regulator, and the Government. In the past, often unavailability of low price domestic gas and low margin structure were the principal reasons. However, the current domestic pricing and gas allocation policy address the concern of availability of cheaper domestic gas for PNG domestic segment. Table 8 presents a sample of cost and margin structure of natural gas supplied to domestic customers. The entity earns a margin of about 16% after covering 76% as supply and distribution cost. Our understanding is that with availability of cheaper domestic gas, adequate coverage for infrastructure cost, and decent margin the entities should not consider serving domestic segment as an obligation. Rather, the entities should consider this as a part of their marketing strategy to fulfil their own commitment. Once the entities realign their orientation, the target of 10 million customers might look achievable.

Table 7 Status of domestic PNG connections in India (as of 30 September 2015)

State	City Covered	CGD Companies	Domestic connection	Share (%)
Delhi /NCR	National Capital Territory of Delhi (Including Noida and Ghaziabad)	Indraprastha Gas Ltd.	601,192	20
Maharashtra	Mumbai, Thane, Mira-Bhayender, Navi Mumbai, Pune, Kalyan, Ambernath, Panvel, and Bhiwandi	Mahanagar Gas Ltd. and Maharashtra Natural Gas Ltd.	854,950	29
Gujarat	Gandhinagar, Mehsana, Sabakantha, Nadiad, Halol, Hazira, Rajkot, Khambhat, Palej, Valsad, Navsari, Surendra Nagar, Ahmedabad, Vadodara, Surat, Ankeleswar, Bhavnara, and Anand.	Sabarmati Gas Ltd., Gujarat Gas Ltd., Hindustan Petroelum Corporation Ltd., Vadodara Gas Ltd., Adani Gas Ltd., and Charotar Gas Sahkari Mandal Ltd.	1,430,566	48
Uttar Pradesh	Meerut, Mathura, Agra, Firozabad, Kanpur, Bareilly, Lucknow, Moradabad, and Khurja	Green Gas Ltd. (Lucknow), Central UP Gas Limited (Kanpur), GAIL Gas Ltd., Sanwariya Gas Ltd., Siti Energy Ltd., and Adani Gas Ltd.	28,784	1
Tripura	Agartala	Tripura Natural Gas Company Ltd.	19,747	1
Madhya Pradesh	Dewas, Indore, Ujjain, and Gwalior	GAIL Gas Ltd. and Avantika Gas Ltd.	4685	0
Rajasthan	Kota	GAIL Gas Ltd.	189	0
Assam	Tinsukia, Dibrugarh, Sibsagar, and Jorhat	Assam Gas Co. Ltd.	28,979	1
Andhra Pradesh/Telangana	Kakinada, Hyderabad, and Vijayawada	Bhagyanagar Gas Ltd.	3748	0
Haryana	Sonepat, Gurgaon, and Faridabad	GAIL Gas Ltd., Adani Gas Ltd., and Haryan City Gas Distribution Ltd.	25,894	1
Total			2,998,734	100

Source Author's analysis based on data collected from PPAC

Table 8 Break-up of retail selling price of PNG (domestic customer) at Gujarat gas

Particulars	In Rs./SCM
Cost of gas	11.92
Supply and distribution cost	9.07
Margin	1.92
Basic selling price	22.91
VAT @ 15%	3.44
Retail selling price	26.35

Source Gujarat gas

In the Round-VI of CGD bidding, 34 geographical areas (GAs) were available for bidding, but only 20 GAs attracted bids and 14 GAs found no bidders. Unfortunately, most of the GAs with higher household population and greater domestic PNG potential did not find any bidders. Table 9 presents status of Round-VI CGD bidding as of 21 January, 2016. The bidding status suggests that the bidders are comfortable investing in Gujarat, Haryana, and Punjab. Even North Gao attracted maximum number of bidders. On the other hand, investors have shown little confidence to invest in CGD projects in Uttar Pradesh and Madhya Pradesh.

5.5 *Direct Use of LNG*

The market for LNG transportation fuel is emerging and relatively immature (PTC 2013).

LNG is a more energy-dense fuel than CNG, (meaning a truck can go farther on one fill up); however, because LNG is a −260 °F liquid, it must be stored in heavily insulated tanks (Helman 2013). Therefore, LNG has become particularly attractive for long-haul or regional commercial trucks in the USA. It can be stored on-board a truck in dual insulated tanks at densities comparable to diesel fuel to provide ranges approaching 600 miles, which are required for profitable long-haul trucking. However, direct use of LNG in transport sector throws more challenges than CNG-driven vehicles. In India, there are no LNG filling stations, which may pose a serious challenge to growth of LNG-driven vehicles in the near future.

The lack of a nationwide LNG fuelling infrastructure has restrained trucking companies from investing in the technology in the USA (PTC 2013). Similarly, the truck manufactures in India may show disinterest in developing and promoting the technology.

There is a strong possibility that LNG fuelling infrastructure may come in the Gujarat, Maharashtra, and Kerala because these states have LNG terminals (Table 10). In India, currently 25 mmtpa of re-gasification facility is available and about 39 mmtpa capacity would be added by 2020, taking the total capacity to 64 MMTPA.

Table 9 Status of round-VI CGD bidding (as of January 21, 2016)

	CGD network	State	Area (Sq. Km)	Inch (km) pipeline	Households	Domestic PNG target by year 5	Bids received
1	North Goa	Goa	1676	650	191,766	9588	7
2	Rewari	Haryana	1519	650	178,043	8902	6
3	Rohtak	Haryana	1749	1342	206,000	10,300	4
4	Panchmahal	Gujarat	5249	1800	446,611	22,331	4
5	Fatehgarh Sahib	Punjab	1146	650	118,100	5905	4
6	Dahej-Vagra Taluka	Gujarat	909	591	21,590	1080	4
7	Yamunanagar	Haryana	1737	650	235,423	11,771	3
8	Rupnagar	Punjab	1370	650	135,635	6782	3
9	Banaskantha	Gujarat	10,668	1800	28,021	1401	3
10	Anand (excluding the area already authorized)	Gujarat	1889	680	253,000	12,650	3
11	Ahmedabad (excluding the area already authorized)	Gujarat	67,629	1800	261,000	13,050	3
12	Patan	Gujarat	5846	1800	267,633	13,382	2
13	Dahod	Gujarat	3636	1309	16,714	836	2
14	Amreli	Gujarat	7477	1800	294,837	14,742	2
15	Saharanpur	Uttar Pradesh	3728	1342	597,656	29,883	1
16	Ratnagiri	Maharashtra	8276	1800	397,115	19,856	1
17	Ramabai Nagar	Uttar Pradesh	3175	1143	329,817	16,191	1
18	Dhar	Madhya Pradesh	8091	1800	21,223	1061	1
19	Bhatinda	Punjab	3357	1209	273,902	13,695	1
20	Ahmadnagar	Maharashtra	17,038	1800	46,501	2325	1
21	Shahdol	Madhya Pradesh	5642	1800	251,452	12,573	0
22	Rewa	Madhya Pradesh	6327	1800	526,065	26,303	0
23	Raebareli	Uttar Pradesh	4593	1653	619,707	30,985	0
24	Nainital	Uttarakhand	4136	1489	191,383	9569	0
25	Mainpuri	Uttar Pradesh	2746	989	313,690	15,685	0

(continued)

Table 9 (continued)

	CGD network	State	Area (Sq. Km)	Inch (km) pipeline	Households	Domestic PNG target by year 5	Bids received
26	Jhabua	Madhya Pradesh	3436	1237	194,157	9708	0
27	Gadag	Karnataka	4673	1682	219,096	10,955	0
28	Etawah	Uttar Pradesh	1328	838	277,527	13,876	0
29	Datia	Madhya Pradesh	2515	905	160,215	8011	0
30	Chitradurga	Karnataka	8465	1800	357,003	17,850	0
31	Bhiwani	Haryana	4637	1669	311,380	15,569	0
32	Baghpat	Uttar Pradesh	1325	650	209,916	10,496	0
33	Auraiya	Uttar Pradesh	2021	728	248,197	12,410	0
34	Amethi	Uttar Pradesh	1956	704	263,647	13,182	0
	Total		209,965	43210	8,464,022	422,903	56

Source Compiled from the website of PNGRB

Table 10 Existing and upcoming LNG terminals in India

Place	State	Company	Status	Existing/planned capacity (mmtpa)	Capacity expansion (mmtpa)
Dahej	Gujarat	PLL	Operational	10	10
Kochi	Kerala	PLL	Operational	5	2.5
Dabhol	Maharashtra	GAIL	Operational	5	
Hazira	Gujarat	Shell	Operational	5	2.5
Dhamra	Odisha	Adani	Upcoming	5	
Ennore	Tamil Nadu	IOCL	Upcoming	5	
Mundra	Gujarat	GSPC	Upcoming	5	
Jafrabad	Gujarat	Swan energy	Upcoming	5	
Haldia	West Bengal	H-energy	Upcoming	4	

Source Compiled from various published sources

According to PTC (2013) estimates, building an LNG station would be costing around $1.8–3.95 million (Rs.122/–268/million) in the USA. The LNG stations would be more expensive than CNG stations. However, the economics may vary from location to location. In India, the cost of building an LNG station may be lower than the projected cost in the USA.

5.6 Others

Development of waterways would provide opportunity for use of natural gas by the boats. Even ships can use natural gas as a fuel. LNG can be used a bunker fuel. Many argue that LNG as a marine fuel is an idea whose time has come, both in terms of reduced emissions (nitrogen oxide, sulphur oxide, and particulate matter) as compared with heavy bunker oil and even low-sulfur marine diesel, and in terms of operating costs. LNG offers the best economics to ship owners and the best environmental benefits to the public (DNV 2014). LNG as a bunkering fuel would create additional opportunity for consumption of natural gas in the locations, such as Kochi, Dahej, Hazira, Dabhol, Dhamra, and Haldia, in India.

6 Challenges for CGD Business Growth

6.1 Demand–Supply–Import

India's net production of natural gas plummeted to 34.5 billion cubic metre (BCM) in the financial year (FY) 2013–2014 from a peak of 51.2 BCM in FY 2010–2011. Increasing demand for natural gas coupled with falling domestic gas production resulted in higher import of liquefied natural gas (LNG). As per the petroleum and natural gas statistics, GoI, during 2012–2013, India imported 13.14 MMT with a value of Rs. 417.31 billion of LNG compared to 8.38 MMT with a value of Rs. 71.97 billion of LNG in 2007–2008. Increasing domestic production of natural gas to meet domestic demand remains a challenging task. We believe that it would continue to remain as one of the biggest challenges for India. On the other hand, import of natural gas in the form of LNG would continue to rise. Therefore, managing cheaper source of LNG could remain a constant challenge for the importers.

6.2 Infrastructure Development

As of 30 September 2015, cross-country gas as pipeline network of 16,065 km with a total capacity of 386.5 MMSCMD primarily caters to the west and north India. The central and eastern parts of the country have limited access to natural gas due to lack of natural gas infrastructure such as LNG terminal and pipeline network. Additional pipeline network of 11302 kms with designed capacity of 468 MMSCMD is under various stages of construction. Within a decade, pan-India natural gas infrastructure can improve natural gas accessibility across the country. With a fully functional national gas grid of about 30,000 km, natural gas can contribute to economic growth of the entire country.

Pipeline infrastructure plays a critical role in evacuating natural gas from the LNG terminal or source of production. Due to delay in building pipeline to evacuate R-LNG from Kochi terminal, Petronet LNG is bleeding about Rs. 400 crores/years. Despite having LNG terminal in Kochi and availability of cheaper spot LNG, customers in the region are not able to receive the benefit. Therefore, it is very important to ensure timely completion of pipeline projects to connect customers with the supplier.

Once the infrastructure is in place, industries, such as fertilizer, power, ceramic, chemical, refinery, commercial, and transport, stand to benefit significantly. The domestic and transport segments will continue to enjoy the advantages of natural gas.

Lack of infrastructure found to be one of the biggest deterrents to natural gas penetration and growth in India. Infrastructure development remains as the one of the biggest concerns because it requires: (i) huge investment and (ii) optimal capacity utilization. Sub-optimal capacity utilization reduces return on investment, often operating assets become unviable. Inadequate return on investment acts as a hindrance to further investment in infrastructure projects.

Infrastructure development for CNG station requires statutory permission from Railway, NHAI, Forest Departments, Irrigation Department, Pollution Control Board, Petroleum and Explosive Safety Organization (PESO), and other relevant organization/departments.

Further, land availability in strategic locations and land prices remain as challenges to build new CNG infrastructure in cities.

6.3 Demand Management

Natural gas usage in Indian cities has been limited primarily due to shortage of supply. Natural gas is a finite resource and domestic availability is limited in India. Natural gas has wider use in domestic, industrial, and commercial set-up. Demand management of the domestic natural gas is a key challenge. Of course, government does rationing of domestic gas through dynamic gas allocation policy. The gas allocation policy often considers the supply constraints of natural gas and requirement of the consuming industry or sector. The gas allocation policy primarily emphasizes on the subsidy outgo from the exchequer and social implications while determining prioritisation and gas allocation. Efficient demand management holds key to sustainable natural gas market developments in the country.

6.4 Supply Security

Ensuring the availability of adequate amount of natural gas at competitive price remains one of the biggest challenges for sustainable growth of natural gas industry

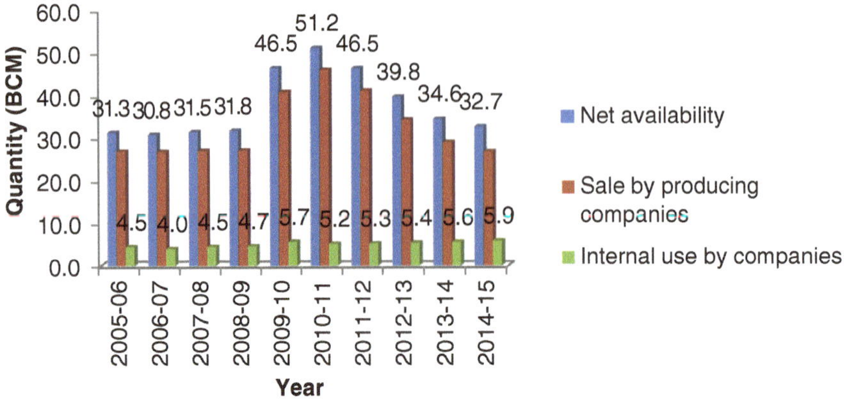

Fig. 3 Domestic natural gas net production trend in India

in India. Many of the natural gas distribution companies argue that the sustainability and viability of CGD business primarily driven by availability of domestic natural gas, because domestic natural gas is cheaper than imported LNG. However, today availability of domestic natural gas is a bigger concern due to the fall in production level from the existing fields and delay in exploration and production of new fields. Figure 3 presents domestic natural gas net production, internal consumption, and sales trend in India. Considering the current situation and future demand–supply scenario, ensuring supply security largely depends on import of LNG or dry natural gas through pipelines. At least for the time being, the domestic and transport segments are having domestic gas supply security. The Government has been working on import of natural gas through transnational pipelines. Turkmenistan–Afghanistan–Pakistan–India (TAPI), which is under construction, would enable India to import about 38 MMSCMD. This would improve gas supply security of north India.

6.5 *Taxation*

Taxation is one of major issue in natural gas business, and it affects the growth of natural gas business in India. Differential taxation structure adopted by State governments affects the final price consumer pay for the same molecule in different states. Non-uniform value added tax (VAT) levied on CNG in various states in India is evident from Fig. 4. Rationalization of VAT rate sounds logical but under the proposed GST bill natural gas kept out of GST. Therefore, the distortion is bound to continue and may have cascading effect. Further, in addition to rationalization of VAT rates, it is prudent to revisit the applicability of excise duty on sale of CNG. The compression of natural gas for supplying to the vehicular

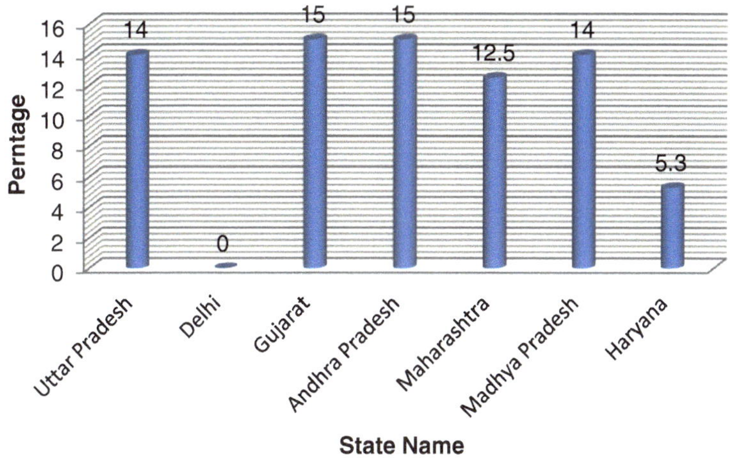

Fig. 4 Value added tax on CNG in various states in India

segment entails change of mass density in order to increase the storability (Mercados 2013). Hence, conversion of natural gas to compressed form is only for the purpose of transportation and should not be considered as manufacturing; thus, excise duty should be exempted from CNG. However, this argument finds no favour from the Government for the time being. The Government should encourage use of more CNG-driven cars. Possibly Government could exempt an Infrastructure cess of 1% (Union Budget 2016) on CNG cars, so that more people buy and drive CNG cars.

6.6 Attracting Private Participation

As of now, natural gas business is primarily driven by public sector undertakings. This business should not be a PSU-centric business (Manglani 2010). For this, PNGRB should come up with a roadmap for the next decade or so. The regulator invites bids for small geographic areas, which practically do not excite large private companies to invest. Returns from small geographic operations hampered by lack of economies of scale. Very often, small independent entities which win the bid but struggle to survive. Market growth and development remains far away from the promise as declared during the bidding process. The private players of course are interested to be part of 'city gas business' evolution in the country. Some of the private entities argue that business risks need to be addressed by ensuring attractive return on investment.

New business models are emerging to address the challenges of risk and return. Now private companies are forming joint venture companies with PSUs to reduce business risk and enhance business viability and success. This is a very good model to capitalize expertise of both private and public sector set-up.

6.7 Price Stability and Uniformity

Price stability is always a big concern for the customers, especially at the time of high international gas price scenario. Frequent changes, especially upward revisions in price of natural gas affect economics and business strategy of industrial and commercial customers. In many industries, natural gas used as a raw material/energy and contributes 40–50% of the total cost of finished products, for those industries upward revision of gas price adversely affects price of the finished products. However, the gas-consuming industrial customers could not pass on the incremental cost to their consumer. This causes undue pressure on bottom line of the natural using industries. Under such circumstances, natural gas users switch back to alternative and competitive conventional energy sources. On the other hand, during the time of gas price reduction, users get the benefit and are not concerned about price stability.

Natural gas price variations may exist across market, and some cases, it may vary from customer to customer. The price variations may be due to sourcing of gas and quality of the gas. Especially, in case of industrial customers, pricing is driven by level of services, daily contractual quantity, and size of the customer, which determine the terms and conditions of the contract. Often industrial customers demand price uniformity, which may be a very challenging task for the entities.

6.8 Customer Management

Customer management received significant emphasis across industries and geographic territories. Extensive studies conducted in the area of customer relationship and customer management. Recently, Nenonen and Storbacka (2014) created a conceptual framework for managing customer assets for improved value capture in the business market context. In the context of natural gas business in India, the industrial/commercial buyers have limited supplier options. Often in the market, there is only one natural gas supplier, so a natural monopoly exists. Whenever natural monopoly exists, customer service could take back seat. However, in natural gas business, both supplier and buyer invest reasonable amount of money to develop infrastructure. Therefore, both parties, especially the supplier should understand the importance of customer management. Better relationship with customer would ensure high returns in the long run.

6.9 Maintaining Growth

Maintaining growth remains a big concern for companies involved in natural gas business. Viability of natural gas business, especially CGD business is primarily dependent on adequate supply of domestic gas at competitive price. Due to lack of large-scale domestic gas findings and lower production from the existing blocks, the future of gas market depends on imported natural gas. Therefore, the natural gas market development will be driven by global factors such as demand–supply, price, and geopolitics.

Under such circumstance, growth would depend on:

- acquiring new customers in the existing market,
- selling more to existing customers,
- geographical expansion of existing market, and
- exploring new markets.

6.10 Coordination with Local Government

As per various Regulations of the PNGRB, the PNGRB authorizes entities to lay, build, and operate CGD network in specific geographical areas. However, the entities need cooperation and permission from the state/local government or government bodies to develop CGD network. Discussions with some of the entities suggest that coordination with the local government found to be increasingly challenging. Some industry experts given an impression that transaction costs for various permissions from the local government/bodies are constantly rising. Often the local government/bodies delay the permission resulting in undesirable delay in project execution resulting in delay in milestone achievement. Such tactics improve the scope of corruption and hamper ease of doing business. The Regulator or the Government should act appropriately to create a single window clearance process for CGD network developments. In addition, most of the permissions should be made available online. This would improve ease of doing business.

7 Developing Sustainable Strategy

Ebinger and Avasarala (2013) argue that a healthy and sustainable natural gas market development in India depends on the pace and scale of reforms that occur domestically. Any sustainable natural gas strategy needs to address core strategic issues such as supply security, pricing, infrastructure, and subsidy. Supply security at desirable and affordable price will continue to remain as the biggest concern at a macrolevel. It is no rocket science to understand that limited domestic reserves

Fig. 5 Action for sustainable gas market developments

coupled with lack of future roadmap for exploration and production would not improve domestic supply for at least next 10 years.

The progress on importing natural gas through pipeline is no better either. Importing natural gas through pipeline from Turkmenistan is going to be reality by 208–2019. However, at the prevailing international gas pricing scenario, the deal of $9/MMBtu that excludes transport cost with Turkmenistan could be a costly affair. Iran–India subsea pipeline to import 31.5 MMSCMD (The Hindu 2015) could be another option to improve supply situation in India.

For the time being, the best option available is to import LNG through by signing long-term contracts or buying from spot markets. However, the bigger question is about affordability and commercial viability of imported LNG. A few years back imported LNG available through long-term contracts considered far more viable than spot LNG. However, the situation has changed; now, spot LNG available at cheaper rate than long-term contracts. Whether spot or long-term, LNG will remain as one of the viable options to meet gas demand in India.

Government should judiciously decide on fuel subsidy and define role of natural gas for developing green economy. The government changed "Clean Energy Cess" to “Environment Cess” on coal and doubled the quantum from Rs. 200 to Rs. 400/ton. This is a tax mechanism to discourage the use of polluting fuel. Sustainable natural gas development strategy heavily depends some of the important factors presented in Fig. 5.

8 Conclusion

Evolving CGD business in India received much need impetus through the reallocation of domestic gas policy by the Union Government. Revised consumer friendly domestic gas pricing policy is a stimulus for CGD business growth in India. So far, the CGD growth has been driven by Gujarat, Maharashtra, and Delhi/NCR. Historically, the state owned companies driving natural gas business in most of the geographic areas. The recent past bidding rounds suggest that private companies are aggressively entering into CGD business. Public–Private partnership through joint venture companies is emerging as a new model for CGD business in the country.

Only winning geographical areas no longer remains the concern, developing market and sustaining market growth become part of strategic bidding. The regulator has increased stakes of failure, so the entities are now showing great intent to succeed. Similarly, as the Government allocating cheaper domestic gas for domestic and transport segments, building desirable pressure to succeed.

The states having natural gas infrastructure become more attractive for companies to invest in new CGD network. High potential for industrial and CNG customers remains as strategic parametre for investing in any geographic area. Future growth of natural gas in a specific geographical area would be driven by the strategic choice made by the company to serve all gas-consuming segments.

References

Ahuja M (2015) Refer a friend for piped gas connection and earn Rs 300 in Indore. Hindustan Times. http://www.hindustantimes.com/indore/refer-a-friend-for-piped-gas-connection-and-earn-rs-300-in-indore/story-GW5QvLVuw2Wv9NqgLBARmL.html. Accessed on 10 Feb 2016

BP Statistical Review (2016) Proven gas reserves history

Business Standard (2014) India and US to join hands for gas hydrates. Available at http://www.business-standard.com/content/b2b-manufacturing-industry/india-and-us-to-join-hands-for-gas-hydrates-114092600811_1.html. Accessed on 11 Feb 2016

DNV (2014) Liquefied natural gas (LNG) bunkering study. http://www.marad.dot.gov/wp-content/uploads/pdf/DNVLNGBunkeringStudy3Sep14.pdf. Accessed on 1 Mar 2016

Doman L (2015) U.S. remained world’s largest producer of petroleum and natural gas hydrocarbons in 2014, EAI, Apr 7

Ebinger C, Avasarala G (2013) Natural gas in India: difficult decisions. In: Harvard University’s Belfer Center and Rice University’s Baker Institute Center for Energy Studies, October

Economic Survey of India (2014–15) Ministry of finance, Government of India
EIA (2016) Henry Hub natural gas spot price (Dollars per Million Btu). Available at https://www.eia.gov/dnav/ng/hist/rngwhhdd.htm. Accessed on 11 Feb 2016
ExxonMobil (2015) The outlook for energy: a view to 2040. http://cdn.exxonmobil.com/~/media/global/files/outlook-for-energy/2015-outlook-for-energy_print-resolution.pdf. Accessed 15 Feb 2016
Gostelow (2016) Asia spot LNG: February JKM below $6/mmbtu to lowest level since May 2010. Available at http://www.platts.com/latest-news/natural-gas/singapore/asia-spot-lng-february-jkm-below-6mmbtu-to-lowest-26339166. Accessed on 11 Feb 2016
Helman C (2013) Should natural gas-powered cars run on CNG, LNG or gasoline? http://www.forbes.com/sites/christopherhelman/2013/06/05/should-natural-gas-powered-cars-run-on-cng-lng-or-gasoline/#c02411f15ef8. Accessed on 10 Feb 2016
Kar SK, Sinha PK, Mishra S (2016) Sabarkantha gas limited: challenges of marketing natural gas. Asian Case Res J 20(01):177–217
Kar SK, Sahu S (2012) Managing natural gas business: a case of Bharat Natural Gas Company Limited. Emerald Emerging Market Studies 2:1–22
Kazmin R (2015) Asia's spot LNG prices rise to $8.20/MMBtu on global demand. Available at http://www.icis.com/resources/news/2015/08/04/9910016/asia-s-spot-lng-prices-rise-to-8-20-mmbtu-on-global-demand/. Accessed on 17 Sept 2015
Manglani S (2010) Developing competitive gas market: current and future challenges. City Gas India roundtable 2010: initiatives and challenges. VIKALPA 35(4)
Mercados Energy Market India (2013) Natural gas vehicle-securing a sustainable future India. Available at http://www.gailgas.com/pdf/Publication/Knowledge%20Paper%20on%20CNG%20Sector.pdf. Accessed on 8 Feb 2016
MOSPI (n.d) Motor vehicles. Available at http://mospi.nic.in/Mospi_New/upload/SYB2015/CH-20-MOTOR%20VEHICLES/MOTOR%20VEHICLE-WRITEUP.pdf. Accessed on 9 Feb 2016
PIB (2015) Target of PNG connections in the next two years. Available at http://pib.nic.in/newsite/PrintRelease.aspx?relid=124342. Accessed on 12 Feb 2106
Platts (2016) The US enters a brave new world as it begins LNG exports. Available at http://blogs.platts.com/2016/02/10/us-brave-new-world-lng-exports/. Accessed on 11 Feb 2016
PTC (2013) Feasibility study for liquefied natural gas utilization for commercial vehicles on the pennsylvania turnpike. Revision 1. Available at https://www.paturnpike.com/pdfs/commercial/LTI_2013_04_Final_Reports.pdf. Accessed on 10 Feb 2016
Sen A (2015) Gas pricing reform in India: implications for the Indian gas landscape. OIES paper NG 96. Oxford Energy Institute
Nenonen S, Storbacka K (2014) Management of customer assets for increased value capture in business markets, Management Decision 52(1):101–121
The Hindu (2015) India, Iran discuss undersea gas pipeline bypassing Pakistan. Available at http://www.thehindu.com/business/india-iran-discuss-undersea-gas-pipeline-bypassing-pakistan/article7958494.ece. Accessed on 3 Mar 2016
The Wall Street Journal (2016) Natural gas prices fall to 17-year low. http://www.wsj.com/articles/natural-gas-prices-fall-to-17-year-low-1456761537 Accessed on 1 Mar 2016
Union Budget (2015–16) The Ministry of Finance, Government of India
Union Budget (2016–17) The Ministry of Finance, Government of India

Issues and Challenges in Development of Efficient Gas Market

Akhil Mehrotra

Abstract India needs energy, which its population can access and afford. Government of India intends to increase energy availability by increasing the share of gas, in the energy mix from 7.1 to 20% by the year 2030. However, this is not possible unless there is a thriving gas market in the country. The current Indian gas market has distortions such as monopoly power, inadequate infrastructure, and lack of transparency. Gas market evolution curve helps understand various phases of gas market, starting from government controlling the gas chain, negotiated market, early wholesale and fully developed gas market. Study of major countries reveals India at negotiated market phase, with aspirations similar to other countries, to move to a fully developed market phase such as the USA and UK. The learning from USA and UK can help to understand factors for such a transition; however, this can only be achieved if challenges such as issue relating to large population below poverty line, lack of integrated planning, continued distortion in all energy markets, unstable fiscal regime, and misalignment between federal and state governments are addressed. Government needs to calibrate its approach to develop gas market in order to increase the share of gas in the energy mix.

Keywords Energy · Gas · Market · Infrastructure · Monopoly

1 Introduction

Energy is a key resource for the development of any country. It is the responsibility, of the government of the country, to make energy, accessible, and available to its citizens, at an affordable price. Different countries have taken different routes for meeting this objective. The route to energy self-sufficiency depends upon factors such as availability of energy sources within the country, stage of its economic

A. Mehrotra (✉)
Mumbai, India
e-mail: akhil3871@gmail.com

S.K. Kar and A. Gupta (eds.), *Natural Gas Markets in India*,
DOI 10.1007/978-981-10-3118-2_9

Table 1 Share of energy sources in the total primary energy consumption (India vs. world average)

Energy source	Percentage share of energy source in total energy consumed for the year 2014 (%)	
	World average	India
Coal	30.03	56.47
Oil	32.57	35.47
Gas	23.71	7.14
Others (nuclear, hydro, renewables)	13.69	8.05

development, and depth of development of its core institutions required for governing such development.

Historically, oil and coal have dominated, the world, as the main source of energy. However, the past few decades have seen a quantum jump in development of gas and renewables as an energy resource. Gas being considered a clean fuel has increased in importance, in the energy map of the world. Worldwide, the consumption of gas has increased fivefold in the last 50 years, increasing from a meager 644 billion cubic meters in 1965 to 3393 billion cubic meters in 2015 (BP 2015).

In India, there are difficulties, in both the availability and accessibility of gas, as an energy source. The share of gas in the total primary energy consumption is only 7.1% for the year 2014. Oil and coal formed the bulk of the energy consumed with a combined share of 85%. Table 1 shows the comparison between the world and India, of the share of various energy sources in the total primary energy consumption (BP 2015).

The government of India ("GoI") has aspirations to augment the availability of gas and increase its share, in the total energy consumption to 20% (from current 7.1%) by the year 2030 (GoI 2013), bringing it nearer to the world average. To meet this target, large investments are required in exploration, production, and transportation of natural gas. To find solution to the problems faced by the gas sector, it is important to first understand the current state and characteristics of the Indian gas market.

2 Current State and Characteristics of Gas Market in India

2.1 History and Structure

The gas market in India can be divided into four distinct regimes/eras.

2.1.1 Nomination Regime (Until 1990)

This regime, as the name suggest, provided hydrocarbon acreages to the national oil companies on nomination basis. In the initial phase, the associated gas, produced from such fields, were flared in the absence of any infrastructure and usage for such gas. The gas was used locally until GAIL was formed in 1984 to set up gas transportation infrastructure. The first high-pressure transportation pipeline, the Hazira–Vijapur–Jagdishpur (HVJ), was fully operational in the year 1991. The price of gas, for the fields under the nomination era, was determined on a cost plus basis, known as administered price mechanism ("APM"). North East of India was provided a discount of 40% on the APM price so determined. While the nomination era ended in 1990, the APM continued to be the price setting mechanism till date.

2.1.2 Pre-National Exploration Licensing Policy (NELP) Regime (1991–1998)

The Director General of Hydrocarbons (DGH) was established, in the year 1993, under the administrative control of the Ministry of Petroleum and Natural Gas (MOPNG) to oversee the exploration and production of hydrocarbons.

In order to attract capital and latest technology in the exploration sector, the GoI, provided certain prediscovered blocks to the international oil companies, based on a production sharing contract (PSC). Panna–Mukta and Tapti fields in the Mumbai high on west coast of India were awarded as part of this regime in the year 1994.

Price of gas for such fields was decided based on the formula provided in the PSC, which is primarily linked to oil indices.

2.1.3 NELP Era (Since 1999)

The first NELP round was initiated on January 8, 1999, offering 48 oil and gas blocks under the PSC regime, on an International competitive bidding basis. Since then nine NELP rounds have awarded 294 blocks to various companies.

The price under the NELP regime is based on the PSC signed between government and the contractor, which envisaged that the contractor follows a price discovery process, on an arm's-length basis for such price discovery, with the approval from the government. However, government has, since determined the price of such gas through devising of formula, with the linkage to various world indices to determine the price of gas at the wellhead. The current formulae devised as part of a report by Dr Vijay Kelkar, recommended that the price of gas be determined through volume-weighted linkages to Henry Hub, National Balancing Point, Alberta and Russian Hubs (Kelkar 2014). This changed the earlier formulae set up, as part of Rangarajan report (Rangarajan 2012), dropping Japan LNG and local LNG import indices.

2.1.4 New Domestic Gas Price Regime (2013 Onwards)

Most of the gas in India, except where the PSC has well-defined formulae, is determined, since the year 2013, by the government, through it new domestic gas price policy. Under the policy, price of gas is determined as per the formulae given by Kelkar, i.e., based on a volume-weighted average of Henry Hub, National Balancing Point, Alberta and Russian Hubs. This, the government believed, would act as a proxy in the absence of a gas market and would provide remunerative returns to the exploration companies based on certain well-established hubs.

However, the mechanism devised by Kelkar remains at best proxy to the market until government takes steps for its development. Rangarajan stated in his report (Rangarajan 2012) since the competitive price for domestic gas does not exist and may not be expected to ensue for several more years, the gas pricing policy will have to be based on the searching from global trade transaction of gas, the competitive price of gas, at the global level.

In the recent, small, and marginal field policy of September 2015, the government has moved away from determining the price of gas and has provided freedom to the producers, for the price of gas produced from such identified small and marginal fields, to be determined by the market. These blocks are to be auctioned as part of the tenth NELP round is a step in the right direction, but this potentially needs to be extended to larger fields also.

2.2 *Governing Framework*

The Indian oil and gas sector is divided into upstream, midstream, and downstream. The upstream sector is controlled by the policies formulated by the MoPNG at the federal level. MoPNG is assisted in its efforts to formulate policies and grant concessions and monitor the exploration and production by Director General of Hydrocarbons ("DGH"). The planning issues are dealt by Petroleum, Planning and Analysis Cell ("PPAC") and safety issues by Oil Industry safety Directorate ("OISD"). All these agencies work under the direction and control of the MoPNG, and hence, the government is the regulator for the sector, thereby taking all decisions. The GoI, in the year 2006, enacted through an Act of Parliament the Petroleum and Natural Gas Regulatory Bill ("PNGRB") to establish an independent regulator to regulate the midstream and the downstream sectors. The PNGRB since then has since formulated regulations to govern the midstream and downstream sectors. In addition, there are various other rules and policies as shown in Fig. 1 which govern the oil and gas sector in India.

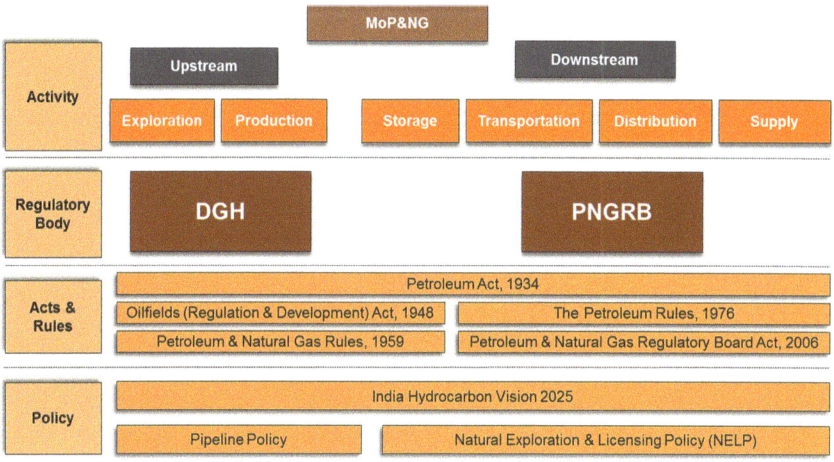

Fig. 1 Oil and gas sector—governing framework

2.3 Key Attributes of Indian Gas Market

The Indian gas market, in spite of opening up of the oil and gas sector for international competitive bidding, and allowing 100% foreign direct investments, has not witnessed the development which was envisaged by the government. This has led to a situation, where there are only few companies exploring the Indian basins, underdeveloped high-pressure cross-country pipeline network and no aggregators/shippers of gas other than the government companies. The key attributes of the Indian gas market can be enumerated as below:

1.0 No independent regulator for the exploration and production sector. DGH is an arm of MoPNG. PNGRB is the midstream and downstream regulator, but does not ave requisite powers to exercise control over the sector. MoPNG still retains power to control the retail segment.

2.0 Unstable price and fiscal regime. International competitive bidding for upstream acreages along with 100% FDI, but price controls at wellhead. Most oil and gas discoveries mired in litigation with the government of India, mostly on interpretation of PSC terms and government approvals.

3.0 Discussion in progress to move from a production sharing to revenue sharing structure. Implemented for identified small and marginal fields.

4.0 Risk reward in favor of government, although the Indian basins are not highly prospective, and the country has not seen large investments in the sector. Revenue sharing mechanism not prudent for acreages in India which are mostly frontier and deep water.

5.0 Domestic gas allocation and pricing determined by GoI, although current PSC provides freedom to determine gas price on competitive arm's-length basis, as approved by GoI.

6.0 LNG import allowed under Open General License (OGL). No terminal currently operating on tolling basis; limiting access to market participants.
7.0 Plan to import gas from resource rich countries such as Turkmenistan, Iran, and Russia, through cross-country pipelines, but geopolitical constraints would delay such projects.
8.0 Transmission pipeline under the development by few large players with major slippages in timeline; large population still not covered through pipeline; open access norm in place. Total high-pressure pipeline stood at a meager 17,000 km as per PNGRB.
9.0 Bundled entities allowed to transport and market gas with accounting separation principle; discussion on timeline for unbundling in progress by PNGRB
10.0 Distribution and marketing segment not developing due to (i) wellhead price regulation, (ii) high LNG prices (although considerably softened since Q1 2015) with no access to terminal capacity to market participants, and (iii) distortions in power, fertilizer, and transport fuel markets
11.0 Challenges in power sector reforms, under domain of state government, and hence not integrated to the federal policy.

Gas market in each country will have its own characteristics. But the basic elements/factors for a market development would remain the same, although these elements may require customization to fit the need of the specific country. Countries such as USA and UK have seen development of gas markets over few decades. Hence, using the learnings from these developed markets, it is important to understand;

(a) The evolution curve of a typical gas market,
(b) The characteristics of various phases within the evolution curve, and
(c) Status of various countries and particularly India on the evolution curve.

This would help us to demonstrate the issues faced by Indian gas market and steps required for addressing these based on the learnings from the key developed markets.

3 Gas Market Evolution Curve and Its Phases

Gas markets can be classified into four to five broad phases based the stage of its evolution (model in Fig. 2 adapted from) (Juris 1998a)

a. **Phase 1: Government controlled gas chain**—In the initial stage of development of any country, the economy is underdeveloped leading to (i) lack of confidence in the institutions of the country and (ii) suboptimal returns on investments for many years. Under such conditions, it is not possible to attract private investment.

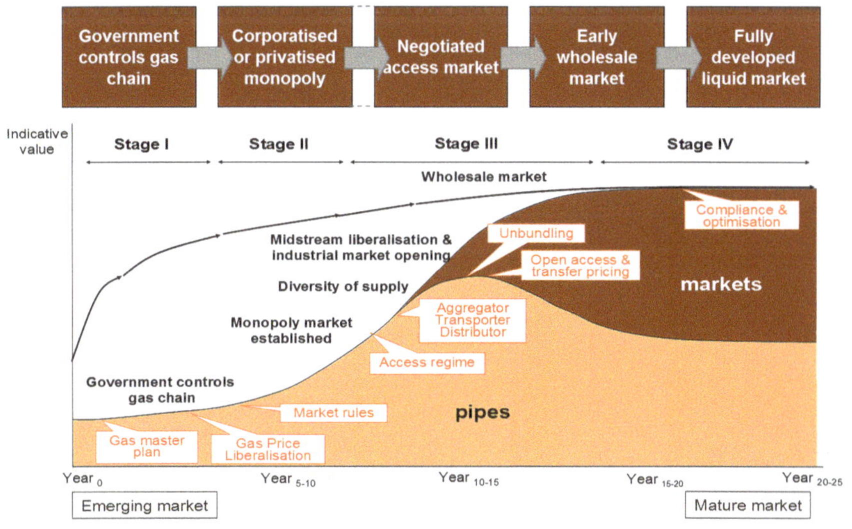

Fig. 2 Phases of gas market evolution

Therefore, in this phase, the role of government is crucial. Government would need to invest through its own institutions/companies to develop confidence and tide over the initial suboptimal return phase. In addition, the government needs to provide opportunity for exploration of acreages and development of transmission infrastructure on nomination basis. Rules and regulations can be framed to provide cost plus returns to such companies.

b. **Phase 2: Negotiated gas market/Corporatized or privatized monopoly**—In this stage, corporatized entities bring focused management; capital is attracted through fiscal/tax incentives. Monopoly rights grant certainty of returns and government still plays a major role in the investment, although parts of the gas chain may become attractive for private investments.
States bifurcate the role of regulation and policy intervention by setting up an independent regulator. Negotiated market allows setting up of large infrastructure, based on large anchor customer. Government intervenes to ensure supply security for the country. New roles develop in the market with marketers and shippers act as large aggregators.

c. **Phase 3: Early wholesale market**—This stage witness standardization of market rules with focus on fair play, transparency, and stability. Regulator has to play an important role since government companies compete with private players. Competition is introduced at the wholesale level while exclusivity is granted at retail level to promote investments in the downstream sector. Physical markets take shape.

d. **Phase 4: Fully developed liquid market**—In this stage, full-scale competition is introduced. Unbundling of transportation and marketing activities takes place, and retail customers are provided choice. Financial markets such as derivatives and secondary capacity trading market develop.

The key drivers and characteristics of each phase of market development are summarized in Table 2.

Table 2 Characteristics and drivers for market development phases

Sr. no.	Market evolution phases	Drivers	Characteristics
a	Government controls the gas chain	• Need for capital and expertise in upstream • Discovery of new fuel source	• Few market players • Monopoly operators • Government controlled supply/demand/pricing • Cost-based pricing sometimes including subsidies • Government acts as an aggregator
b	Corporatized or privatized monopoly-negotiated gas markets	• Desire to raise finance • Encourage network expertise • Lack of market-based gas pricing inhibits new upstream development • Demand is greater than supply (need for LNG and potentially liberalizing E&P terms) • Supply security concerns	• Regulated supply prices (mostly on cost plus basis) and/or limited interfuel competition •Administered alternate fuel pricing • Can involve a degree of vertical disaggregation (e.g., separately owned monopoly distribution franchise buying from an integrated transmission/supply business) • Parts of the value chain become attractive to private and foreign investments • Negotiated access regime large customers need to anchor network expansion • Rapid demand growth + unwinding of subsidies • New roles: shipper + aggregator + initial midstream marketers • Establishment of regulator
c	Early wholesale markets	• Enactment of early government market rules (access, transparency, unbundling) • Belief in market (security of supply + efficiency)	• More market participants • Contract diversity • All large customers have choice of supply • Force capacity release of pipeline ullage, used to service small-/medium-sized customers • Initial market rules • Price transparency emerges • Established midstream marketers
d	Fully developed liquid markets	• Market development dependent on – Diversity of supply – Market framework (removal of dominance)	• Full unbundling, transparency, full open access • Bidding used to ration tight capacity and encourage investment • Full market for commodity (derivatives, secondary capacity market, etc.)

4 Mapping of Indian Market on the Evolution Curve

Based on the characteristics of each stage, key countries were mapped on the gas market evolution curve (IEA 2013; Allan Consulting Group 2005; Baker Institute Study 2005; Jain et al. 2011; Council of European energy regulators 2011; Chawla 2011). India, as shown in Fig. 3 is currently struggling to move from a negotiated market stage to the early wholesale stage, with some government interventions (price caps at wellhead) pulling it back toward government control stage. The desired state is the fully developed liquid market where countries such as USA and UK are currently operating. These markets are fully developed and attract investments based on economic signals emanating from the market. The review of literature on various countries also clearly brings out that most countries are aspiring to move to the stage where USA and UK have reached after spending decades in reforming the policies and learning from their mistakes.

To understand each of the above phases better, these have been depicted via model (Juris 1998a) as given in Fig. 4.

Model 1 shows a fully integrated gas industry where there is monopoly across the gas chain and the flow of gas and the contracts follow the same linear path. Model 2 shows a negotiated market where competition is introduced in production, but transportation and distribution is on a negotiated basis. Model 3 introduces a wholesale market place and introduction of market players such as traders and suppliers. Physical flow of gas remains the same, but new contractual path emerges. Model 4 shows a development of financial markets and full retail competition.

In addition to the above mentioned, it has been observed that various countries have either followed a wholesale or retail competition model during the beginning of the market opening phase. Hence, it would be important to understand the concept of various forms of gas market.

Forms of gas market

Juris (1998a) has illustrated the principles of wholesale and retail gas markets. Based on the characteristics and need of market, liberalization can take place either in the wholesale or retail market or in both. USA has followed a wholesale liberalization approach while UK was more on retail competition approach, but gradually both are opening up market in other segment also. Andrej Juris describes both the market as below.

Wholesale and retail gas markets: Purchases of natural gas for further resale take place in the wholesale gas market. Purchases of natural gas for end use take place in the retail gas market. Wholesale transactions are all those concluded among the producers, traders, suppliers, and pipeline and distribution companies; retail transactions are those between suppliers and end users. The structure of wholesale and retail markets is important for the efficiency of pricing in them. Strong

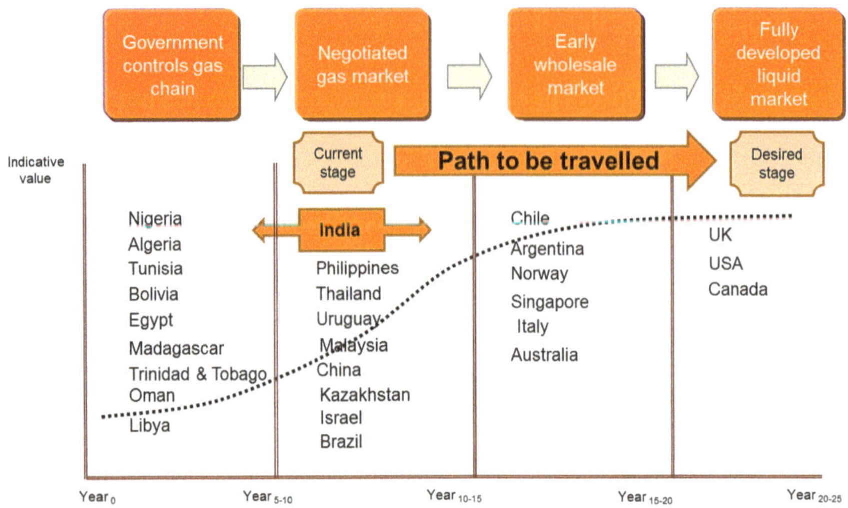

Fig. 3 Mapping India on the evolution curve

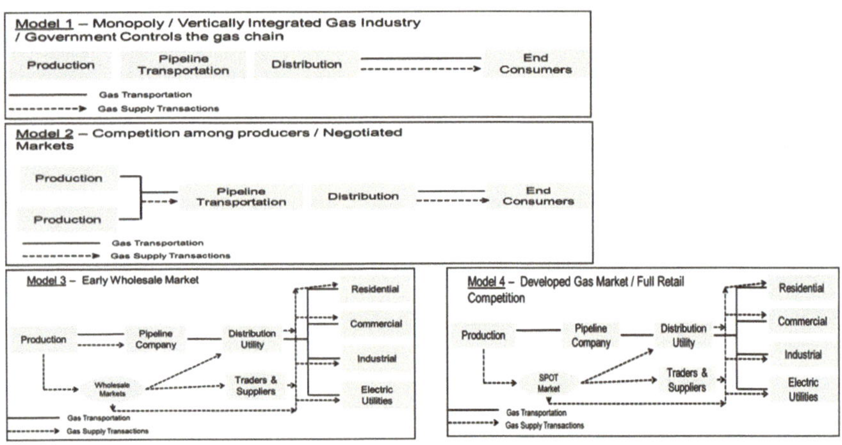

Fig. 4 Model depicting various phases of evolution curve

competition in these markets increases the economic efficiency of decentralized pricing and reduces the need for price regulation, while a lack of competition raises concerns about market power and price efficiency and calls for price regulation.

A large country like India can probably first open the market to wholesale competition and then gradually move to open up retail market.

In addition to the wholesale and retail markets, gas markets can have various subcharacteristics such as physical and financial markets similar to an equity market and also a transportation capacity market.

In natural gas market, natural gas is traded as a commodity, separated from transportation services, in the form of gas contracts. In addition, transportation services are traded as separate contract. Hence, when we talk of gas markets, it needs to be understood that gas markets can comprise of the following:

(1) Physical and financial gas market and
(2) Transportation markets both primary and secondary (capacity release).

Figure 5 represents the current and future state of gas market in terms of (a) market structure i.e., gas transportation and (b) gas supply, i.e., contractual obligation.

Present state in India is where GAIL acts as a nominee of government of India to aggregate all domestic and major imported gas which it then supplies thru contracts (bundled or otherwise) to large consumers and city gas distributers. The supply molecules are either allocated by Government of India or parties enter into a bilateral contract for purchase of spot or term gas mainly linked to Japanese Crude Cocktail (JCC) or Brent Crude.

In the target stage, a market mechanism should take shape (wholesale markets) where most of the transactions for sale and purchase take place at the market. Thus, market can have various forms and shapes such as hub-based trading, bulletin board, etc., which is the matter of further research.

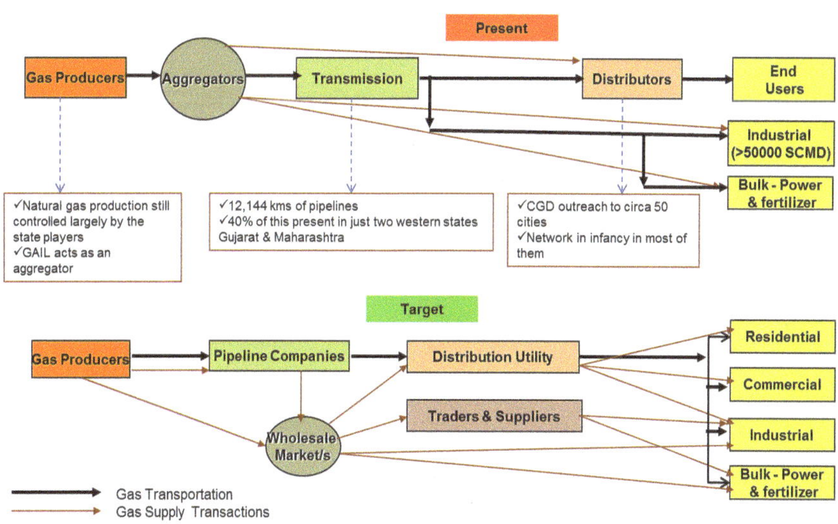

Fig. 5 Current and future state of gas market

5 Characteristics of an Efficient Market

An allocation of resource that maximizes the sum of consumer and producer surplus is said to be efficient. Consumer surplus equals buyer's willingness to pay for a good minus the amount they actually pay for it and it measures the benefit buyers get from participating in a market. Producer surplus equals the amount sellers receive for their goods minus their cost of production, and it measures the benefit sellers get from participating in a market (Mankiw 2007).

The equilibrium of supply and demand maximizes the sum of consumers and producers surplus. That is the invisible hand of the market place leads buyers and sellers to allocate resources efficiently. Markets do not allocate resource efficiently in the presence of market failures such as market power or externalities (Mankiw 2007).

Pareto efficiency (or sometimes just efficiency) occurs when no possible reorganization of production or distribution can make anyone better off without making someone else worse off. Under conditions of allocative efficiency, one person's satisfaction or utility can be increased only by lowering someone else's utility (Samuelson 2010).

Information on market and system operations and capabilities at all stages of the supply chain should be publically available and frequently updated for a reliable, competitive, and efficient gas market (MCE 2004).

Gas on gas competition (determination of gas price through forces of demand and supply) is the soundest of all mechanisms (for gas price determination) when free trade prevails in the gas markets (Rangarajan 2012).

Therefore, the competitive markets such as USA and UK where prices of gas are determined through forces of demand and supply with no or minimum government interference can be termed as efficient gas markets. We would use the US and UK markets as benchmark to draw learnings for the Indian market.

6 Evolution of USA and UK Markets

In order to understand the prerequisites for the Indian gas market, to evolve, from the current stage of oligopoly and monopoly to a fully developed stage, the evolution of most developed market of United States and United Kingdom needs discussion. Market in USA has seen major developments over last 70 years starting from the passage of the Natural Gas Act in 1938 till date. The evolution (year wise) of US market (Pierce 2004; Juris 1998b; Smead 2013; Kim 2014; Vany 1994; Herbert 1996; GAO 2006) is presented in Table 3.

Similarly, the UK market has evolved over last 40–50 years. The preliberalized UK market had a single-dominant player (BG) which required government and

Table 3 Evolution of gas markets in USA

Year	Key development/evolution
1938	• Passage of Natural Gas Act to prevent exercise of monopoly power over interstate pipelines • The Act established federal jurisdiction over natural gas companies operating in interstate commerce • The Act placed the authority to regulate the gas industry with Federal Power Commission (FPC) later reorganized as Federal Energy Regulatory Commission (FERC)
1950s–1960s	• Post-world war economic boom. No wellhead price regulation. Plentiful supply of natural gas. Major expansion of interstate pipeline network • Supreme Court ruling of 1954 brought in rate regulation for all gas producers (provided FERC jurisdiction over wellhead prices), having major impact on producer revenues. However, pipeline expansion accelerated by low commodity prices • The above order created major problems in creating very high demand for such gas even more than unregulated oil and coal
1970s–1980s	• Natural gas market bifurcated in interstate and intrastate markets; with low wellhead gas prices leading to severe shortages in interstate market while intrastate market having market clearing prices for all production resulting in ample supplies • Natural Gas Policy Act of 1978 lifted wellhead price ceiling for all new gas resulting in surge in exploration and production. Gas starved interstate pipeline sign long-term unregulated contracts resulting in take or pay liabilities. First US spot gas market created • FERC order 380—Relieved the utility purchasers from any contractual obligation to the pipeline for minimum bills for system supply they elected not to take • FERC order 436—provided open access regime on pipelines (voluntary basis). Enabled development of gas shippers. No trading of gas took place till this time. Producers sold gas to pipeline owners which in turn sold to local distribution companies and large industrial consumers
1990s	• Natural Gas Wellhead Decontrol Act of 1989 eliminated wellhead price regulation on all gas removing distortion from the market. Unbundling of transportation and supply on interstate pipelines removed impediment on open access • 1992—FERC order 636—provided for ownership unbundling on interstate pipelines. Owners cannot hold capacity for own use. Encouraged formation of market centers (hubs) on pipeline intersection points. Capacity release and secondary trading of capacity allowed
Beyond 2000	• FERC order 637—removes price ceiling on short-term capacity release and allows peak and off-peak rates • FERC hackberry decision—In a significant shift in its policy, FERC allowed LNG terminals to operate without any open access requirement and charge market-based rates. The regulator now believes that the new non-open access policy would encourage development of new terminal and hence provide alternate source of supply • However since then due to shale gas boom, most LNG terminals are converting into export terminals • Order requiring interstate and major intrastate pipelines to post on a daily-basis capacity, schedule, and actual flow basis (promoting transparency)

regulatory interventions as outlined in Table 4 (Stern 1997; Simmonds 2004; Calliope 2009; Melling 2010; Thuber 2011; European federation of energy trade 2004) to create a market.

Table 4 Evolution of gas market in UK

Year	Key developments/evolution
1948s	• Gas available before 1948 was synthetic gas manufactured from Coal (or town) gas, and the market was run primarily by county councils and small private firms • Gas Act 1948 nationalized the UK gas industry. Came in effect in May 1949 when 1046 privately owned companies were merged into 12 area gas boards or councils organized geographically • The above marked the beginning of publicly owned vertically integrated monopoly for downstream supply of gas • UK remained isolated from continental Europe in the absence of any interconnector
1960s	• UK became first European country to import LNG (from Algeria) driven by supply diversification strategy • 1966—Post-discovery of natural gas in UK continental shelf, UK supply was converted from town gas to natural gas • High-pressure transmission system was built as a means to supply natural gas to regional boards
1970s	• UK facing stagflation; from oil price shocks of 1978 • Gas Act 1972—Gas councils were renamed the British Gas Corporation • British Gas extended monopsony powers and continued to enjoy monopoly power on supply of gas • Gas prices from domestic production was negotiated at wellhead • Downstream prices based on the cost of the gas, transmission and distribution cost, and a profit margin
1980s	• 1982—Oil and Gas (Enterprise) Act creates the possibility of competition. Pre-emption rights of BG on gas purchase removed, but still it was not practically possible for entities to purchase, transport, and sell gas • 1986—Gas Act passed; first gas regulator (OFGAS) established. British Gas privatized, formation of British Gas Plc. Large customers given open access on pipes (>25,000 therms/annum) • 1988—Monopolies and Merger Commission (MMC) recommends that BG cannot contract 90% of any new gas and should publish tariff for industrial and commercial customers
1990s	• 1990—First contract for transportation signed (quadrant gas) • 1992—Office of Fair Trading (OFT) review reduces thresholds of customers open for competition to 2500 therms/annum • 1992—Gas release program commences (target to reduce share of British Gas to 40% by 1995) • 1993—BG's share of non-residential market falls to 70% • 1993/94—MMC report recommends demerger of BG's trading activity from transportation • 1994–95—Surplus of supply over demand; spot market evolves • 1994—British Gas was again reorganized into British Gas (trading—gas supply, service, and retail) and Transco (National Transmission System—NTS and storage)

(continued)

Table 4 (continued)

Year	Key developments/evolution
	• 1995—Gas Act 1995 passed to create a fully liberalized gas market. Paved way for competition in residential market. Separate licensing system established for gas transportation, gas shippers, and gas suppliers • 1996—Network code passed. Rules and procedures for third part access established including daily balancing • Feb 1997—British Gas demerged into Centrica plc (trading) and BG plc (E&P and transportation) • 1998—Opening of complete residential market to competition by April 1998 • UK—Belgium interconnector became operational, allowing UK gas to be exported to Europe and vice versa • The National Balancing Point (NBP) was established as the national, central trading point or hub
2000s	• The Utilities Act 2000 merged the regulatory office of the electricity and gas regulator into a single regulatory office, the Office of the Gas and Electricity Markets (Ofgem) • Also it put an end to the exclusivity in transportation licensing and allowing a single license for Great Britain

7 Learnings from the US and UK Markets

The USA and UK have had their own journey from a developing market to an efficient market stage. Based on the learnings from the market evolution from USA and UK, the following can be enumerated as the key elements for development of any gas market.

(1) A strong and independent regulator is must for a vibrant market; US and UK regulators have made mistakes but have made course corrections as their markets evolved, as evident from Tables 3 and 4.

(2) Promoting infrastructure development key for any market to develop. USA has 4.92 lakh kilometers of wide diameter, high-pressure pipelines, and 30 major market hubs including Henry Hub in Louisiana. In comparison, India can boost of just 17,000 km of pipeline and no trading hub.

(3) Prices and rates of gas at source of supply need to be deregulated for attracting investments (wellhead for producers and LNG terminals for imported gas). The success of US market is evident from the fact that after undergoing numerous policy changes, they have today more than 20 major gas producers (in addition to 6800 independent producers). India in comparison has just handful of producers.

(4) Limited control over regas terminals helps higher investment. Higher control-deterred investment.

(5) Open and transparent access to cross-country pipelines is must for any competition. Open access in India exists, but it cannot have its impact without adequate coverage.

(6) Unbundling of ownership of pipeline with shipping activity prerequisites to remove hurdles on open access.
(7) Promoting transparency (no information asymmetry in the market) is must for helping develop markets which are efficient.

8 Factors Effecting Development of Gas Market

Literature review has provided the list of key variables for development of gas markets across the world. The variables are listed below for reference.

- Existence of cross-country gas grid,
- Existence of sufficient number of import terminals,
- An independent regulator,
- Access to infrastructure—robust access code,
- Unbundled transportation and marketing segments of gas sector,
- An independent system operator for pipelines,
- Transparency in regulated businesses,
- Bulletin Board for information sharing,
- Stable long-term policy from the government,
- No wellhead price regulations,
- Entry/exit system for pipeline tariffs,
- Existence of physical hubs and financial markets,
- Secondary trading in pipeline capacity,
- Efficient downstream (power and fertilizer) markets,
- Presence of unconventional-like shale/CBM, and
- Presence of gas imports through pipelines.

The above factors are similar to the enablers listed from the learnings from the USA and UK and hence can be considered while deciding the (i) the need for development of gas market; (ii) list factors that would help development of such gas market; and (iii) issues and challenges India may face to develop its market.

9 Need for Development of Gas Market in India

Based on the study of various world markets, the current state of Indian gas market and the process of evolution of various markets, the following are the key reasons why the Indian gas market needs reforms.

(a) Contribution of gas in primary energy consumption of India is projected to increase from current 8.7 to 20% in the next 15–20 years. To achieve this, supply from domestic and imported sources has to increase fivefold in the next

2 decades. Any decrease in supply has to be either met through costly crude imports and coal imports leading to huge drain on Forex reserves. Increased usage of coal will have its own environmental affects.

(b) 12th–13th plan document envisages doubling of per capita energy consumption from 0.5 toe presently to 1.8 toe by 2030. This would require huge investments in gas chain to the extent of $45 bn in the next 10 yrs. Investments can only be attracted through right policy measures and price signals.

(c) The Indian gas market is still at the early stage of development in the evolution phase (as per path traversed by UK/USA), and there is no comprehensive work on the path to be traversed from current stage to a fully liquid and efficient gas market. Uniqueness of Indian market with respect to its domestic consumption (no heating load), available natural resources, and political realities do not allow learning of US and UK markets to be applied directly.

(d) Government has provided ad hoc remedies by artificial linkage of gas price to various indices which may not reflect the true economic value of gas in India. GoI acknowledges that this is an interim measure but does not provide mechanisms to move toward the end state.

(e) Different arms of government like PNGRB work in isolation and have brought in measures like open access to help develop markets, but the same will not be sufficient and a more joined-up approach is required.

(f) World over, power is the main consumer of gas, proving it an anchor load. Indian power sector and fertilizer sector reforms are integral to the development of gas markets.

(g) Gas prices trading at the developed hubs, lowers that GoI ad hoc price fixation indicating a need to move toward market may be more economical.

(h) Study of price formulations across the world reveals that the share of gas on gas competition in price formulation is around 38%, which proves the point that world is moving toward gas market liberalization.

Ensuring that India has well-developed and functioning gas markets that would deliver reliable, efficient, and competitively priced energy is vital to India's economic development.

10 Issues and Challenges

It is not that the factors outlined in the chapter, for development of gas market, are hidden from the decision makers in the Indian government. However, in spite of having knowledge of how to develop the Indian gas market, the real challenge is in executing such measures. We would discuss below some key challenges and steps which can be taken to resolve them.

(a) Large population living below poverty line, thereby making governments nervous in implementing reforms which impact the masses. Government should work toward expediting direct subsidy transfer to address this challenge.
(b) Attracting capital to invest in upstream and midstream sector to resolve the basic issue of availability and accessibility. This would require a stable fiscal regime, an independent and strong regulator, and incentives in the initial years to attract large capital.
(c) Lack of integrated planning among the various government institutions. The reformed NITI Aayog should help integrate, planning of various energy resource institutions, and guide them toward achieving the national goal of self-sufficiency and making available energy to all.
(d) Misalignment between federal and state policies. State plays an important role in development of any sector especially in terms of land access for any development and labor norms to sustain investments. States have also been reluctant to reform electricity sector due to political compulsions. Higher central assistance for direct subsidy transfer can help states take bold decisions. Land and labor reforms will remain important for such development.
(e) Putting the cart before the horse, Indian policy makers have taken steps to open up the gas sector with measures like open access and plan for unbundling of midstream sector. Such measure is meaningless in the absence of pipeline to transport gas. Policy measures to help attract large investment in gas transportation pipeline are a prerequisite before opening up the market.
(f) Monopoly power and lack of transparency. Indian government has allowed monopoly powers to government companies for too long with no attempt to promote transparency. Breaking of monopoly power and promoting transparency would be important in development of markets.
(g) Distortions in the energy market, the key downstream markets do not sell energy at market prices. Any reform has to span all energy markets in the absence of which the reforms in gas sector would be difficult to implement.

The answers to the above challenges are difficult. Therefore, the government needs to calibrate the opening of the gas market, first by making gas available and accessible with limited competition and then introducing full-scale competition to make it affordable. Financial market will follow the physical market.

References

Allan Consulting Group (2005) Options for development of the Australian gas market—Report to the ministerial counsel on energy standing committee of officials
Jain A et al (2011) Natural gas in India—an analysis of policy. Oxford Institute of Energy Studies
Baker Institute Study (2005) The geopolitics of natural gas—James A Baker III Institute for public policy of rice university and Program on energy and sustainable development at STANFORD university

BP (2015) BP statistical review of world energy. BP retrieved from bp.com/statistical
Calliope W (2009) The evolution of gas industry in the UK. International Gas Union (IGU)—gas market integration task force paper
Chawla A (2011) Open and competitive mechanism for allocation, pricing and utilization of natural resources
Council of European energy regulators (2011) CERC vision for European gas target model, CERC
European federation of energy trade, E (2004) Expanded version of principles of gas market development EFET
GoI (2013), Indian Petroleum and natural gas statistics, Ministry of petroleum and natural gas, Government of India
Herbert JH (1996) US natural gas markets—how efficient are they. Energy policy, page 1–5IEA (2012), Gas pricing and regulation—China's challenges and India's experience. International Energy Agency
IEA (2013) Developing a natural gas trading hub in Asia—obstacles and opportunities, International Energy Agency
GAO (2006) Natural gas, Factors affecting prices and potential impact on consumers, GAO, Government accountability office, an investigating arm of US congress
Juris A (1998a) World Bank Paper, The emergence of markets in the natural gas industry
Juris A (1998b) The world bank group, Development of competitive natural gas markets in United States
Kelkar V (2014) Roadmap for reduction in import dependency in the hydrocarbon sector by 2030. MoPNG, GoI
Kim T (2014) United States natural gas markets, contracts and risks: what lessons for the European Union and Asia pacific natural gas markets. Energy Policy
Mankiw GN (2007) Principles of micro economics, Thomson Learning
MCE (2004) Statements on principles of gas market development, Australia, MCE
Melling AJ (2010) Natural gas pricing and its future—Europe as the battleground. Carnegie Endownment
Pierce RJ (2004) Reconstituting the natural gas history from well head to burner tip, Energy Law J 57–112
Rangarajan C (2012) Report of the committee on the PSC mechanism in petroleum industry, Government of India
Samuelson AP (2010) Economics, Tata McGraw Hill
Simmonds G (2004) The UK Gas Industry 2003/4. Centre for the study of regulated Industries (CRI)
Smead R (2013) NG market notes, Navigany Energy practice publication
Stern JP (1997) The British Gas market 10 years after privatization: a model or a warning for the rest of Europe. Energy Policy 25(4):387–392
Thuber Mark C et al (Feb 2011), The policy tight rope on gas producing countries, stimulating domestic demand without discouraging supply. Pacific Energy Summit. Jakarta Indonesia
Vany, AD (1994). Natural gas industry transformation, competitive institutions and the role of regulation. Energy Policy, pg 755–763

Demand Assessment and Design Aspect of City Gas Distribution Network

Husain Ahmad

Abstract Natural gas is a promising policy element in achieving equitable, balanced, and sustainable economic growth by widening its user base beyond conventional industries. The City Gas Distribution is what we call the "last mile" in entire gas value chain. Gas supplies usually are at low/medium pressure to the residential/domestic, commercial, industrial consumers, and this is called as pipe natural gas and is compressed to very high pressure to increase the volumetric efficiency in order to make it compatible for the usage automobile sectors as compressed natural gas.

Keywords Demand assessment · Pipe natural gas · Compressed natural gas · Distribution network · CNG station · CNG kit · CGD technical initiatives

1 Introduction

Natural gas is cleanest burning fossil fuel among the hydrocarbons group, enhancing its desirability as a fuel option in a carbon-constrained environment. The Indian Energy Basket has a share of natural gas touching approximatey 9% as against 24% of the world's average.

The most significant use of natural gas is in City Gas Distribution (CGD). This sector comprises of following segments:

- Domestic—for use in households;
- Commercial—for use in eateries, inns, hospitals, offices, etc.;
- Industrial—for use in the small-scale industries including power generation; and
- Transport—for use in three-wheelers, four-wheelers, buses, etc.

H. Ahmad (✉)
EPIC (Engineering, Procurement, Installation and Commissioning) Company, Doha, Qatar
e-mail: husainsinbox@gmail.com

S.K. Kar and A. Gupta (eds.), *Natural Gas Markets in India*,
DOI 10.1007/978-981-10-3118-2_10

The utility of natural gas in Indian households have gained momentum expressly when multifarious use of natural gas in the houses was recognized. The gas can be used for cooking in the kitchen, generating power, heating water, air-conditioning and refrigeration, and fueling the automobiles. The model of a one-switch solution in dwellings is gaining impetus. This notion provides that a household could be independent gas consumers to meet its all energy necessities at one go which signifies that once the gas supply is switched on, all the aforementioned services can be utilized simultaneously which are gas operated, certainly selectively or requirement based.

In the industrial and commercial segment, the city gas is very useful especially in the applications such as cookery in eateries and hotels, air-conditioning, and power generation.

The efficacy of natural gas in the vehicular segment has a say to an extent in curbing the pollution intensity due to exhaust emissions.

Natural gas supply to industries in and around a city significantly helps in limiting pollutants in the city apart from providing ease of operation and fuel handling to the industries. Specifically, CGD projects propose ways to improve its accessibility by means of huge allocation of gas through piped natural gas (PNG) to domestic segment and compressed natural gas (CNG) to automobile/vehicular segment. PNG provides a possibility to broaden the benefits of a new fiscal policy to masses. CNG is evenly vital to strengthen this objective as it is coupled with improving our health and quality of life by providing a clean and pristine atmosphere.

On heat equivalent basis, natural gas costs significantly lesser than liquefied petroleum gas (LPG), petrol, and diesel. Natural gas also decreases maintenance costs of the vehicle. A surplus benefit is that contrasting to liquid fuels is that gas cannot be contaminated or drained off from an automobile. Though, certain economic support or incentives may be requisite to stimulate consumer for switchovers from conventional ways.

Emissions from CNG vehicles are lot lesser than from petrol or diesel vehicles. Natural gas, primarily methane contains less carbon atoms than any other fossil fuel and as a result produces smaller quantity of carbon dioxide emissions during oxidation. CNG vehicles do release fugitive methane which is a greenhouse gas, but the small rise in methane emissions is counterbalanced by a considerable lessening in Carbon Dioxide emissions. Owing to low specific gravity, CNG, contrasting to liquid fuels, dispels into the environment in case of occurrence of any accident. CNG fuel systems are completely closed systems, putting off any spills or evaporative losses of the system gas. Also, it is nonhazardous or noncorrosive in nature and does not pollute ground water.

Worldwide, CGD is an area which has sprung up together with the gas industry growth. In India, this sector would get the necessary thrust in the coming years. It has to be remembered that the development of CGD network has developed in proximity to major transmission networks.

1.1 Benefits of PNG and CNG

1.1.1 PNG has quite a lot of distinctions to its credit of being an eco-friendly fuel, as PNG is directly disseminated at customer's location through pipelines; hence, there are no problems related to fuel handling, refilling, and changing of gas cylinders with a distributor, also no storage is required which leads to secure, easy and safe handling.

- **Consistent and reliable supply**
  PNG provides the ease of ensuring continuous and adequate delivery of PNG at all times without calling for any space to store.
- **Convenience**
  The household customers have to themselves do booking of conventional cooking gas cylinder in order to get it refilled; additionally the consumers have to wait for the delivery of cylinder. Moving on to PNG causes this complete process to be superfluous. PNG also eradicates the droning routine of inspecting conventional cooking gas cylinder for any doubted leakage or it being underweight, at the juncture of delivery.
- **Safety**
  If the combustible mixture of natural gas, which is primarily of methane and air, is lesser than 5%, and is more than 15%, it does not catch fire. This narrow inflammability range makes natural gas as one of the most dependable and safest fuels across the world.
  In case of leakage, LPG owing to higher specific gravity settles at the bottom while natural gas being lighter than air, just goes up, and disperses into the atmosphere.
- **Economy with PNG**
  In various segments of CGD network, i.e., domestic—households, industrial, and commercial, the customer pays upfront for the price LPG cylinders or any other fuels; however in case of PNG, billing in general is done on fortnight/month/quarter basis after the utilization of gas by the consumer. As an effect, there are savings on account of outgo of funds for the Industrial and commercial segment and deferment of everyday expenditures in case of domestic sector.
- **Billing**
  The customer has to pay only for the amount of PNG consumed. No pilferage is likely to happen with PNG as billing is prepared in line with the reading of the meter. A distinctive element is that the user pays only after consuming the gas. Furthermore, there are no bare minimum utilization charges, i.e., there shall not be any bill, if there has not been any consumption. The customer pays the bills based on the exact amount of consumption by him as shown by the meter setup at its location.

- **Environment friendly**
  Natural gas helps in ameliorating the quality of air, particularly when used instead of polluting energy sources. Its combustion results in far lower release of carbon monoxide, reactive hydrocarbons, carbon dioxide, and almost no distinctive releases of sulfur dioxide as compared to combustion of other fuels.

1.1.2 CNG is a smart and reasonably priced alternate for fleet vehicles, trucks used for waste disposal, delivery vehicles, etc. CNG helps a consumer to save money on fuel, limits emission intensity, and extends the life of a vehicle by lowering the maintenance cost.

- **Reduced Fuel Cost**
  CNG offers almost a 50% cost savings over other conventional fuels.
- **Environmentally Friendly**
  CNG burns cleaner than other fossil fuels as it contains appreciably less pollutants than conventional liquid fuels. CNG generates approximately 20–30% fewer of greenhouse gas emissions and 95% of tailpipe emissions lesser than petroleum-based products. Further vehicles running on CNG produces no evaporative emissions as CNG fuel systems are completely sealed.
- **Reduced Maintenance Cost**
  CNG does not contain lead; hence, there is no fouling which extends the spark plug life. CNG do not dilute or contaminate the crankcase oil, so the time period between oil change and overhaul is also increased. Also CNG does not react with the metals due to which life of pipes and mufflers also increases. All the said factors reduce the maintenance expenses while increasing by and large life of the engine.
- **Performance Advantages**
  CNG is better than liquid-based fuels as its octane rating is of approximately 120. CNG vehicle undergoes reduced knocking and zero vapor-locking circumstances. CNG vehicles have superior start even under harsh weather conditions as natural gas is already in a gaseous state.
- **Safety Advantage**
  CNG storage tanks are safer than liquid tanks decreasing the probability of accidental release. If released, CNG scatters rapidly into the air rather than settling on the ground, hence reducing the likelihood of fire or ground contamination. Besides, CNG emits modest to no carbon emissions during refueling. In addition, CNG is less likely to autoignite owing to its high autoignition temperature, i.e., approximately 540 °C and a narrow range of 5–15% of inflammability. The aforesaid ignition temperature and restricted flammability sphere makes accidental fire very unlikely.

1.2 Current Scenario

At present, India has approximately 14,789.82 km of existing natural gas pipeline infrastructure. Of these, the major ones are the Hazira–Vijaipur–Jagdishpur–Dahej–Vijaipur HVJ/VDPL of 4658 km, Dahej–Vijaipur (DVPL)–Vijaipur–Dadri (GREP) Up gradation DVPL2&VDPL of 1118.73 km, GSPL's High Pressure Gujarat Gas Grid network of 2120.62 km, Dabhol–Bangalore pipeline of 1004.14 km, and the Dadri–Bawana–Nangal pipeline of 810.43 km. As of date, there are 26 companies operating CGD networks in 64 Geographical Areas (GAs) across 26 States of India. Out of which 59 GAs are authorized by the Petroleum and Natural Gas Regulatory Board under various provisions of the PNGRB Act, 2006, and the Regulations thereof. As of November 2015, the average natural gas consumption by CGD sector has been approximately 17 MMSCMD.

There were a total of 3,033,168 PNG consumers as of December 2015. Though there has been a more than threefold growth in the past six years, the industry still remains small. Among states, Gujarat has the largest base, followed by Maharashtra and Delhi (Fig. 1).

As of September 2015, there were about 21,925 commercial customers and 6044 industrial customers. While there were approximately 1026 CNG stations which served approximately 2.43 Million vehicles in the country.

With the Government of India's plans to increase PNG connections to 10 million households in the next four years in order to reduce subsidy and import bill, natural gas demand for CGD sector is likely to go up progressively (Business Standard 2015), due to the addition of gas networks in various new cities, price benefit of CNG, and increased use of PNG in domestic, industrial, and commercial segments. Environmental concerns will further set in motion the use of natural gas, particularly in the transport and industrial sectors. Currently, sixth round of bidding is going on wherein 34 GAs are being considered for authorizations of CGD networks.

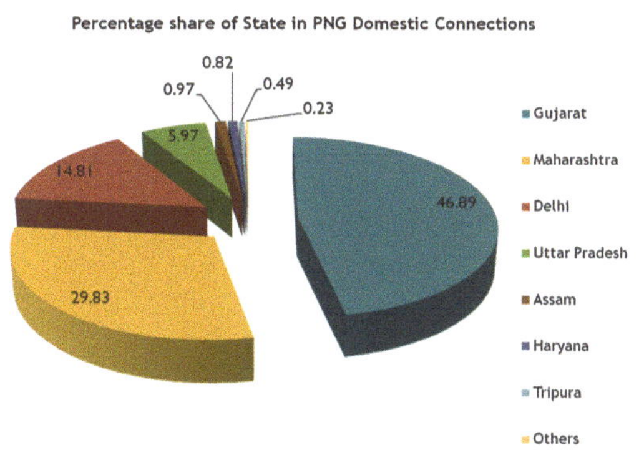

Others States

States	Percentage share in the country
Madhya Pradesh	0.10%
Andhra Pradesh	0.09%
Telangana	0.03%
Rajasthan	0.01%
Others	0.00%

Fig. 1 Percentage share of state in PNG domestic connections. *Source* Adapted from Petroleum Planning and Analysis Cell, Ready Reckoner, November 2015

2 Demand Assessment and Design Aspects of City Gas Distribution Network

The most vital component of a CGD network is the distribution part. Gas transportation and distribution are coupled with a fixed infrastructure using public land. This service cannot easily be changed for small number of customers once they are linked to it.

Across various countries' gas transportation is a heavily regulated business as the pipelines exhibit extensive technical economies of scale, i.e., given adequate gas volumes, larger the pipeline is built, lower the tariff requisite to produce certain net return on the outlay as the pipeline grows in capacity. Its costs increase less than linearly while output increases exponentially. Gas transportation involves substantial capital investments, implying considerable barriers to entry.

2.1 Demand Assessment for a Geographical Area (GA)

The demand may be anticipated for any GA for the below-mentioned four customer segments namely,

- Domestic,
- Commercial,
- Industrial, and
- Transport (CNG).

The overall demand possible for any natural gas consumer segment has to be understood in terms of the following three levels:

- **Demand Potential**: By and large demand potential that a customer segment holds;
- **Realizable Demand**: The demand that can be really achieved after discounting for the constraints such as geographical terrain, nonwillingness of end consumers to switch; and
- **Demand Penetration**: The phasing of realizable demand on a year-on-year basis which is catered by CGD.

2.1.1 Approach for Demand Estimation

The approach for demand estimation used for the different user segments has been detailed below:

(a) **Domestic segment—Demand estimation methodology**
The demand arising out of the domestic user segment has to be estimated following the methodology described below:

Obtain the number of households for the given GA as per the last census. The household's population has to be considered for demand estimation as against the overall population.

- Increasing the total population to current level based on composite population growth rate.
- Obtaining the number of households consuming the LPG as a percentage of households.
- For the purpose of demand estimation, the extent of LPG penetration may be considered to arrive at a realistic level of demand that can be captured through the method of replacement of LPG by natural gas. The remaining households are anticipated to be consuming other fuels such as kerosene and wood. These households have a low likelihood of switching to natural gas and therefore may not be considered for the purpose of demand projection.
- Computing the number of LPG-consuming households that can be realistically projected to be converted to natural gas. A conversion factor has to be considered for domestic demand estimation arising out of the households due to the possible substitution of LPG by natural gas. Hundred percentage conversion of all the households consuming LPG to usage of natural gas is not always practically feasible given the various factors such as variations in the geographical terrain and the level of eagerness of the users. Therefore, conversion factors of 65–80% may be considered to capture the extent of conversion of households from LPG to natural gas usage.
- Computing household demand based on the average cylinder consumption per household. The consumption pattern of cooking gas in households suggests that on an expected value of 0.4–0.6 standard cubic meter per day (SCMD) of gas consumption is likely per household.
- Estimate the realizable natural gas demand emerging out of the households and project the demand. The natural gas demand probable for a given GA may be projected by applying the expected value of gas consumption per household (0.5 SCMD) to the number of eligible households considered for demand estimation.

(b) **Commercial Segment—Demand Estimation Methodology**

The demand arising out of the commercial establishments should be estimated following the methodology as detailed below:

- Obtaining alternate fuel consumption data based on the primary study. Alternate fuel consumption data have to be collected for the commercial units covering the entire GA. The fuel consumption details include the consumption levels of LPG/petrol/diesel and other competitive fuels used in various types of commercial units as identified in Table 1.
- Converting alternate fuel consumption into equivalent natural gas consumption. Natural gas demand potential has to be established based on the

Table 1 Commercial unit type

S. no.	Commercial units type
1	Hotel/restaurant
2	Lodging/boarding
3	Office
4	Hospital and nursing home
5	Home
6	School and college hostels
7	Bakery/cinema hall/multiplex, etc.

alternate fuel consumption for various commercial units. All fuels may be considered 100% convertible to natural gas use, except for diesel and solid fuels. The utilization of diesel may not be converted to corresponding to natural gas equivalent demand as diesel is used typically used for power generator sets.

- Extrapolating the demand in order to account for the demand arising out of the commercial units not surveyed. The expected value of natural gas consumption for each commercial unit category has to be considered to arrive at the total demand potential or the commercial user segment. As targeting this entire demand is not expected to be practically achievable due to various factors such as lack of willingness to convert on the part of some users or difficult geographical terrain, the realizable demand has been measured at 90% of the total demand likely estimated for given GA.
- Arriving at projections of gas consumption for the upcoming years. The demand arising from the commercial units present in the given GA may be projected based on a higher growth rate than historical population growth rate assumed for the households as the level of urbanization is anticipated to happen at a higher rate compared to the population growth rate.

(c) **Industrial Segment—Demand Estimation Methodology**

The demand arising out of the industrial units should be estimated on the following methodology:

- Establishing the demand based on alternate fuel consumption in the given GA. Most of the Industrial units today use fuels like furnace oil, diesel, naphtha, petcoke, and other fuels. Industrial customer has an incentive in replacing aforesaid fuels completely or partially by natural gas due to economic/environmental causes and convenience. This makes all the industrial units consuming alternate fuels (liquid), a potential consumer of natural gas supplied through the CGD network. The various types of commercial units that may be identified for the assessment are briefed in Table 2.
- Converting the alternate fuel consumption into equivalent gas consumption. The alternate fuel consumption may be converted into equivalent gas consumption based on the possible conversion of industries from alternate

Table 2 Industrial unit types

S. no.	Industrial units type
1	Agro processing/food processing/tobacco processing
2	Plastics products
3	Automobile part/marine products
4	Electrical/electronics/ceramics products
5	Glass/paper/rubber-based products
6	Cement/sugar factory
7	Cold storages
8	Chemicals/pharmaceuticals/cotton yarn spinning/other metals
9	Steel sponge iron units engineering/fabrication, etc.

fuel to natural gas. A conversion of 100% may be considered from the usage of all the alternate fuels to natural gas except for diesel.

- Extrapolating the demand in order to account for the demand arising out of the industrial units not surveyed. The natural gas demand arrived for the industrial units using the alternate fuel consumption data may be further extrapolated to consider the possible natural gas demand for the industrial units present in the given GA but not surveyed. As targeting this entire demand is not likely to be practically possible due to factors such as lack of willingness to convert on the part of some users or difficult geographical alter rain, the realizable demand may be considered at 90% of the entire demand potential estimated for given GA.
- Arriving at projections of gas consumption for the upcoming years. The demand projection for the next years has been arrived at based on the target growth rate in line with the various Government policies.

(d) **Transport (CNG) Segment—Demand Estimation Methodology**
The demand arising out of the CNG user segment may be estimated following the methodology as detailed below:

- Obtaining vehicle (by type) data for all the areas individually comprising the GA. The district-level data (available from RTO) can be acquired for the number of motor vehicles within different categories such as three-wheelers, taxis, buses, and private vehicles.
- Obtaining vehicle (by type) data for entire GA and projecting the data to current levels based on historical growth rates.
- Estimating the percentages for maximum conversions possible under different categories of vehicles. Auto rickshaw (three-wheeler)/taxis and buses fall under the public vehicles category and usually convert to natural gas as typically Government mandates all public transit to switch to CNG once it is available. The conversion for the buses and three-wheeler/taxis segment may be measured at 100% (differs from states to states). Conversion levels in private vehicle (cars) category may be considered as about 50% as no mandate is applicable to them for conversion.

Table 3 Per day gas consumption of various category of vehicles

S. no.	Type of vehicles (kg/day)	kg/day	SCMD
1	Three-wheelers	3	3.9
2	Buses	65	84.5
3	Taxi	6	7.8
4	Cars	4	5.2

- Estimate the per day gas consumption of each category of vehicles. Based on the industry standards, assumptions with respect to average gas consumption, mileage, average run per day for each vehicle segment have been taken. The following Table 3 provides the assumptions for projecting the demand from the CNG segment.
- Projecting the future demand based on growth rate of future vehicle population. Future demand has been projected for the next years assuming a growth rate that is in line with the historical growth rate in the number of registered vehicles in the given GA.

2.2 *Design Aspects of CGD*

City or local natural gas distribution network, i.e., CGD network means an interlinked network of gas pipelines and associated equipment utilized for shipping natural gas from a bulk supply high-pressure transmission main pipeline to the medium pressure distribution grid and then to the service pipelines to facilitate supply of natural gas to domestic/households, industrial or commercial locations, and CNG filling stations located in a specific GA.

2.2.1 General Provisions

The selection of design for CGD network should be based upon the gas attributes, requisite flow rates, operating pressures, and the environs. The following parameters should be considered while carrying out the design of a pipeline network:

- Estimated load,
- Peak hour flow,
- Maximum operating pressure,
- Minimum network pressure,
- Maximum permissible velocity,
- Specific gravity,
- Compressibility factor,
- Temperature, and
- Spare capacity and future requirements.

2.2.2 Network Design Requirements

The design of CGD networks and its elements may be such that it safeguards:

- Supply of gas at fixed volume into the system, which fluctuates in pressure amid predetermined upper and lower limits in the network.
- Delivery of gas at a steady pressure at customer's end.
- The designer shall take care of the need for protection against failure of any equipment and offer adequate redundancy to make sure that the supply is secured against such malfunctions.

Facilities and equipment forming part of CGD networks may be designed allowing range of pressures and flow rates requisite in various parts of the system

- Quality of gas as well as cleanliness concerning both solid and liquid particles,
- Metering necessities,
- Noise control, and
- Protection against corrosion.

Essential calculations should be carried out to authenticate structural integrity and stability of the pipeline for the combined effect of temperature, pressure, bending and pipeline route profile, exterior loads, and other ecological parameters as applicable, throughout all stages of work from safe installation to safe operations.

The aforesaid calculations may also consider but not restricted to the followings:

- Stability and buoyancy control of pipeline in the areas prone to floods/submergence,
- River crossing to be set up by trench less technique, wherever soil data are favorable for such systems, and
- Damage potential for steel pipeline from risks related to earthquake, if applicable.

2.2.3 A typical CGD Network

A typical CGD network should comprise of one or more or all of the followings:

(a) **Primary Network**: The part of City Gas Distribution CGD network that works at pressure more than 100 psig (7 bar) and lower than 711 psig (49 bar). A pipeline which forms part of this network called Gas Main or Distribution Main or Ring Main is designed to make sure constant supply of gas from one or more City Gate Stations (CGS) to supply gas to the secondary gas distribution network or service lines to bulk consumers (Oil industry Safety Directorate 2013). The uppermost velocity in the pipeline network should be limited to 100 ft/s (30 m/s) at once after pressure regulating instrument.[1] As far as, primary network should be fed through more than one City Gate Stations/sources of supply.

(b) **Secondary Network**: The part of City Gas Distribution network that works at a pressure lower than 100 psig (7 bar) and more than 1.5 psig (100 mbar) (Oil industry Safety Directorate 2013). Pipelines which form part of this network are called as low-pressure distribution mains. These low-pressure distribution mains are designed to ensure uninterrupted and unfailing supply to tertiary network or to industrial consumers through service lines and are generally constructed by thermoplastic piping (MDPE) and connects District Pressure Regulating Station or District Regulating Station to different service regulators at domestic/households, industrial, and commercial customers. The system is usually designed for utmost flow velocity of 100 ft/s (30 m/s).

(c) **Tertiary Network**: It is the part of network that works at pressure lower than 1.5 psig (100 mbar). Pipelines which form part of this network to service pressure distribution mains are planned to make sure uninterrupted gas supply to service lines. A service pressure distribution system consists of Service Lines, Service Regulators and meter set assemblies at consumer's location which are constructed using an arrangement of Medium Density Polyethylene and Galvanized Iron/copper tubing components.

2.2.4 City Gate Station

A CGD entity takes its contracted gas from its supplier at the entry point of its own system which is called as CGS, from where it brings it to end users at the necessary volumes, in the requisite form at the required time.

The point where charge transfer of natural gas from natural gas trunk pipeline to the CGD network takes place is called as CGS. This point is often referred as Pressure Regulating Station or City Gate Measuring.

It is the entry point of natural gas from a sub-transmission high-pressure gas pipelines to a city, hence the name.

A CGS generally has following major facilities:

- Isolation valve,
- Knockout drum,
- Cartridge filters,
- Pressure regulators,
- Slam shut valve,
- Mass flow meter,
- Flow control valve,
- Cold venting facility,
- Pressure safety valve,

[1]IGE/TD/3—Steel and Pipelines for Gas Distribution.

- 3 nos. of tap-offs,
- Odorization unit, and
- Control room with UPS, battery bank, Electrical Distribution Board, and SCADA provision (Fig. 2).

Isolation valves are provided to isolate the system when considered necessary. At CGS, the gas first enters the knockout drum where the liquid particle (moisture) is removed. Two cartridge filters (1 + 1) are provided to remove all the dust and foreign particles greater than 5 micron in size.

One pressure regulator in each stream (1 + 1) with one slam shut valve upstream of regulator is provided to regulate the gas pressure so that the gas supply to main grid line remains constant. The regulators are fail open type and usually gas actuated. High and low indication is also provided to indicate malfunction of regulators. In the event of high pressure at downstream of pressure regulator, the upstream slam shut valve with an alarm for closure. The set pressure of slam shut valve is kept slightly higher than set pressure of pressure control valve. The set pressure of standby regulator is kept slightly lower than the set pressure of the operating regulator. The standby regulator is closed under normal operation and becomes operative automatically in the event of failure of the operating regulator.

Mass flow meter of preferable turbine type is provided to measure mass flow rate. Flow, temperature, pressure, and density are telemetered to SCADA.

As far as possible, CGS should be installed at the outside boundary of the inhabited area. The CGD operator should try to have more than one CGS in order to tackle supply security. Platforms and crossovers should be provided for ease of operation and maintenance of equipment and piping where necessary. Provision should be made for venting, purging, and draining all parts of pipeline works and equipment that may have to be secluded during the construction or maintenance period.

All vents are supposed to be routed to a protected area and in a mode that gas aired out is blown away from the adjacent buildings. Height of vent should be at least 3 m above the working level. Distance between vent and boundary wall/fence should be less than 5 m. Gas detectors should be installed at strategic locations which is able to detect any gas leakage.

Fig. 2 A view of CGS

2.2.5 Odorizing Facility

Natural gas supplied through CGD Network should have a distinct odor, strong enough to distinguish its existence in case of leakage.

As predetermined quantity, equivalent to maximum 12.5 PPM ethyl mercaptan, of any other odorant may be dosed into the gas stream.

The odorant typically consists of sulfur-based chemical compounds, and a combination of two odorants (30% of tetra-hydro-thiophane and 70% of tetra-butyl mercaptan) provides a better odor impact than a single odorant, in particular if there is significant background odor such as in a mildly polluted environment.

One odorizing plant on upstream side of the distribution system is suggested as odorant compounds are dangerous organic chemicals and highly flammable in nature, a reduced number of odorant plants should provide a safer environment. The odorant sulfur compounds do not harm a natural gas transmission or steel distribution system because high elevated temperatures are essential to convert these compounds into acid.

The odorizing equipment should be located separately at CGS. Also, there should be a minimum safety distance of 1.5 m in the area of the odorizing and rest of the facilities at the station in order to make easy maintenance and personnel movement. The odorant unloading should be done in a safe way. Precautions for handling odorant should be prominently displayed.

Provision should be made for eye wash or emergency shower near odorant handling and injection systems. Use of personal protective equipment such as face shield, mask, rubber hand gloves, gumboot, and safety goggles. for handling of odorant spillage must be ensured.[2]

2.2.6 Steel Pipe Grid

The steel pipe grid is designed to cover the entire GA and to maintain required pressure at all points in the grid so that sizes can be optimized. Steel grid of different sizes (12″, 10″, 8″, 6″, etc.) may be taken into account to cater the demand of CNG stations and to supply gas to District Pressure Regulating Station or District Regulating Station for domestic, commercial, and industrial customers. Also steel pipe thickness calculation should be done in accordance with ASME B31.8 Clause No. 841.1[3]

$$T = (P * D)$$
$$(\overline{2SFET})$$

[2]IGE/SR/16—Storage of Odorant and Use of Odorant Systems.

[3]ASME B 31.8—Gas Transmission and Distribution Piping Systems (ASME 3 Park Avenue, New York, NY USA 10016.

where

T Nominal wall thickness (inch)
P Design pressure (psi)
D Nominal outside diameter of the pipe (inch)
S Specified minimum yield strength (psi)
F Design factor
E Longitudinal joint factor
T Temperature derating factor

Consideration should also be given to the loading because of followings factors while choosing nominal wall thickness as per ASME B 31.8 as appropriate:

- Overburden loads,
- Dynamic and seismic loads,
- Cyclic and vibratory loads Internal pressure fluctuations,
- Geotechnical, and
- Support and thermal effect of the pipeline on soil properties.

However, minimum thickness of pipe as allowed by aforesaid standard should not be below 6.4 mm, regardless the grade of the pipe material (see Footnote 3).

2.2.7 District Pressure Regulating Station or District Regulating Station

A station situated within authorized region for CGD network having isolation, pressure regulating, and overpressure protection systems is called as Distribution Pressure Regulating Station (DPRS) or District Regulating Station (DRS).

- DRSs are installed at various strategic locations in the network to supply PNG to domestic, commercial, and industrial consumers,
- Twin-stream regulating system with hot standby stream,
- Regulating from inlet pressure of 18.5 bar (g) pressure to outlet pressure of 1.5–4 bar (g), and
- Downstream distribution through PE pipeline.

Situated at various demand centers for domestic/commercial customers usually comprises of:

- Gas filter;
- Heater (if compulsory);
- Pressure reduction skid comprising active and[4] monitor regulators arrangement with at least 50% redundancy with stream discrimination arrangement,

[4]A combination of two regulating devices in series whose pressure settings are stepped so as to permit one, i.e., active, ordinarily to control the outlet pressure and the other, i.e., monitor, to take control of the active devices in the event of failure are called as monitor and active regulators.

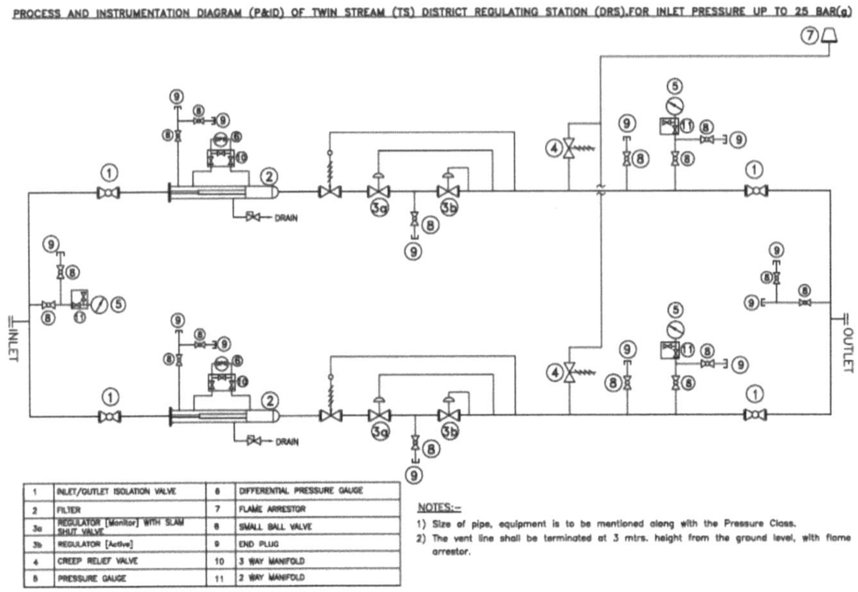

Fig. 3 A sample process and instrumentation diagram of district pressure regulating station or district regulating station

including[5] slam shut valve for over and under pressure fortification with additional feature of under pressure protection if requisite; and

- inlet and outlet isolation valves.

A process and instrumentation diagram (P&ID) (Fig. 3) and a general arrangement drawing (GAD) (Fig. 4) have been given for better understanding.

DPRS facilities can be positioned on top of ground or below ground. If DPRS is found underground, the complete facility has to be either inside a enclosed chamber with a provisions for admission of authorized personnel or be a buried in a air proof and sealed module type with proper cathodic protection. For underground setups, the vent line should be concluded at a minimum height of 3 m on top of the ground level.[6]

Individual Pressure Regulating Station (IPRS) Situated at the location of an individual consumer and having setups alike DPRS though, monitor regulator may or may not be present. Metering facilities may or may not be components of this station.

[5]A valve which is designed to close quickly in the event of an abnormal pressure whether excess or low, being detected at a selected point in a gas system is called as slam-shut valve.

[6]IGE/TD/13—Pressure Regulating Installations for transmission and Distribution Systems.

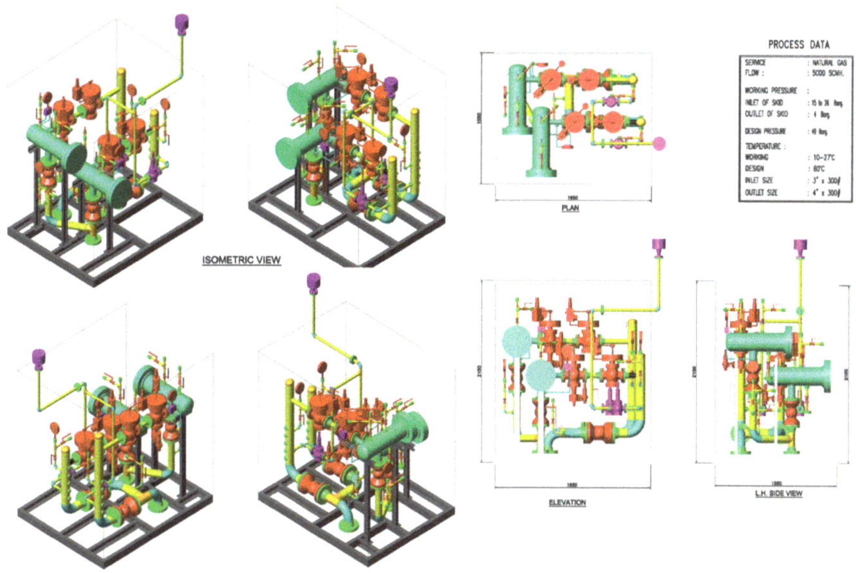

Fig. 4 A sample general arrangement drawing of district pressure regulating station or district regulating station

IPRS are usually positioned above ground. DPRS/IPRS installed on top of ground. It is provided with appropriate security fencing as per requisites of statutory bodies.

2.2.8 Medium Density Polyethylene Network

For supplying gas to domestic, commercial, and industrial segments, the pressure of piped gas is reduced to distribution pressure in DPRS, which feeds gas to Medium Density Polyethylene Network (MDPE) distribution network. The domestic consumers are fed from the MDPE distribution network through service lines till the domestic location where regulating and metering facilities are provided. Industrial and commercial consumers are also supplied gas from distribution network through service lines up to the industrial and commercial consumer's premises from where regulating and metering facilities are arranged by individual industrial and commercial consumer.

The routing of pipeline of different diameter, i.e., 180, 125, 90, 63, 32, and 20 mm of MDPE may be considered.

Recapitulation of MDPE pipes:

- Technical Specification: ISO 4437.
- Material grade and color: Internationally approved resins of PE 80 grade of yellow color.
- Minimum required strength of PE 80-grade pipe: 8 MPa.
- Pressure class: SDR 9 (diameter 20 mm) and SDR 11 (diameter 32, 63, 90, 125, and 180 mm).
- Operating pressure: 4 bar (g).
- Operating temperature range: −10 to +40 °C.
- Density: 930 kg/m^3.
- Tensile strength at yield elongation: 15 MPa.
- Elongation at break: More than 350% up to pipe size of 125 mm diameter, 500% for pipe sizes more than 125 mm diameter.

Following are few advantages of MDPE pipes:

- Sustainable design life of PE pipes (50 years) as compared to steel pipeline (30 years).
- Avoidance of nondestructive techniques (NDT) in building premises, which is very critical.
- Size of trench is less in case of laying PE pipe as compared to steel.
- Easier and speedier joining techniques to ensure leak tight joints by employing electrofusion techniques.
- More productivity, i.e., reduction in installation time (15 min in case of PE against 4 h in case of steel), in that way lesser hassle for the public.
- Reduced number of joints, hence safer and leak free system.
- Less time is devoted in repairing PE damages as compared to steel damages.
- High performance (globally proven leak free system).
- More elasticity, coil ability, ductility, etc.
- Low density (low weight, high strength to weight ratio).
- High resistance to corrosion.
- Low heat conductivity (small thermal loss).
- Smooth surfaces (low pressure losses due to low pipe friction).
- Simple to transport, handle and lay, and longer life.
- Good squeeze off properties.

2.2.9 Sectionalizing Valve Station

Sectionalizing valve (SV) is installed to aid maintenance of installation and equipment and permit instant segregation in the event of damage. The position of the valves is decided based on the risk associated with maintaining the gas supply and to ease repair and maintenance, alteration, testing and commissioning, etc. In no case, distance between two valves should exceed:

(i) 3 km for steel pipelines and
(ii) 1 km for MDPE pipelines of diameter more than 63 mm.

For MDPE pipelines below diameter of 63 mm, the space between valves should be based upon risk assessment. Such valves are also provided on each riser, upstream of each meter. The SV station should consist of the followings:

- Sectionalizing valve,
- Tap-off point on upstream for future provision, and
- Cold venting facility.

SV should be reduced bore ball valve with extended stem. 2″ tap-off for 4″ line should be made available for future use. Provision should be made for safe blow down of gas into the atmosphere. A vent pipe of minimum 3 m height above ground should be provided and located at a suitable distance from the valve assembly. Also a plot size of 5 m × 5 m (minimum) is requisite for installing the facilities.

2.3 Design Aspects of CNG Station

CNG means mixtures of hydrocarbon gases and vapors, comprising primarily of methane in gaseous form, which has been compressed to a pressure of 200–250 kg/cm^2 (g) for use as a vehicular fuel. CNG has been identified as one of the alternate fuels to liquid fuels, which are adding environmental benefits. CNG is being used internationally with the established success as an automotive fuel.

Besides, installation of conversion kit in an automotive system, two major installations must be erected, i.e.,

- Compressor station either for online operation or to act as a mother station and
- Refueling station either with online facility or with moveable cascade.

These facilities may be either colocated in the petrol/diesel retail outlets or at dedicated stations.

Safety at these installations is most vital factor considering the operating settings such as pressure and storage of other flammable materials.

2.3.1 Different Types of CNG Stations

(a) **CNG "Mother" Station**

- CNG setup (equipment and facilities) directly interlinked to a natural gas pipeline and having a compressor proposed primarily to fill mobile cascades for "daughter" stations. Such facilities, further act as "mother" station also fills stationary cascades for CNG dispensing into vehicles (Oil industry Safety Directorate 1998).

- Additionally, to its own sales, it should be able to feed 2–3 daughter stations.
- Sizing of compressor is carried out taking peak hourly demand and taking into consideration a factor of additional 10% to take care of the lower output due to various technical reasons.
- It is ideal to go for more than one compressor at each mother station for better reliability.
- Depending upon electric power availability, decision may be taken to either go for gas engine-driven compressor or electric motor driven or to go for a mix of both, if numbers of compressors are more than one.
- If the mother stations have the facility to feed buses, separate storage system, and tubing up to dispensers need to be considered.

(b) **CNG "Online" Station**

- CNG setup (equipment and facilities) directly interlinked to a natural gas pipeline and having a compressor mainly to fill stationary cascades for dispensing CNG to vehicles. In case the "online" station has enough space to accommodate mobile cascades filling, it can be used to act as mother compressor station (Oil industry Safety Directorate 1998).
- Online station may have the facility to feed small vehicles alone or small vehicles and buses.
- Sizing of compressor are carried out taking peak hourly demand and considering a factor of additional 10% to take care of the lower output due to various technical reasons.
- For better reliability, it is better to have more than one compressor.
- Storage cascade capacity and design considerations are same as those for mother stations.

(c) **CNG "Daughter Booster" station**

- The daughter Booster station have only Booster compressor and are placed away from the grid line. The gas in the Daughter Booster station is delivered through mobile cascades filled from mother station.
- Booster stations are for dispensing CNG to small vehicles (taxis/cars, autos, etc.)
- Booster stations are installed at those locations where it is not feasible to lay pipelines.
- Booster compressors of variable suction pressure are installed at the daughter Booster Stations.
- One booster compressor of average capacity of 250 Sm^3/h is adequate for two dispensers (two arms).

- Besides dispensing CNG from mobile cascade, Daughter Booster Station can have its own fixed storage cascade for dispensing CNG when there is no mobile cascade at the station.

2.3.2 Major Equipment Used in CNG Stations

The systems and components of CNG facility(s) are required to be certified for CNG use and marked accordingly by applicable statutory authority or its agent. The various components of the CNG facilities which require such certification/approval are stipulated as below:

- Compressors,
- Dispensers,
- Cylinders,
- Cylinders/cascades,
- Pressure relief devices including pressure relief valves,
- Pressure gauges, and
- Valves, hose, etc.

The piping and fittings thereon, up to the battery limit of CNG installation must be in line with ANSI B 31.8 or equivalent.

(a) **Compressor**: Compressor should be designed for use in CNG service, for the pressures and temperature to which it may be subjected under normal operating conditions conforming to API 618/API 813 or equivalent standard. Flameproof electric motor and associated fittings should also conform to IS: 2148 suitable for class I division I group II area. Compressor should be fitted with the following minimum devices:

  - Pressure relief valves so as to prevent pressure surge against the predecided set point.
  - Switch for high discharge temperature shutdown.
  - Switch for high-inlet, interstage and discharge pressures shutdown.
  - Switch for low lube oil pressure shutdown.
  - Switch for low cooling water flow.
  - A remote isolation knob for emergency shutdown should be made accessible with manual reset at control panel.

  Compressor may be provided with the following legible and permanent markings in the installed facilities:

  - Name of the manufacturer,
  - Model number,
  - Serial number, month, and year of manufacture,
  - Approval number of certificate,
  - Capacity rating usually standard cubic meter per hour,

- Operating speed usually in revolution per minute,
- Requisite driving power usually in kilowatt,
- Maximum supply pressures usually in bar (g),
- Minimum supply pressures usually in bar (g), and
- Maximum outlet pressure usually in bar (g).

A compressor along with its all fittings must be tested for conformity of relevant statutory provisions appropriate for CNG use by a competent authority before to installation.

(b) **Dispensers**

CNG from compressor/cascade is dispensed to natural gas vehicles such as cars, three-wheelers, buses, through dispensers.

- **Bus Dispenser**: To meet the requirement of bus filling at mother stations, single-arm bus dispenser, each of capacity 75 kg/min may be used.
- **Car Dispenser**: To meet the requirement of car and three-wheeler filling at mother and daughter stations, double-arm-type car/auto dispensers, each with a capacity of 15 kg/min may be used.

Main Features of Dispenser

- Dispensers have to be installed on an appropriate foundation considering the bare minimum safety distances and other related aspects. Dispensing units are required to be safeguarded against probable harm by automobile movement in the filling stations.
- The flexible hoses installed/connected to the dispenser are supposed to be mechanically and electrically uninterrupted. The design, material, and construction of hoses should be appropriate for CNG and must withstand not safe working pressure.
- The dispensing unit must be of a kind approved by the PESO/statutory authorities.
- The car dispenser should have two arms and bus dispenser should have one arm for dispensing the gas.
- "Coriolis" true mass flow metering system or equivalent with necessary sensor and electronics should be provided with provision of liquid crystal backlit display for night viewing which will show unit price of CNG in Rs = net price/kg, quantity of gas sold in kg and total sale in Rs = net price. A tamperproof totalizer may be provided.
- Program logic controller-based sequencing software and controller including hardware along with ball valves "associated with pneumatic actuation" for dispensing of gas have to be provided.
- Two CNG flexible electrically conductive twins fill and vent hoses with two Natural Gas Vehicle-l Type-2, Class-A fill nozzles with captive vent including 3-way vent should be provided.
- All dispensers are required to be earthed.

(c) **Cylinders and Cascades (CNG Storage System)**

Cylinder: "Gas cylinder" or "cylinder" means any closed metal container having a volume exceeding 500 ml but not exceeding 1000 L intended for the storage and transport of compressed gas, including any LPG container/CNG cylinder fitted to a motor vehicle as its fuel tank but not counting any other such container fitted to a special transport or undercarriage and comprises of a composite cylinder, still, the water capacity of cylinders used for storage of CNG, nitrogen, compressed air, etc., may exceed 1000 L up to 2500 L provided the diameter of such cylinder does not exceed 60 cm (Gas Cylinder Rules 2004).

- The cylinders and its fittings must be designed, manufactured, tested together with hydrostatic stretch test at a pressure in accordance with IS:7285 part 1 and Gas Cylinder Rules 2004 taking into account the maximum allowable operating pressure of 250 kg/cm^2 (g).
- The cylinders have to be permanently and legibly labeled as "CNG ONLY" in letter at least 25 mm high in contrasting color in a position which is evident after installation.
- The cylinder should be fabricated from steel. Though, cylinders with assorted materials may be considered after the establishment of its suitability and approval of the PESO.
- The cylinders must be re-examined and hydro tested after every five years as per the relevant statutory guidelines.
- In order to protect the cylinders from atmospheric corrosion, it may be painted white to reduce solar heating effect.

Cascades

Cascades is a group of gas cylinders with a total water capacity not exceeding 4500 L, contained within a length of 5.5 m, a height of 1.6 m, and a width of 1.2 m in case of cylinders kept vertical, or 5.5 m, 1.6 m, and width of 1 cylinder up to 2 m in case of horizontal cylinders. Either arrangement used may be interlinked by high-pressure piping so as to form a singly gas storage unit. Cascades are used to store the CNG at high pressure, to absorb the surge of reciprocating compressor, frequent start and stop of compressor and to supply additional gas when dispensing rate is more than compressor capacity. Compressor will take start, if pressure in cascade falls below 210 kg/cm^2 and run till it reaches at pressure 255 kg/cm^2. The cascade can supply gas to bus as well as car/auto dispensers.

Main Specification Features

- Cascades are made of group of cylinders fixed with structural steel frame having facility of lifting and placement.
- The cylinder and their neck threading should be designed as per IS:7285-1988 and IS:3224-1979 respectively and approved by Chief Controller of Explosives.

- The cylinder shutoff valve should be with fusible disk confirming to requirements of IS:3224 or CCOE approved.
- All end connections for quick release couplings, PG, valves, and fittings of cascade should be within tamperproof enclosure. These should be on one side of cascade for ease of operation.

CNG Storage System (Static)

- The cascade having horizontal cylinders and sited parallel to other cascade, cylinder fittings should be arranged so that they do not face cylinder fittings of other cascade.
- Cylinder installed horizontally in a cascade should be separated from another cylinder in the cascade by a distance of not less than 30 mm.
- Cascade with horizontal cylinders should have the valves fitted on the same side within the cascade opposite to the refueling point and arranged in a manner that any gas leakage is discharged upward.
- Cascade/bulk units should be installed on a firm, compacted, well-drained foundation. The base possibly could be in the form of a platform with the raised edge at 2 m from frontage and sides of the cascade forming a curb up to which vehicles should be allowed. The cascade should be firmly fixed to shun, floating in case flood is probable.
- Gas storage facility should be protected from the consequences of the weather by a roof or canopy designed to facilitate the dispersion of free or escaped gas and should not permit gas to be trapped.

Adequate means should be provided to prevent the flow or accumulation of flammable or combustible liquids under containers for instance by grading, pads, or diversion curbs.

CNG Storage System (Mobile)

- Only dedicated trailer, truck, or any other vehicle to be utilized for transportation of CNG storage units. The storage units should have lugs especially built-in for lifting purpose and in no case magnetic device to be utilized as an alternative.
- The vehicle with the cascade thereon may be located with easy entrée and exit on a low raised area or hard flattened ground, which should extend to at least another one meter on all sides and this platform or hard ground should be under a light roof or canopy as described above.
- The trailers/vehicle carrying CNG should be made immovable by application of brake and wheel choke prior initiation of filling or dispensing operation.

(d) **CNG Piping**

- All rigid piping, tubing, fittings, and other piping components should conform to the recommendations of ANSI B 31.3. All the elements of piping should be designed for the full range of pressures, temperatures, and loading to which they may be subjected with a factor of safety of at least 4 based on the minimum specified tensile strength at 20 °C.

- Gaskets, packing, and any other materials used should be compatible with natural gas and its service conditions.
- All the piping and tubing should have bare minimum turns with enough provisions for expansion, contraction, jarring, vibration, and settling. Exterior piping may be either buried with suitable corrosion protection or installed 30 cm. above the ground level with supports and protection against mechanical and corrosive damage.
- Rigid pipelines should have welded joints between their respective components.
- All the piping and tubing should be able to endure a pressure up to that of safety relief device and subsequently tested after assembly. The testing should be done by using less reactive/inert gas like nitrogen. However, in case natural gas is being utilized as a testing medium, appropriate safety measures are obligatory to be adhered to.

(e) **Values**

A minimum of four shutoff valves should be fitted between the gas storage unit and vehicle refueling filling nozzle as briefed below:

- Every CNG storage unit should have a quick action isolation (usually ball type) valve in the steel supply pipe right away adjoining to such storage unit to allow isolation of individual storage unit. These valves should be surrounded by fencing of storage unit.
- The master shutoff valve equipped with locking facility in closed position should be established in steel outlet pipe outside, yet in proximity to the gas storage unit so as to cut off all downstream equipment from the gas storage unit. This valve should be located outside the fencing.
- A fast action emergency and isolation shutoff valve must be installed next to dispensing unit with easy access and to remain closed until the refueling is not done.
- An automobile refueling shutoff valve should be installed for each flexible vehicle refueling hose to manage the refueling operation. It should have provisions related to vent which allows the flow of the remaining high-pressure gas after refueling is done.

These valves along with all other essentials of piping should be apt for the full range of temperature and pressure to which they may be exposed to.

(f) **CNG Hoses**

- Internally braided, electrically uninterrupted, nonmetallic and metallic hoses which are anticorrosion and apposite to the natural gas usages shall be deployed for CNG service in the downstream of isolation shutoff valve and emergency switch.

- The flexible hoses and their connections are appropriate for high-pressure and high-temperature service settings estimated with a burst pressure of as a minimum four times the maximum operating pressure.
- These flexible hoses along with their connections should be tested after assembly and before using, no less than twice the operating pressure.
- These flexible hoses must have permanent markings signifying the name identification of the manufacturer, operating pressure, and other details for usage with CNG.

(g) **Pressure Gauges**

- Each CNG bulk storage unit including every cascades or bulk storage tank ought to be provided with an apposite pressure gauge directly mounted on it.
- The CNG storage unit should have an opening not greater than diameter of 1.4 mm at the connection point where pressure gauge is installed.
- The pressure gauge should have dial graduated to read approximately double the operating pressure but in no case less than 1.2 times the pressure at which PRV is decided to operate.

(h) **Pressure Relief Device**

- Safety relief devices should be in compliance with the requisites of OISD-STD-132. It may comprise of burst disk/safety relief valve.
- Safety relief devices should be installed with unhindered full-size discharge to a safe site on bulk tanks and cylinders in the vertical position with right kind of rain caps. These devices should have their outlet arranged in such a way that in case of emergency a high-pressure gas escapes from it does not directly hit on operators/persons who are in close vicinity.
- Cylinder should have safety relief devices installed in accordance with the requisite statutory rules and regulations.
- Piping should be safeguarded by safety relief devices as per the OISD Standard 132.
- These valves should have a locking facility to prevent any tempering.
- The safety relief devices must have manufacturer's permanent marking indicating following details:
  - Set pressure to start discharge usually in bars.
  - Discharge capacity usually in cubic meter per min.

(i) **SS Tubing**

Stainless steel (SS 316 grade or SS316 L grade) tubing generally of sizes 1/2″, 3/4″, or 1″ should run in underground concrete masonry trenches for conveying of compressed gas from compressor to priority panel to dispenser and priority panel to stationary cascade to dispenser.

(j) **Electrical Equipment**

All electrical wiring and equipment, and other facility located in hazardous area Zone I and II should be as per relevant statutory provisos.
The earthing at the facility, protection against ignition due to static, lightning, and stray currents has been detailed in OISD Standard 110 and OISD Standard 137.

2.3.3 Layout and Interdistances

- Interdistances among various facilities/equipment such as dispensers, compressors, and storage cylinders/cascade should be as per Tables 4 and 5 (Gas Cylinder Rules 2004). A sample of layout is given at Fig. 5.
- Safety distances not suggested above should be as approved by PESO on case to case basis after taking into account all the impelling factors.
- Whenever inside a petrol/diesel service station, the CNG storage and dispensing facilities should be positioned in a secluded area not getting disturbed due to the vehicular movement on the drive way and not coming within the risk-prone areas of liquid petroleum facilities. The CNG facility should not be situated below the electric power lines.
- The fencing may be restricted up to the dispensing unit to prevent barrier in the driveway, if the requisite clear space is available thereafter within the service station location.

2.3.4 Safety at Vehicle for Refueling

- The vehicles should have authenticated CNG kit installed as per the guidelines of Ministry of Surface Transport, by an authorized workshop and such workshop should issue a fitness certificate to the vehicle for its appropriateness for

Table 4 Minimum interdistances required from buildings and outer boundaries compressed natural gas storage units

Total capacity of compressed natural gas storage units (liter)	Minimum required distance from buildings and boundaries (meter)
1	2
Up to 4500 L	2.5
4500–10,000 L	4
10,000–100,000 L	10

Note If on the side(s) toward the boundary of the storage, the clearance is not available as mentioned above, the same may be compensated to 2 m if a 4 H-FRR RCC wall of suitable height and length covering the storage units is constructed at the boundary and suitable clear space is available on the other side of the wall

Table 5 Interdistances between various facilities/equipment of natural gas handling at CNG refilling stations

Distance from various equipments/ facilities (in meters)	1	2	3	4	5	6	7
1. CNG compressor	–	3	2	3	6	6	Table 4 (Min-3)
2. CNG dispensing unit	3	–	2	4	6	4	Do
3. Storage cascade	2	2	–	Table 4	Table 4	Table 4 (Min-6)	Do (Min-4)
4. Outer boundary wall/chain-linked fencing	3	4	Table 4	–	6	4	Do
5. Petrol/diesel dispenser	6		Table 4 (Min-6)	6	–	6	Do
6. Vent of petrol/diesel underground storage tanks	6	4	Table 4 (Min-4)	4	6	–	6
7. Filling point of petrol/diesel	Table 4 (Min-3)					6	–

Note
Distances stipulated as "–" should be any safe distance required for operational suitability
A CNG storage cascade having total capacity not exceeding 4500 water liters can be installed on top of the compressor superstructure. The assembly shall have clearance distance of 3 m around and also from the dispensing unit. However, this distance can be reduced up to 2 m in accordance with the Note stipulated in Table 4

CNG use. Such certificate should be all the time carried by the driver of the vehicle.

- Each vehicle running on CNG fuel should display "CNG" labels prominently in compliance to Central Motor Vehicle Rules.
- Vehicle driver is supposed to carry the record showing the fitness details of the essential parts of the system equipped in the vehicle for CNG.
- The cylinder along with the complete assembly mounted on the vehicle must be as per the Gas Cylinder Rules and be subjected to hydraulic test in any case once in every five years.
- The vehicle refueling must be done by as killed and duly certified operator.
- Before refueling the gas, operator of the CNG dispensing unit must check followings:
  - Vehicle driver is carrying updated Fitness Record.
  - No smoking, naked flame, or any other source of fire within 6 m of the refueling point is present.
  - No leakage in the CNG fuel facility is reported by the vehicle driver.

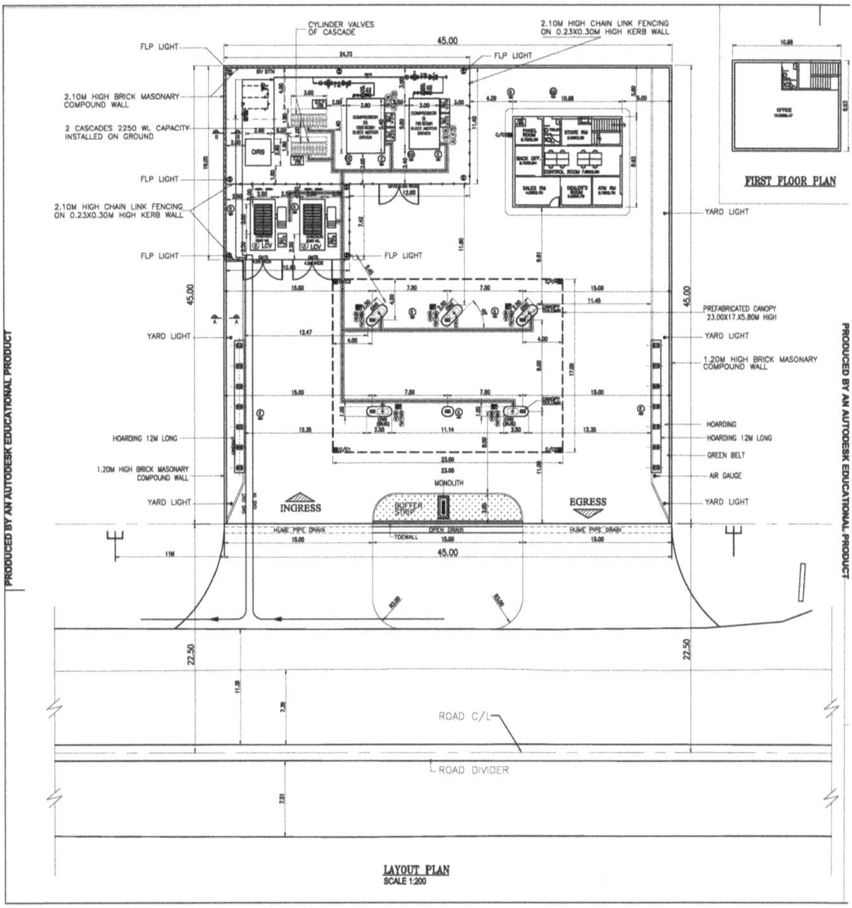

Fig. 5 A sample layout of CNG station

- The fuel connection is in fine condition and matches the dispensers filling nozzle.
- No passenger should remain inside the vehicle.
- Hand brake is firmly applied, the vehicle parked in gear or in "P" with automatic transmission.
- The vehicle engine is switched off.
- Detailed safety measures and procedure (also in local language) of refueling have to be displayed near the dispensing unit and shave to be strictly adhered by the operator.
- The operator should not leave the vehicle unattended during refueling.
- The refueling in the vehicle should not be done in excess of maximum allowable operating pressure at standard temperature.

- Warning signage such as "FLAMMABLE GAS," "STOP VEHICLE," "NO OPEN FLAME PERMITTED," and "NO SMOKING," must be exhibited near dispensers and compressor areas.

2.3.5 Fire Protection

- Fire-fighting facilities need to be cautiously examined and planned considering the accessibility of municipal fire tenders, etc.
- All approaches to dispensers, compressors, gas cascades, and work places should be free from obstruction in order that they are readily reachable in case of emergency.
- The electrical installations must be inspected by a competent electrical inspector as per relevant rules and adherence should be made as stated during the inspection. Proper documentations must be done for all periodical and routine examinations.
- The flameproof attributes of electrical equipment must be checked via visual checks, and as suggested by manufacturer.
- There shall be no unauthorized additions or alterations of the service station whether temporary or permanent in nature.
- All electrical maintenance should be carried out by skilled technician and under the supervision of authorized and competent person.

3 Various Major Components Used in CNG Kit of Vehicles

3.1. **Cylinder**: It is used to store CNG at an operating pressure not exceeding 200 bar (g). It is equipped with a shutoff valve and a protection burst disk.

3.2. **Vapor bag**: It is fitted on to the cylinder. It is mainly used to cover the valves and piping of the cylinders and is vented out of the vehicle to the open atmosphere.

3.3. **High-pressure pipe**: It hooks up the refueling valve to the pressure regulator and cylinder valve.

3.4. **Refueling valve**: It is used to refueling the cylinder.

3.5. **Pressure regulator**: This has a solenoid valve to shutoff gas delivery to the engine cylinder. The CNG stored at a high pressure is brought to slightly above the atmospheric pressure by this regulator.

3.6. **Gas–Air mixture**: Particularly designed to suit each model of the vehicle.

3.7. **Petrol solenoid valve**: It is used to stop petrol supply to the engine cylinder when the vehicles run on CNG.

3.8. **Selector switch**: Mounted on the dashboard, selector switch enables the driver to select either CNG mode or petrol mode.

4 Technology Initiatives in CGD

(a) **Social media initiatives**: A twitter account, Facebook account, and SMS facility may be opened for approaching the public and spreading the awareness.

(b) **Mobile app for PNG and CNG customers**: User-friendly mobile app may be available for download on Android play store providing the below salient features:

Features for PNG users

- Customer can log complaint-linked issue by choosing the type of complaint from the drop-down menu.
- Customers can find their current and previous PNG bills make online payments and also find locations wherever bills can be done.
- Customer can themselves provide meter readings for the period of billing cycle by uploading a photograph of their meter display.
- In regard to the customers' safety and to restrain fraudulent events, customers can validate the meter reader by feeding his employee ID displayed on his identity card.

Features for CNG users

- Tracing nearest CNG stations.
- Details related to car models available in CNG variant or CNG Kit suppliers and hydro testing centers, instructions on CNG vehicle maintenance and repair et al.

(c) **Mobility solutions**: Image of actual meter reading by a meter reader is transferred via cell phone to billing server and the photograph of actual reading is printed in the invoice. It provides comfort to domestic consumer on accurate meter reading. For a collecting agency man visiting domestic customer's home, account history is available on cell to view. Customer can pay to the agency man at his home, and receipt generated through live billing system is printed on portable printer.

(d) **GIS**: Geographical information system is a computer-based system used for capturing, storing, querying, and analyzing pipeline network information data.

(e) **SCADA**: SCADA provides online monitoring and controls to CNG outlets, commercial and industrial customers, and taking daily sales information with no or minimum human intervention.

(f) **E-Payment to suppliers**: By using ERP and Web-based banking platforms, payments are directly credited into bank accounts of suppliers. No need of cheques so no hassles, total transparency and maintaining high standards of corporate governance.

(g) **GPRS/GPS**: Vehicle tracking system for safe and efficient management of CNG cascade carrying vehicles.

References

Business Standard, 28th March 2015, UrjaSangam: PM urges nation to give up LPG subsidy

Gas Cylinder Rules (2004) Petroleum and Explosive Organization, India

Oil industry Safety Directorate (1998) OISD Standard-179, safety requirements on compression, storage, handling & refueling of natural gas for use in automotive sector

Oil industry Safety Directorate (2013) OISD Standard-226, natural gas transmission pipelines and city gas distribution networks

Petroleum Planning & Analysis Cell, Ministry of Petroleum and Natural Gas (2015) Ready Reckoner, Snapshot of India's Oil & Gas data

Part IV
Legal, Regulatory Developments: Impact on Natural Gas Market in India

Petroleum Exploration and Licensing Policy in India

T.P. Rao and Sanjay Kumar Kar

Abstract This chapter lucidly brings out policy developments in the area of petroleum exploration in India. Authors discuss historical perspectives, existing policy frameworks and recent developments and set futuristic directions. Findings suggest that the current policy fails to attract desired level of investment and retaining confidence of investors. Therefore, to bring back petroleum exploration and production on track, the Government recently approved Hydrocarbon Exploration and Licensing Policy (HELP) in March 2016. The Government is strategically moving away from cost-sharing model to revenue-sharing model with marketing and pricing freedom for crude oil and natural gas produced by contractors under HELP. The new policy regime is expected to attract more investment to boost exploration and production of oil and gas from conventional and unconventional sources. Further, the HELP is designed to improve bidding for designated areas throughout the year in a very transparent manner.

Keywords Exploration policy · Bid evaluation · Work programme · Fiscal package

1 Introduction

The story of oil in India began in the dense jungles, swamps, damp, and undulated terrain of Brahmputra Valley, Assam in the middle of nineteenth century. Just seven year after famous "Drake Well" drilled in Titus Ville, Pennsylvania, USA (1859); India discovered and drilled its first oil well in Jaipur, Upper Assam, in 1866. No

T.P. Rao (✉)
NELP, DGH, Noida, India
e-mail: tprao1959@yahoo.co.in

S.K. Kar
Department of Management Studies, Rajiv Gandhi Institute of Petroleum Technology, Jais, India
e-mail: skar@rgipt.ac.in; karsanjay1@gmail.com

S.K. Kar and A. Gupta (eds.), *Natural Gas Markets in India*,
DOI 10.1007/978-981-10-3118-2_11

technologies deployed to find the evidence of oil rather an elephant carrying logs provided clues for availability of oil.

W.L. Lake of Assam Railway and Trading Co. (AR & T Co) started Digboi well No. 1 in September 1889, and the drilling of 662 feet well was successfully completed in 1890. The discovery of the "Digboi Oil field" in Upper Assam was a landmark in the history of oil industry in India.

In 1955, the Government of India decided to harness the oil and natural gas resources available in the various regions of the country. To achieve this objective, an Oil and Natural Gas Directorate was set up in 1955, as a subordinate office under the then Ministry of Natural Resources and Scientific Research. Subsequently in August 1956, the Directorate was raised to the status of a commission with enhanced powers for efficient and effective operations. Further to provide statutory authority, in October 1959, the Commission was converted into a statutory body by an act of the Indian Parliament, which enhanced powers of the commission further and ONGC became an autonomous body. On 18 February 1959, Oil India Private Ltd. (OIL) was incorporated and registered as a Rupee Company with the BOC holding two-third of the shares and GOI one-third. On 27 July 1961, the Government of India share holding in the company increased from one-third to one-half, and GOI and BOC become equal partners in OIL. On 1 January 1962, OIL formally came into being at its new head quarters at Duliajan.

After liberalization in 1991, ONGC was reorganized as a limited company under the Companies Act, 1956 in February 1994. Today, ONGC has grown into a full-fledged horizontally integrated petroleum company. ONGC has a fully owned subsidiary, ONGC Videsh Ltd. (OVL) that is responsible for acquisition of foreign oil and gas assets and exploration and exploitation of potential petroleum reserves in those blocks. ONGC has also acquired 72% stake in MRPL with full management control of the 9.69 tonne, state-of-the-art refinery.

To address increasing demand of petroleum products, the Government recognized the role of higher domestic petroleum production. As a result, a strategic shift from allocation of blocks from nomination to competitive bidding emerged. In 1998, New Exploration Licensing Policy (NELP) was launched, and 48 exploration blocks were offered under round 1. In 2000, 25 exploration blocks were offered under NELP round II. In 2002, 27 exploration blocks offered under NELP round III. In 2003, 24 exploration blocks were offered under NELP round IV. In 2005, 20 exploration blocks were offered under NELP round V. In 2006, 55 exploration blocks were offered under NELP round VI. In 2007, 57 exploration blocks were offered under NELP round VII. In 2009, 70 exploration blocks were offered under NELP round VIII. In 2010, 34 exploration blocks were offered under NELP round IX. Nine bidding rounds have been completed so far, and Tenth round was planned but did not happen.

Once again the Government pushed the petroleum exploration and production policy reforms with lot of intent, and in March 2016 launched a comprehensive "Hydrocarbon Exploration Licensing Policy" aimed at increasing investor confidence and reducing trust deficit between the contractors and the Government.

2 Historical Perspective

2.1 *Pre-new Exploration and Licensing Policy (Pre-NELP)*

Two national oil companies, namely Oil and Natural Gas Corporation Limited (ONGC) and Oil India Limited (OIL) carried out petroleum exploration and exploitation activities in India from 1956 to 1980. The sector was opened to the private sector in 1980s for few foreign oil companies for exploration to augment exploratory effort, mainly in offshore. With the launch of liberalization policy by the Government of India in 1991, the sedimentary basins were opened to domestic private companies. During 1991–1995, 5 Rounds of bidding (4th–8th) for exploration blocks were carried out. National Oil Companies (NOCs), designated as Licencees, were liable for payment of royalty. Private/JV Companies were exempted from payment of Royalty and/or Cess, and Government/NOCs had carried in rights.

During 1992–93, 2 rounds of discovered fields were held. During these rounds, signature and production bonus were biddable. In mid-size fields, NOCs hold 40% working Interest and balance 60% by Pvt. Joint Ventures (JVs) where as in small size fields, and 100% is held by Pvt. JVs. Royalty and Cess at fixed rates is payable by companies on basis of its participating interest (PI).

Figure 1 presents growth exploration and production activities in India. It is evident that initiatives taken in the 1990s resulted in expanding the producing basins and increasing the number of E&P companies in India.

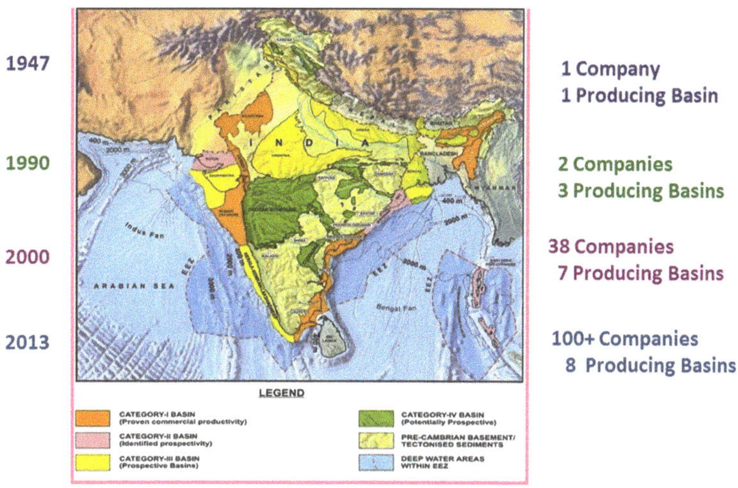

Fig. 1 Growth of exploration and production activities in India

2.2 New Exploration Licensing Policy (NELP)

In the backdrop of slow progress in terms of petroleum exploration and production activities in the country, New Exploration Licensing Policy (NELP) was formulated during 1997–1998 by the Government of India with Directorate General of Hydrocarbons (DGH) as a nodal agency. The NELP was expected to address key challenges including creating level playing field for both public and private sector, attracting higher investment and attracting foreign players with experience and expertise. The change in policy regime expected better channelization of resources, reduce market failure, remove information asymmetry, and increase efficiency in the hydrocarbon sector.

The NELP Policy was operationalized in 1999. The NELP was an evidence of India's commitment towards liberalization, privatization and globalization with focus on domestic market developments.

The NELP regime not only promoted spirit of competition between public and private companies, but also created opportunities for collaboration. As a result companies begun to realize strength of complementary resources and deployed them for better results.

The NELP has been a landmark event in the growth of the upstream oil sector in India.

The development of E&P sector has been significantly boosted through this Policy of Government of India, which brought major liberalization in the sector and opened up E&P for private and foreign investment, where 100% Foreign Direct Investment (FDI) is allowed. NELP provides a level playing field to the private operators either Indian or Foreign, by giving them the same fiscal and contract terms as applicable to NOCs for offered acreages. Apart from providing a level playing field to all the exploration and production (E&P) companies, such as private, foreign, and National Oil companies (NOCs), the policy was aimed at attracting much needed investments as well as infusion of new technologies in E&P sector for bringing more and more unexplored area under the exploration and increasing oil and gas production in the country. Table 1 shows increase in crude production after 2009–10, whereas natural gas production shows lot of variations.

Initially, the foreign and Indian private companies invested with a lot of enthusiasm and interest. However, owning to lacuna in the NELP the later rounds

Table 1 Crude oil and natural gas production trend in India (2007–08 to 2013–14)

Year	Crude oil production (MMT)	Natural gas production (BCM)
2007–2008	34.118	32.417
2008–2009	33.508	32.845
2009–2010	33.690	47.496
2010–2011	37.684	52.219
2011–2012	38.090	47.559
2012–2013	37.862	40.679
2013–2014	37.788	35.407

of NELP received minimal or no interest from foreign companies. Even the Indian companies started to withdraw their interest.

2.3 Features of NELP

Under the New Exploration Licensing Policy (NELP) for exploration of oil and natural gas, the Government of India announces offer of exploration blocks from time to time. Companies are invited to bid for exploration blocks on offer. Companies may bid for one or more blocks, singly or in association with other companies, through an unincorporated or incorporated venture. A company interested in Indian exploration acreages can enter exploration arena either through direct participation in bidding process, as operator or a consortium partner, or by acquiring participating interest in existing acreages held by companies.

Some of the attractive features of the terms offered by the Government are as follows:

- The successful bidder is required to enter into a Production Sharing Contract (PSC), which is based on a Model Production Sharing Contract (MPSC).[1]
- There is one exploration phase of 7 years for on-land and shallow water blocks with initial exploration period of 4 years and subsequent exploration period of 3 years.
- There is one exploration phase of 8 years for deep water blocks and frontier area blocks with "initial exploration period" of 5 years and subsequent exploration period of 3 years.
- Mandatory and committed work programme are to be completed in the initial exploration period. Operators have option to relinquish entire area after completion of minimum work programme (MWP) or retain the block by committing to carry out drilling of one well per year in case of on-land and shallow water blocks or one well in 3 years in case of deep water blocks.
- Up to 100% participation by foreign companies is permitted.
- At the time of signing contract, at the time of discovery or at the time of production no bonus is required to be paid by the contractor, which is a practice in many countries.
- There is no mandatory State participation.
- There is no carried interest by National Oil Companies (NOCs).
- There is no customs duty on imports, which are made for petroleum operations.
- Income Tax Holiday for seven years from start of commercial production of "Mineral Oil".

[1]For more features of the Model Production Sharing Contract please visit the link available at http://www.dghindia.org/pdf/MODEL%20PRODUCTION%20SHARING%20CONTRACT(MPSC).pdf?, accessed on 3 February, 2016.

- Biddable cost recovery limit up to 100% is permitted.[2]
- Option to amortize exploration and drilling expenditures over a period of 10 years from the first commercial production.
- Sharing of profit petroleum with Government of India based on biddable pre-tax investment multiple achieved by the contractor is biddable.
- Royalty for on-land areas is payable at the rate of 12.5% for crude oil and 10% for natural gas. For shallow water offshore areas, royalty is payable at the rate of 10% for both crude oil and natural gas; whereas for deep water offshore areas (beyond 400 m iso-bath), royalty is payable for both crude oil & natural gas at the rate of 5% for the first seven years of commercial production and thereafter at the rate of 10%.
- There is a Fiscal stability provision in the contract i.e. the offer of profit share made to the Government is firm throughout the contract. For increased profit either due to larger discoveries based on geological surprises or due to crude oil/gas price hike, no additional revenues are required to be shared.
- Contractors are permitted to assign their share to any eligible company with the approval of the Government, which effectively means Contractor can quit in the middle of contract period if suitable contractor is willing to take the entire stake.
- Arbitration and Conciliation Act, 1996, based on United Nations Commission on International Trade Law (UNCITRAL) model, applicable.
- Predetermined Liquidated Damages (LD) are specified upfront for unfinished minimum work programme (Table 2).
- One time Bank Guarantee (BG) at the rate of 7.5% of the value of total committed work programme, is required to be provided. In some of the countries the same is as high as 50%.
- A nominal bid bond at specified rate to encourage serious bidders and discourage non-serious bidders. The rates are given below:

  (i) Deepwater block ₹ 2.0 million
  (ii) Shallow water block ₹ 1.5 million
  (iii) On-land block ₹ 1.0 million
  (iv) On-land Type-S block ₹ 0.5 million

Table 2 Rate of predetermined liquidated damages

	US$		
	On-land	Shallow water	Deep water
Exploratory well (per well)	1,000,000	3,000,000	6,000,000
3D seismic surveys (per sq. km)	5000	1500	1500
2D seismic surveys (per line km)	2500	1000	1000

Source: DGH (http://petroleum.nic.in/docs/rti/MPSC%20NELP-IX.pdf p. 17. Accessed on 11 May 2016)

[2]http://petroleum.nic.in/docs/rti/MPSC%20NELP-IX.pdf accessed on 10 May, 2016.

The bid bond will be released on signing of PSC for the block. If the PSC is not signed within 90 days after the award of the block, the bid bond will be forfeited.

2.4 *Biddable Terms of NELP*

Companies are required to bid for:

- Work programme commitment
- Percentage of annual production sought to be allocated towards cost recovery
- Profit petroleum share offered to Government of India (GOI) at the lowest Pre-Tax Investment Multiple (PTIM) (1.500 and below) and the highest Pre-tax Investment Multiple (PTIM) (3.500 and above).

2.5 *Bid Qualifying Criteria (BQC)*

Under the NELP the Government set following qualifying criteria for bidders:

- Payment of Tender Fees, by bidding company or any member of the consortium, by way of purchase of the requisite Data Package of the block to be bid as the case may be on or before bid closing date
- The bidder must be a company singly or in association with other companies, through an unincorporated or incorporated venture
- The designated operator for on-land blocks (except onshore S-Type blocks) and shallow water blocks is required to obtain non-zero score on one out of the three sub-criteria of technical capability apart from non-zero score on operatorship experience
- The designated operator for Deep water blocks is required to obtain non-zero in technical capability parameters on an aggregate basis i.e. the total score of the designated operator on account of Acreage Holding, Operatorship Experience, Average Annual Accretion of Proved Reserves (1P) and Average Annual Production taken together should be more than zero
- Bidders are required to confirm to carry out the mandatory work programme given for the blocks
- The net worth of the bidding companies should be equal or more than its participating interest value of the minimum work programme and mandatory work programme
- In case a parent company provides financial and performance guarantee, the financial capability of the parent company is considered for evaluating the financial capability of a bidding company

2.6 Bid Rejection Criteria (BRC)

- A bid without documentary proof of payment of Tender Fees by way of purchase of the requisite Data Package of the block
- A bid not submitted in "Format for Submission of Bids" covering all the information/details listed therein including Bid Bond
- A bid not confirming to carrying out the mandatory work programme during the initial exploration period
- A bid with any assumptions/deviations which are inconsistent or not complying with the contract terms
- In case a bidding company or each of the company constituting consortium does not furnish the annual report including the audited annual accounts for the latest completed year and a Certificate of net worth from company's statutory auditor(s) based on the latest audited annual accounts certifying the net worth of the bidding company.

2.7 Bid Evaluation Criteria (BEC)

A. *Net worth of the company*

(a) The net worth of the bidding companies should be equal or more than its participating interest value of the minimum work programme and mandatory work programme.

(b) In case a bidding company, either bidding alone or as a consortium happens to be the best ranked bidder for two or more blocks, the net worth of the company is required to be equal to or more than its participating interest in the value of minimum work programme commitment for all such blocks. In case, the company's net worth is less than its participating interest in the value of minimum work programme commitment for such blocks, the bids will be considered in order of priority given by that company in their bids for respective blocks.

B. *Technical capability of the proposed Operator*

(a) For the on-land Type-S blocks, only work programme and fiscal package are considered for bid evaluation. Technical capability is neither a pre-qualification criterion nor a bid evaluation criterion.

(b) For the on-land and shallow water blocks, technical capability is a pre-qualification criterion. The designated operator has to score non-zero on one out of the three sub-criteria of technical capability apart from non-zero score on operatorship experience.
Table 3 presents the sub-criteria of Technical Capability (only for pre-qualification and nil weightage for bid evaluation).

Table 3 Sub-criteria of technical capability

Sl. no.	Sub-criteria
(i)	Acreage holding petroleum exploration licence (PEL) (sq. Km.) (total on-land, shallow and deep waters)
(ii)	Operatorship experience (experience of operatorship in oil and gas exploration and/or development and/or production in the last consecutive 10 years)
(iii)	Average annual accretion of proved reserves (1P) during last 5 years (MMBoe) (total on-land, shallow and deep waters)
(iv)	Average annual production (O + OEG) for the previous 5 years (MMBoe) (total on-land, shallow and deep waters)

Source Bid invitation, DGH

(c) For Deep water blocks, technical capability is an evaluation criterion. Evaluation of technical capability of the proposed operator is based on experience as operator in terms of number of years, acreage holding (PEL), average accretion of proved reserves (1P), average annual production and drilling of exploratory wells. The evaluation parameters and points for deep water blocks are presented in Table 4.

(d) Where a parent company endorses the bid and undertakes to provide financial and performance guarantee, the technical parameters of the parent company of the designated operator are considered for evaluation of the bid.

C. *Work Programme*

(a) Only the committed biddable work programme by the bidding company (ies)/consortium is considered for evaluation purposes. Work programme consists of the following:

- 2D Seismic surveys
- 3D Seismic surveys
- Exploratory Wells

(b) Some 2D seismic work is prescribed (mandatory work programme) for blocks where no data is available previously. Such mandatory work programme is not considered for evaluation. The work programme bid and mandatory work programme are required to be completed in the initial exploration period.

D. *Fiscal Package*

(a) Profit petroleum share offered to Government of India by the bidder at the lowest pre-tax Investment Multiple (PTIM) (1.500 and below) and the highest PTIM (3.500 and above) along with offered annual cost recovery limit is taken into account for evaluation of Fiscal Package.
Table 5 presents summary of the evaluation criteria/points adopted for evaluating bids under the NELP rounds. It is clear that technical capability is very important component of bidding system in case of deep water blocks.

Table 4 Evaluation parameters and points for deep water blocks

Sl. no.	Sub-criteria	Max. point	Scale	
			Parameter (point) minimum	Parameter (point) maximum
(i)	Acreage holding (sq. km.) petroleum exploration licence (PEL) beyond 400 m. bathymetry	3	(0) 0	(3) 50,000
(ii)	Operatorship experience (years) (experience in years of operatorship in oil and gas exploration and/or development and/or production in the last consecutive 10 years beyond 400 m bathymetry)	3	(0) 0	(3) 10
(iii)	Average annual accretion of proved reserve (1P) during last 5 years (MMboe) >400 m bathymetry	4	–	(4)
(iv)	Average annual production of O + OEG (MMboe) during last 5 years >400 m bathymetry	4	(0) 0	(4) 50
(v)	Average annual production of O + OEG (MMboe) during last 5 years up to 400 m bathymetry from its share of total acreage held either as operator or non-operator	4	(0) 0	(4) 150
(vi)	Drilling of deep water exploratory wells beyond 400 m bathymetry during last 5 years	2	(0) 0	(2) 4
(vii)	Companies bidding in consortium/partnership with Indian company <u>or</u> companies bidding alone (For consortium with Indian company(ies) one of the partners should have deepwater production beyond 400 m bathymetry with minimum PI of 10% and be designated as operator.)	5	Production beyond 400 m and up to 1000 m (3 <u>OR</u> 2) or production beyond 1000 m bathymetry (5 <u>or</u> 4) (a) Operator with production of O + OEG (MMboe) during last 5 years beyond 400 m and up to 1000 m bathymetry bidding singly: 2 points (b) Operator with production of O + OEG (MMboe) during last 5 years beyond 400 m and up to 1000 m bathymetry bidding in consortium with Indian companies: 3 points (c) Operator with production of O + OEG (MMboe) during last 5 years beyond 1000 m bathymetry bidding singly: 4 points	

(continued)

Table 4 (continued)

Sl. no.	Sub-criteria	Max. point	Scale	
			Parameter (point) minimum	Parameter (point) maximum
			(d) Operator with production of O + OEG (MMboe) during last 5 years beyond 1000 m bathymetry bidding in consortium with Indian company(s): 5 points	

Source Production sharing contract, DGH

Table 5 Summary of the evaluation criteria/points adopted for evaluating bids under the NELP rounds

Criteria/points	NELP		
	On-land type-S block	Other on-land blocks and shallow water blocks	Deep water blocks
Technical capability	Nil	Qualifying	25
Work programme	50	50	25
Fiscal package	50	50	50
Total	100	100	100

Source Production sharing contract, DGH

2.8 Basin Information Availability

Dockets containing basin information where the blocks on offer for exploration and production are available with Directorate General of Hydrocarbons for purchase. These Dockets provide valuable information on regional and local geology, status of exploration activities, hydrocarbon potential and a brief write-up on block(s) on offer. The CD-ROM of Basin Information Dockets is available for purchase. The Data Packages of the blocks on offer, prepared by DGH, are available in digital mode. The digital data set comprising seismic data, navigation data and well-log data for the individual blocks are available for purchase. A detailed blockwise price list for Data Packages is also available for reference.

2.9 Process of Offer and Award of Blocks Under a NELP

- Initially, larger exploration areas are identified in different sedimentary basins based on perceived hydrocarbon prospectivity and availability of geoscientific data. These blocks are carved out from the acreage which is not awarded to any of the companies and which is with the Government of India. Some of relinquished areas are recycled and further offered after integration of old and new data generated. Some new areas are also opened up for exploration.

- These larger areas are forwarded to the Ministry of Petroleum and Natural Gas (MOPNG) to obtain clearances from Ministry of Environment and Forests (MOEF), Ministry of Defence (MOD), Ministry of Home Affairs (MHA), Ministry of External Affairs (MEA), Department of Space (DOS) and Survey of India (SoI) to exclude the restricted areas.
- Subsequent to the inter-ministerial clearances, exact blocks are carved out after leaving indicated or demarcated restricted areas.
- The Notice Inviting Offers (NIO) which consists of Bid Evaluation Criteria (BEC), Terms and conditions and Model Production Sharing Contract is made available to prospective bidders free of cost.
- Geological and geophysical data that have been acquired earlier are populated in the Blocks. Such data is made available for viewing to the bidders free of cost, and this enables bidders to analyze the prospectivity of the areas prior to bidding.
- The bidders can purchase the Data Packages. Data Package includes processed output of 2D/3D Seismic data and drilled well data.
- In addition to the above, DGH has carried out seismic surveys in some areas through Speculative surveys mode and these data are also available for purchase from the respective companies.
- Additional data such as Well completion reports, Composite logs, Master logs and VSP data, are also available for purchase.
- To propagate and promote NELP, road shows are conducted at Mumbai, London, Houston, Calgary, Singapore, Brisbane and Perth.
- Physical Data viewing rooms are opened at NOIDA, London, Houston, Calgary and Perth.
- Four to five months period is provided between offer of blocks and bid closing date.
- Bid evaluation is done objectively as per the Bid Evaluation Criteria (BEC) approved by Empowered Committee of Secretaries (ECS). The award is scrutinized and recommended by ECS, and Cabinet Committee on Economic Affairs (CCEA) approves the award.

3 Present Scenario

India's 26 sedimentary basins have estimated area of 3.1 million square kilometre (sq. km), out of which, 1.35 million sq. km area is in deep water (water depth >200 m), 0.4 million sq. km area is in shallow water (water depth up to 200 m) and 1.35 million sq. km area is on-land.[3]

[3]http://www.dghindia.org/EandPAcreages.aspx accessed on 11 May, 2016.

Subsequently, area beyond 400 m bathymetry has been considered as Deep water area, under NELP. At present, around 3,81,600 sq. km area is held under Petroleum Exploration Licences in 18 basins by national oil companies such as Oil and Natural Gas Corporation Limited (ONGC), Oil India Limited (OIL) and Private/Joint Venture companies.

During NELP I–IX, a total of 360 blocks were offered for bid, and 282 blocks received 750 bids. The government awarded 261 blocks with 1,500,957 sq. km area and signed PSC for 254 blocks. Table 6 presents details of the nine NELP bidding rounds held in India.

During the NELP-VIII, round 32 PSC signed which is 22% lower than NELP-VII round (Table 7).

During the NELP-IX round, 19 PSCs contracts signed with a maximum of 9 blocks from Cambay Basin. E&P activities for deep water blocks constantly falling. For example, under the NELP-XI, PSCs for 1 deep water block signed compared to 8 and 11 deep water blocks signed under NELP-VIII and NELP-VII, respectively. This is an indication that investors are not willing to take undue risk. Possibly investor are looking for suitable incentives to invest in relatively risky deep water exploration and production activities.

Table 8 presents a brief summary name of the blocks, basins and company/consortium signing PSCs with the Government.

Under various rounds of NELP, a maximum of 144 blocks (57%) were awarded to public sector undertakings (PSUs), followed by Indian private (70 blocks) and foreign (40 blocks) operators (Fig. 2).

So far, the following exploration activities have been carried out by the contractors in the awarded blocks under NELP.

(i) Acquisition, Processing and Interpretation (API) of 3,53,523 Line Kilometre (LKM) of 2D seismic and 2,53,118 Sq Km (SKM) of 3D seismic data.
(ii) Drilling of 543 exploratory wells.
(iii) A total of 131 hydrocarbon discoveries, comprising of 47 oil and 84 gas discoveries, have been made in 42 blocks.
(iv) In-Place volume accretion of 935 Million Metric Tonnes (MMT) of oil and oil equivalent of gas (O + OEG) has been realized as on 01. 04. 2014. The corresponding recoverable reserves are to the tune of 548 MMT (O + OEG).
(v) NELP bidding rounds have attracted many private and foreign companies in addition to Public Sector Undertakings (PSUs). Before the NELP, a total of 35 exploration and production (E&P) companies (5 PSUs, 15 Private and 15 Foreign) were working in Nomination and Pre-NELP regime. After the conclusion of nine rounds of NELP bidding, the total number of companies increased to 117 (11 PSUs, 58 Private and 48 Foreign Companies as Operators and Non-operators/Consortium Partners).[4] Major central PSUs such as ONGC and Oil India; state PSU such as GSPCL; private companies such as Reliance Industries, Jubilant and Essar; and Foreign Companies such

[4]http://www.petroleum.nic.in/docs/exp.process.awardprocess2015.pdf accessed on 11 May, 2016.

Table 6 Summary of NELP bidding rounds in India

Parameter	NELP-I	NELP-II	NELP-III	NELP-IV	NELP-V	NELP-VI	NELP-VII	NELP-VIII	NELP-IX	Total
No. of blocks offered	48	25	27	24	20	55	57	70	34	360
No. of blocks bid for	28	23	24	21	20	52	45	36	33	282
No. of bids received	45	44	52	44	69	165	181	76	74	750
No. of blocks awarded	25	23	23	21	20	52	44	34	19	261
No. of PSC signed	24	23	23	20	20	52	41	32	19	254
Area awarded (sq. km)	228,472	263,050	204,588	192,810	113,687	306,331	112,988	52,603	26,428	1,500,957

Source DGH (2014–15), Kar and Sinha (2014)

Table 7 Name of blocks and company/consortium awarded under NELP-VIII

Sl. no.	Deep water block (24)	Basin	Awarded to company/consortium
1	MB-DWN-2009/1	Mumbai Basin	Cairn Energy India Pty Ltd. Cairn India Ltd.
2	KG-DWN-2009/1	Krishna Godavari	British Gas Exploration. And Production. India Ltd., Oil India Ltd. and Oil and Natural Gas Corporation Andhra Pradesh Gas Infrastructure Company
3	AN-DWN-2009/1	Andaman and Nicobar	Oil and Natural Gas Corporation and Oil India Ltd.
4	AN-DWN-2009/2	Andaman and Nicobar	Oil and Natural Gas Corporation and Oil India Ltd.
5	AN-DWN-2009/3	Andaman and Nicobar	Oil and Natural Gas Corporation and Oil India Ltd.
6	AN-DWN-2009/5	Andaman and Nicobar	Oil and Natural Gas Corporation and Gujarat State Petroleum Corporation Ltd.
7	AN-DWN-2009/13	Andaman and Nicobar	Oil and Natural Gas Corporation, NTPC Ltd., GAIL India Ltd. and Gujarat State Petroleum Corporation Ltd.
8	AN-DWN-2009/18	Andaman and Nicobar	Oil and Natural Gas Corporation, Oil India Ltd. and GAIL India Ltd.
	Shallow off. (28 blocks)		
9	GK-OSN-2009/1	Gujarat-Kutch	Oil and Natural Gas Corporation, Gujarat State Petroleum Corporation Ltd., Adani Welspun Exploration Ltd. and Indian Oil Corporation Ltd.
10	GK-OSN-2009/2	Gujarat-Kutch	Oil and Natural Gas Corporation, Adani Welspun Exploration Ltd. and Indian Oil Corporation Ltd.
11	MB-OSN-2009/3	Mumbai Basin	BHP BILLITON
12	MB-OSN-2009/6	Mumbai Basin	BHP BILLITON
13	MB-OSN-2009/7	Mumbai Basin	BHP BILLITON
14	CY-OSN-2009/1	Cauvery	Bengal Energy International Inc.
15	CY-OSN-2009/2	Cauvery	Oil India Ltd. and Oil and Natural Gas Corporation
16	KG-OSN-2009/1	Krishna Godavari	Oil and Natural Gas Corporation, Andhra Pradesh Gas Infrastructure Company and NTPC Ltd.
17	KG-OSN-2009/2	Krishna Godavari	Oil and Natural Gas Corporation and Andhra Pradesh Gas Infrastructure Company
18	KG-OSN-2009/3	Krishna Godavari	Cairn Energy India Pty Ltd. and Cairn India Ltd.

(continued)

Table 7 (continued)

Sl. no.	Deep water block (24)	Basin	Awarded to company/consortium
19	KG-OSN-2009/4	Krishna Godavari	Oil India Ltd., Oil and Natural Gas Corporation, Andhra Pradesh Gas Infrastructure Company and NTPC Ltd.
	On-land blocks (18)		
20	AA-ONN-2009/1	Assam-Arakan	Jubilant Oil and Gas Pvt. Ltd., Jubilant Energy (KHARSANG) Ltd. and Jubilant Offshore Drilling Pvt. Ltd
21	AA-ONN-2009/2	Assam-Arakan	Jubilant Oil and Gas Pvt. Ltd., Jubilant Energy (KHARSANG) Ltd. and Jubilant Offshore Drilling Pvt. Ltd.
22	AA-ONN-2009/3	Assam-Arakan	Oil and Natural Gas Corporation and Oil India Ltd.
23	AA-ONN-2009/4	Assam-Arakan	Oil India Ltd. and Oil and Natural Gas Corporation
24	VN-ONN-2009/3	Vindhayan	Oil and Natural Gas Corporation
25	CB-ONN-2009/1	Cambay	Esveegee Steel (Gujarat) Pvt. Ltd.
26	CB-ONN-2009/2	Cambay	Esveegee Steel (Gujarat) Pvt. Ltd.
27	CB-ONN-2009/3	Cambay	Harish Chandra (India) Limited
28	CB-ONN-2009/4	Cambay	Oil and Natural Gas Corporation and Gujarat State Petroleum Corporation Ltd.
29	CB-ONN-2009/5	Cambay	NTPC Ltd.
30	CB-ONN-2009/6	Cambay	Harish Chandra (India) Limited
31	CB-ONN-2009/7	Cambay	Esveegee Steel (Gujarat) Pvt. Ltd.
32	CB-ONN-2009/8	Cambay	JAY Polychem (India) Ltd. and JAY Polychem (Pte) Ltd.

Source Compiled from published sources and DGH

as British Gas, British Petroleum, Cairn Energy, ENI, Santos and BHP Billiton participated in various NELP rounds.

As on 31 March 2013, a total of 477 wells drilled under various NELP rounds. NELP Roundwise well drilling statistics presented in Fig. 3.

One of the major objectives of NELP was to attract investments in E&P sector. To a greater extent, NELP rounds achieved the objective of attracting more investment to the sector. NELP committed exploration investment statistics reveals that a total of about US$11.73 billion of exploration investment committed by the contractor. The details of roundwise committed investment for exploration are presented in Fig. 4. The contractors actually invested about $14.25 billion for exploration and $9.42 billion[5] for carrying out development activities. As of

[5]http://www.petroleum.nic.in/docs/exp.process.awardprocess2015.pdf accessed on 11 May, 2016.

Table 8 Blocks awarded to company/consortium during NELP-IX

Sl. no.	Block name	Basin name	Awarded to company/consortium
1	MB-DWN-2010/1	Mumbai offshore	BG-BHP Billiton
2	GK-OSN-2010/1	Gujarat-Kutch	ONGC-OIL-GAIL
3	GK-OSN-2010/2	Gujarat-Kutch	ONGC and GAIL
4	MB-OSN-2010/2	Mumbai offshore	**OIL-HPCL-BPRL**
5	AA-ONN-2010/1	Assam-Arakan	**Prize Petroleum-ABG Energy**
6	AA-ONN-2010/2	Assam-Arakan	OIL-ONGC-GAIL and EWP
7	AA-ONN-2010/3	Assam-Arakan	OIL, ONGC and BPRL
8	VN-ONN-2010/1	Vindhyan	Deep Energy, LLC and KGN Industries Limited
9	VN-ONN-2010/2	Vindhyan	Deep Energy LLC—Deep Natural Resources Limited and Safal WSB Energy Private Limited
10	CB-ONN-2010/1	Cambay	ONGC
11	CB-ONN-2010/3	Cambay	Deep Energy LLC and KGN Oil and Gas Private Limited
12	RJ-ONN-2010/2	Rajasthan	Focus Energy Limited and Birkbeck Investments Limited
13	CB-ONN-2010/4	Cambay	PRATIBHA OIL AND NATURAL GAS PVT. LTD.
14	CB-ONN-2010/5	Cambay	Pan India Consultants and Frost International Ltd.
15	CB-ONN-2010/6	Cambay	**ONGC-IOC**
16	CB-ONN-2010/8	Cambay	**GAIL-BPRL-EIL-BFIL and MIEL**
17	CB-ONN-2010/9	Cambay	**ONGC**
18	CB-ONN-2010/10	Cambay	SANKALP OIL and NATURAL RESOURCES LTD.
19	CB-ONN-2010/11	Cambay	**GAIL-BPRL-EIL-BFIL and MIEL**

Source Compiled from published sources and DGH

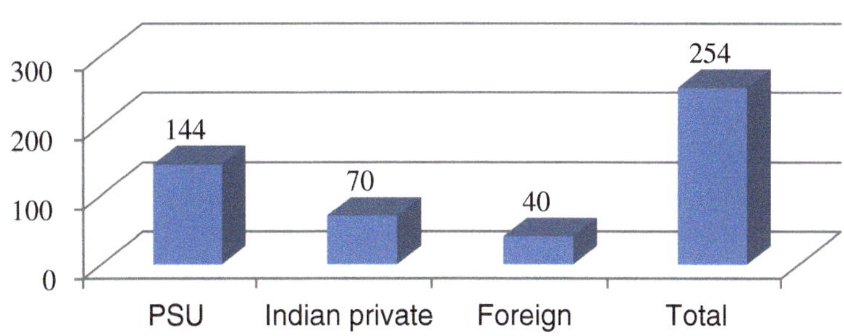

Fig. 2 Blocks awarded to various types of operators during the NELP rounds in India

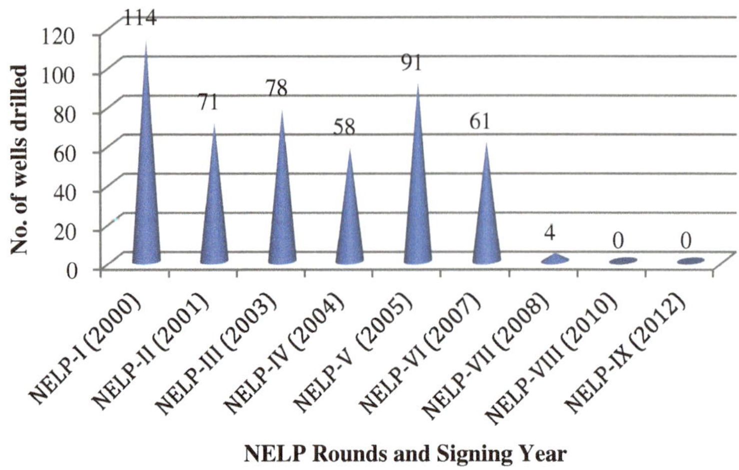

Fig. 3 Number of wells drilled under NELP (As on 31 March, 2013). *Source* Authors analysis based on data available on the website of DGH (2016)

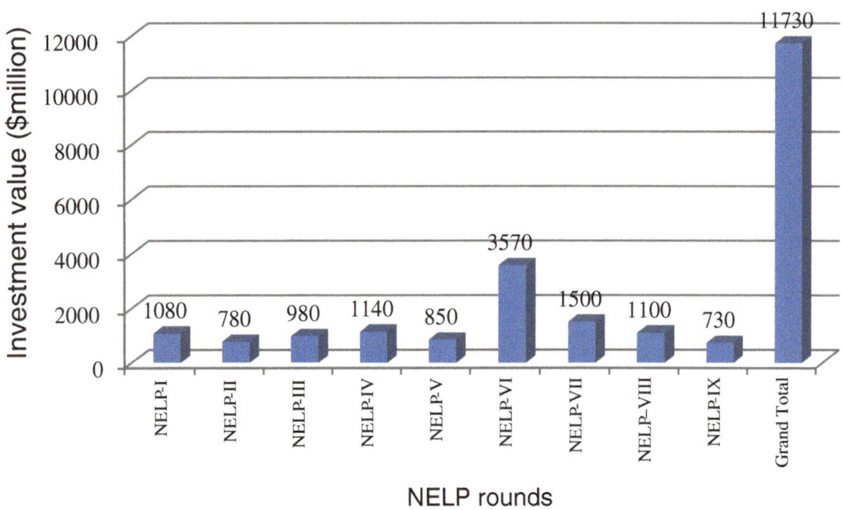

Fig. 4 Investment commitments during various NELP rounds in India

30 June, 2014, actual exploration investment was about 1.2 times higher than the committed exploration investment. Under the NELP rounds, the sector attracted cumulative total investment of $23.67 billion for exploration and developmental activities.

NELP regime has proved to be a success so far in terms of accelerating the pace of exploration in the country, deep water oil and gas production, infusion of technology and investment in E&P Sector, number of discoveries, etc.

The exploratory works relating to seismic survey and exploratory drilling in many of the NELP blocks are still going on in accordance with Production Sharing Contract (PSC). Once these activities are undertaken, there are likely to be more hydrocarbon discoveries. Further, the development of existing discoveries will also increase oil and gas production in the country.

Following other benefits have accrued under NELP:

(i) Increasing participation of private companies in NELP bidding rounds.
(ii) India has joined the list of global deepwater players.
(iii) India has been recognized as one of the desired destination for E&P activities.
(iv) The earlier perception about low prospectivity of Indian Sedimentary Basins has been changed globally.
(v) Offering of "S" type on-land blocks from NELP–VII onwards, having area of less than 200 sq. km, has enabled many new and non-E&P companies to bid and win the blocks, thereby helping to develop in-country expertise in E&P sector

4 Hydrocarbon Exploration Licensing Policy

Initial NELP rounds were quite successful; however, the latest rounds were far below the expectation. Due to various reasons including pricing mechanism of natural gas, contentious cost recovery policy and its interpretation; the Government explored the possibility of introducing more investor friendly exploration policy in the country. To further the development of E&P sector, the Union Cabinet, chaired by the Prime Minister Shri Narendra Modi, approved the Hydrocarbon Exploration and Licensing Policy (HELP) in March 2016. Key features of HELP[6] are as follows:

(a) uniform licence for exploration and production of all forms of hydrocarbon including unconventional oil and gas resources such as Coal bed methane, Shale gas/oil, Tight gas, Gas hydrates and any other resources to be identified in future which fall within the definition of "Petroleum" and "Natural Gas" under Petroleum and Natural Gas Rules, 1959,
(b) an Open Acreage Licensing Policy, which allows E&P companies submit an initial expression of interest for blocks on offer,
(c) easy-to-administer revenue-sharing model compared to complicated production sharing and cost recovery linked payment model and
(d) marketing and pricing freedom for the crude oil and natural gas produced.

[6]PIB (2016)

The new policy regime is designed to:

- enhance domestic oil and gas production to improve energy supply security
- bring substantial investment in the sector
- generate sizable employment
- enhance transparency and reducing administrative discretion, thereby increasing participation of global and domestic energy companies

The HELP is designed to improve bidding for designated areas throughout the year in a very transparent manner. Generally higher risks and costs involved in exploration and production from offshore areas. Therefore, for encouraging greater participation in offshore exploration and production activities in the country, provisions for lower royalty rates (Table 9) for such areas have been created under HELP. Introduction of graded system of royalty rates with decreasing order from shallow water to deep water and ultra-deep water is intended to incentivize investments in risky but equally rewarding projects.

At the same time, royalty rate for on-land areas have been kept intact so that revenues to the state governments are not affected. On the lines of NELP, custom duty on all machineries, plants, equipment, materials and supplies to petroleum operations will be exempted on blocks awarded under the HELP. Further, no Cess will be levied on crude oil. In addition, HELP increases exploration phase for onshore areas from 7 to 8 years and for offshore from 8 to 10 years. The contractor will be allowed to carry out exploration activity throughout the contract period at its own risk and cost.

HELP offers much debated and desired marketing freedom to the producers to sale their crude oil and natural gas produced from news allocated blocks. This is in tune with Government's policy of "Minimum Government—Maximum Governance".

The HELP guidelines would be applicable to future and existing discoveries which are yet to commence commercial production as on 1.1.2016. But this would not be applicable to existing discoveries which are under arbitration for gas price.

Table 9 Rate of royalty under NELP & HELP

Blocks	Duration	NELP royalty rates		HELP royalty rates	
		Oil (%)	Gas (%)	Oil (%)	Gas (%)
On-land		12.5	10	12.50	10
Shallow water		10	10	7.50	7.50
Deepwater	First seven years	5	5	Nil	Nil
	After seven years	10	10	5	5
Ultra-deep water	First seven years	5	5	Nil	Nil
	After seven years	10	10	2	2

Source Ministry of Petroleum and Natural Gas, India (http://www.petroleum.nic.in/docs/HELP.pdf accessed on 11 May 2016, http://www.petroleum.nic.in/docs/imppolicy.pdf accessed on 11 May 2016)

5 Future Ahead

Government of India approved Hydrocarbon Exploration Licensing Policy (HELP) to replace NELP and CBM regime. Now new policy enables E&P operators to explore and extract all hydrocarbon resources covered under the Oilfields Regulation and Development (ORD) Act, 1948, and Petroleum and Natural Gas (PNG) Rules, 1959 under one Petroleum Exploration Licence/Petroleum Mining Lease. Figure 1 presents progressive evolution of exploration and production regime in the country (Fig. 5).

The uniform licence enables the contractor to explore conventional and unconventional oil and gas resources including CBM, shale gas/oil, tight gas, gas hydrates and any other resource to be identified in future which fall within the definition of "Petroleum" and "Natural Gas" under PNG Rules, 1959.

Under Uniform Licensing Policy (ULP), a Revenue-Sharing model based on an incremental production-based sliding scale combined with a fixed, price-sensitive scale is envisaged. The contractor is required to pay biddable Government share of revenue (net of royalty) as per Revenue-Sharing Contract (RSC). Revenue, net of royalty, in case of on-land (including CBM) blocks and total revenue in case of offshore blocks will be shared between the contractor and the Government, based on the average daily production in a month for oil and gas, using a sliding scale calculation methodology.

It is envisaged that companies are required to bid for the Government share of revenue, as per the production–price matrix provided. The production and price bands will be pre-determined and specified in the bid document for each block to enable the bidders to bid appropriate revenue shares for each cell in the matrix.

ULP ensures exploration and production of all kinds of "petroleum" and "natural gas" under one licence in the awarded acreage. This removes impediments on

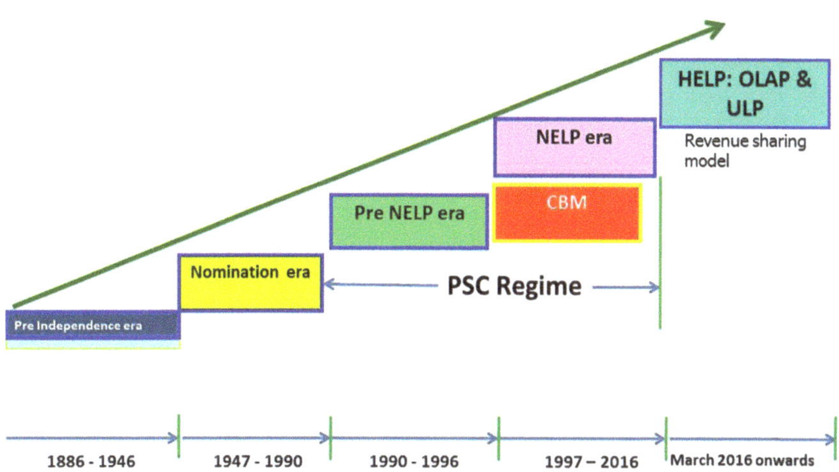

Fig. 5 Future of E&P policy in India. *Source* Authors analysis

account of multiple operators, thus, expedites exploration. Contractor will have incentive in keeping the costs low and maximize production. This is in line with the Government's broad objectives of efficiency in oil field operations and optimizing scarce hydrocarbon resources.

The new system is simple and transparent with easy-to-monitor parameters of production and price. Management Committee (MC) will have more time to focus on monitoring of technical aspects for effective exploration and optimal exploitation of reservoirs, as monitoring of budget and expenditure will not be required, since no profit computation is involved.[7] The Revenue-sharing model addresses gold plating and windfall profit. The Revenue-Sharing model would create "Win-Win" situation for the Government and the Contractor. The regulator and the Government certainly could save time and money by reducing conflicts and legal issues.

Option of bidding the revenue share at various production levels and oil price tranches allows the bidder to factor in the fiscal terms of contract in view of the market conditions. There is no minimum government share prescribed, and the bidder is free to bid any non-zero share.

Under HELP, the blocks will be awarded based on a maximum of 100 points allocated to following biddable components:

- revenue share offered to the Government (maximum 50 points)
- committed minimum work programme (maximum 50 points)

The new model a royalty—tax regime driven by commercial production-based payment. Government share arrived at through competitive bidding increases with the output and price rise, for the government take.

The HELP will lead to simple and transparent administration with easy-to-monitor parameters of production and price. The new system enables both the contractor and Government to generate revenue from the date of commercial production. In the new system "Government Take" would be progressive with respect to field sizes and with respect to prices. Contractor will be able to explore all types of hydrocarbon within the awarded acreage. Under HELP, there is no provision of cost recovery, so the Contractor will be allowed to carry out exploration in mining lease area. New policy addresses windfall profits accruing to Contractor in case of price upheaval.

References

DGH (2014–15) Hydrocarbon exploration and production activities 2014–15. Available at http://www.dghindia.org/pdf/2014-15.pdf. Accessed on 3 Feb 2016

DGH (2016) Bidding rounds. Available at http://www.dghindia.org/NELPRounds.aspx?tab=0. Accessed on 3 Febr 2016

[7]http://eac.gov.in/reports/rep_psc0201.pdf accessed on 11 May, 2016.

Kar SK, Sinha PK (2014) Ensuring sustainable energy security: challenges and opportunities for india. Oil Gas Energy Law Intell 12(4)

PIB (2016) Hydrocarbon Exploration and Licensing Policy (HELP) 10-March-2016 14:50 IST http://pib.nic.in/newsite/PrintRelease.aspx?relid=137638. Accessed on 12 Mar 2016

Downstream Petroleum and Natural Gas Regulatory Developments in India

Sanjay Kumar Kar

Abstract The downstream regulatory environment is passing through developmental stages in India. This chapter covers historical and recent downstream regulatory developments in India. A sincere attempt is made to discuss about vision, functions, and operations of the Petroleum and Natural Gas Regulatory Board (PNGRB). This chapter highlights all the important regulations developed since the inception of the PNGRB and authorizations granted to entities as per the regulations. This chapter discusses the challenges faced by the regulator to develop efficient natural gas market in India. It is concluded that the regulator may be empowered to regulate the entire petroleum sector.

Keywords PNGRB · Natural gas · Regulations · Downstream · Efficient

1 Introduction

Regulations, also called administrative laws or rules, are the primary vehicles by which the federal government implements laws and agency objectives (Dudley and Brito 2012).[1] These specific standards or instructions governs what individuals, businesses, and other organizations can or cannot do.

Regulation is a binding legal norm created by a state organ that intends to shape the conduct of individuals and firms.[2] The state organ, the regulator, may be any legislative, executive, administrative, or judicial body that has the legal power to create a binding legal norm.[3] Normally regulations were developed and administered by government or its agencies or independent regulators with adequate

[1]Dudley and Brito (2012).
[2]Baron (1989).
[3]Orbach (2012).

S.K. Kar (✉)
Department of Management Studies, Rajiv Gandhi Institute of Petroleum Technology, Jais, India
e-mail: skar@rgipt.ac.in; karsanjay1@gmail.com

S.K. Kar and A. Gupta (eds.), *Natural Gas Markets in India*,
DOI 10.1007/978-981-10-3118-2_12

authorities as assigned by the competent body. Often the regulators may be independent regulatory body with no interference from the political system.

Regulations are required for competitive market development and effective delivery of products and services. Often regulations desired to remove monopoly[4] and bring fair competition for the benefit of all stakeholders. The Office of Information and Regulatory Affairs, USA, estimates the cost of major regulations issued from 2001 to 2011 at \$43–\$67 billion per year and the benefits at \$141–\$700 billion per year (Dudley and Brito 2012). This is a clear indication that regulations are better for the society, business, and economic development.

In the literature, regulations are divided into two main categories: social regulations and economic regulations. Social regulations address issues related to health, safety, security, and the environment, whereas the economic regulations are often industry-specific.

The "public interest" theory of regulation[5] recognizes that markets are very efficient at allocating scarce resources to their best uses. However, there are many possibilities of suboptimal use of scarce resources resulting in inefficiency and market failure. In other words, market failure is a situation where scarce resources are not put to their highest valued uses.[6]

Dudley and Brito (2012) highlight that the concept of market failure comes from the economic paradigm of a perfect competitive market, characterized by (1) private decisions, such that individual choices do not affect the welfare of others; (2) private goods; (3) market participants who are price takers; and (4) perfect information about price and quality.[7]

The market failure may fall into one of the following four categories:

- Externalities: externalities may occur due to action of one party resulting in uncompensated costs or benefits on another party. Under such circumstances, underutilization or undue exploitation of resources may occur.
- Public goods: public goods are those for which the cost of providing an additional unit is negligible and excluding users is costly (violating the "private goods" condition).
- Monopoly power: the presence of monopoly power in a market allows a firm to control prices, so resulting in market distortion. In the case of monopolistic market condition, the consumer is likely to pay more than competitive market price. In the past, monopoly power has been seen in infrastructure like natural gas and petroleum product pipelines business.
- Asymmetric information: asymmetric information may result in inefficient allocation of resources, thereby creating chances for market failure.

[4]https://www.djei.ie/en/Publications/Publication-files/Forf%C3%A1s/Sectoral-Regulation.pdf accessed on 9 May 2016.

[5]Richard (1971).

[6]Johan de Hertog (2010).

[7]Theories of regulation, Why do we regulate? Dudley and Brito (2012).

Fig. 1 Classification of regulation

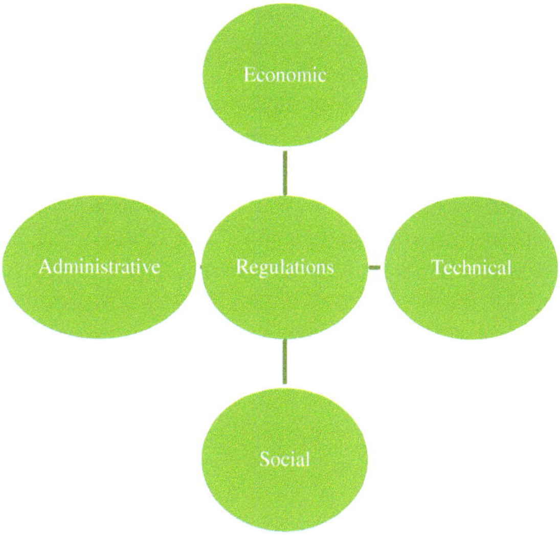

Further, regulations can be categorized as economic, technical, social, and administrative (Fig. 1).

(i) Economic Regulations: these regulations are meant for regulating market decisions such as pricing, competition, and entry/exit. These regulations affect entry, exit, and operations of entities.

(ii) Technical Regulations: these regulations regulate the technical aspects, which are distinct and unique to the sector. The technical regulations designed to improve technical standards of products or services, thereby reduce the chances of product or service failure and enhance customer satisfaction.

(iii) Social Regulations: these regulations are designed to protect public interest such as health, safety, and environment. In the absence of such regulations, environmental externalities and market failure may be rampant.

(iv) Administrative Regulations: administrative formalities through which government collects information and intervenes in individual economic decisions. Often the administrative regulations protect citizens from unauthorized encroachment on their rights and interests and make public servants accountable to people.[8]

Regulations may have positive or negative effect on business operations. For example, Greenstone et al. (2012) findings suggest that stricter air quality regulations are associated with a roughly 2.6% decline in total factor productivity (TFP) in the USA. Greenstone et al. (2012) suggest that the regulations governing ozone

[8]http://www.lawnotes.in/Administrative_Law accessed on 9 May 2016.

have particularly large negative effects on productivity, though effects are also evident among particulates and sulphur dioxide emitters.

Traditionally, government interventions need to correct market failures. However, over a period, it is observed that government interventions are not timely or yield limited positive results. Therefore, government initiated many reform measures to set up regulators to ensure competitive and level playing fields for all the stakeholders. The regulators are required to deliver decisions with transparent, consultative, and participatory approaches. Regulators such as Securities and Exchange Board of India (SEBI), and Telecom Regulatory Authority of India (TRAI) have established their credentials in their domain.

India started developing regulatory institutions with the introduction of reforms in 1991. Singh and Mitra (n.d) argue that the regulatory environment that has developed over a period does not seem homogeneous across sectors. Over the last two decades, many regulatory reforms are carried out by the government. Such regulatory reforms have been very much beneficial for the investors, consumers, and government. Some of the sectors immensely benefited are telecommunication, financial services, insurance, and petroleum and natural gas.

It is to be noted that not all sectors have regulators, so the government continues to perform the desirable regulatory function in those sectors without regulators. In addition, Competition Commission is established through Competition Act 2002 to enforce national competition laws. Competition authorities have a regulatory mandate over competition issues that cover all the sectors of the economy. The sector-specific regulators and the Competition Commission share a common goal of protecting and enhancing social/economic welfare (CCI 2012). The Competition Act, 2002, envisages[9] prohibition of anti-competitive agreements as well as abuse of dominant position (Table 1).

There are very many sectors with many complexities and operational challenges. These sectors need sector-specific regulators for better development and fair competition. The Competition Commission of India (CCI) may not be equipped with sufficient technical knowledge to handle the emerging challenges of all sectors. There is a provision for CCI getting knowledgeable experts to handle issues pertinent to any specific sector. The sector-specific regulators deploy the preventive measures, whereas the CCI deploys curative measures. So, the sector-specific regulators can complement the CCI for efficient functioning of the sector and the economy. There has to be an interface between the sector regulators and the CCI mechanism to avoid duplicity of efforts.

[9]http://www.cci.gov.in/sites/default/files/cci_pdf/competitionact2012.pdf accessed on 9 May 2016.

Table 1 Comparison between sector-specific regulator and competition authority

Sector-specific regulator	Competition authority
• Tells businesses "what to do" and "how to price products"	• Tells businesses "what not to do"
• Focuses upon specific sectors of the economy E.g. Petroleum and Natural Gas Regulatory Board (PNGRB) for Petroleum and Natural Gas Sector	• Focuses upon the entire economy and functioning of the market
• E.g. ante-sector-specific regulators create provisions and develop regulations to avoid potential behavioural issues among the stakeholders	• E.g. post-addresses behavioural issues after problem arises
• Focus upon orderly development of a sector to ensure fair competition, level playing field, and consumer welfare	• The competition authority prevents practices having adverse effect on competition, promotes and sustains competition in markets, protects the interests of consumers, and ensures freedom of trade carried on by other participants in markets
• Sectorial regulators are usually more appropriate for access, infrastructure development, network tariff, reducing entry barrier, prevention of discriminatory tariff, and opening up market, and resolving price issues	• Competition legislation is usually more appropriate for affecting conduct and maintaining competition
• Often determines fair return for the industry	• Prohibition of abuse of dominant position

Source PNGRB, Competition Commission of India, Telecom Regulatory Authority of India

2 Downstream Regulatory Developments in India

Regulation and policy: the under-developed natural gas market in India faces multifaceted challenges including policy framing, implementation, and market regulation. Limited availability of domestic natural gas along with domestic gas pricing influences most of the policy decisions. Domestic gas prices still attract critic's attention very regularly. Of late, the final price chargeable to consumer attracted attention of various stakeholders such as marketing entities, regulator, government, and finally the judiciary. As the natural gas market in India is under development stage, gas infrastructure developments and wider penetration requires more invest. Greater participation of private companies expected to complement the public sector companies to develop the market.

Deloitte (2013) observed that equal treatment to private/foreign players, transparent and fair bidding, market driven pricing and only need-based interventions from state are certain aspects, which are considered by investors to make investment decisions. The gas market needs to be fiercely competitive, therefore need for sector-specific regulator and stronger regulations. The Petroleum and Natural Gas Regulatory Board (PNGRB) Act, 2006, is a step in the right direction.

2.1 Historical Perspective

Over the years, the government has framed various Acts to provide directions, regulatory framework, and operating guidelines for the petroleum and natural gas industry in the country. Some of the Acts are pre-independence and some are post-independence. Some of the prominent Acts are presented below to build on historical perspective on regulatory developments in the country.

- The Petroleum Act,[10] 1934
- The Oil Fields (Regulation and Development) Act,[11] 1948
- The Petroleum and Natural Gas Rules,[12] 1959
- The Petroleum Pipelines Act,[13] 1962
- The Oil Industry Act,[14] 1974
- The Petroleum Rules,[15] 1976
- The Petroleum Rules,[16] 2002
- Petroleum and Natural Gas (Amendment) Rules,[17] 2003
- Petroleum and Natural Gas Regulatory Board Act,[18] 2006
- Petroleum and Natural Gas (Safety in Offshore Operations) Rules,[19] 2008
- Petroleum and Natural Gas (Amendment) Rules,[20] 2009
- The Petroleum (Amendment) Rules,[21] 2011

The Petroleum Act, 1934, was enacted to control issues related to import, transport, storage, production, refining, and blending of petroleum. As per this Act, "petroleum" means any liquid hydrocarbon or mixture of hydrocarbons and any inflammable mixture (liquid, viscous, or solid) containing any liquid hydrocarbon.

The Oil Fields (Regulation and Development) Act, 1948, was enacted to provide regulations of the oil fields and development of mineral oil resources. Under this Act, mineral oils include natural gas and petroleum.

---

[10]http://www.petroleum.nic.in/lrbody.htm accessed on 5 May 2016.

[11]http://petroleum.nic.in/docs/exp/circulars%20&%20notifications66.pdf accessed on 5 May 2016.

[12]http://www.petroleum.nic.in/pngrules.pdf accessed on 5 May 2016.

[13]http://www.petroleum.nic.in/docs/ref/Petroleum&MineralPipelines(P&MP)Act,1962andAmendmentsthereto.pdf accessed on 9 May 2016.

[14]http://www.petroleum.nic.in/oilind74.pdf accessed on 9 May 2016.

[15]http://www.petroleum.nic.in/lrrules.htm accessed on 9 May 2016.

[16]http://www.petroleum.nic.in/Rules2002.pdf accessed on 9 May 2016.

[17]http://www.dghindia.org/pdf/PNG_AMENDMENT_RULES_2003.pdf accessed on 9 May 2016s.

[18]http://www.petroleum.nic.in/regbill.pdf accessed on 9 May 2016.

[19]http://www.petroleum.nic.in/rule.pdf accessed on May 2016.

[20]http://www.petroleum.nic.in/gaz2009.pdf accessed on 9 May 2016.

[21]http://www.petroleum.nic.in/amd2011.pdf accessed on 9 May 2016.

The Oil Fields (Regulation and Development) Act, 1948 empowers the Central Government to make:

- regulations for granting or prohibiting mineral leases of mineral oil or any area;
- the fixing of the maximum and minimum rent payable by a lessee, whether the mine is working or not;
- the maximum or minimum area and the period for which any mining lease may be granted and the terms on which leases in respect of contiguous areas may be amalgamated; and
- rules for the conservation and development of mineral oils.

The Petroleum and Natural Gas Rules, 1959, provided regulatory framework for domestic exploration and production of oil and gas.

The natural gas sector is under transformation stage and smooth transition from liquid fuel to natural gas-driven economy is underway. Growth of natural gas can be ascribed to many global and domestic factors. Among the domestic factors, increased production of domestic gas due to some large-scale discoveries is considered as one of the most important factors. The New Exploration and Licencing Policy (NELP) led to big natural gas discoveries in KG basin in the East Coast of India. This resulted in the development of new infrastructure and consumer base in various demand centres in the country.

Some of global factors like availability of relatively cheaper natural gas in the international markets especially in the Middle East and North Africa influenced India's decision to import natural gas. Therefore, the need for developing necessary infrastructure in terms of LNG receiving terminal and liquefaction facility emerged. This resulted in the development of LNG terminal in the west coast of the country.

Owing to the emerging petroleum and natural gas market developments and competition, need for a regulator is felt. This resulted in set-up of the Petroleum and Natural Gas Regulatory Board (PNGRB) through Petroleum and Natural Gas Regulatory Board Act, 2006. The President of India has given assent to The PNGRB Act, 2006, on 31 March 2006 and was published in the official gazette on 3 April 2006.

Major objectives of PNGRB include ensuring fair trade and competition, protection of consumer's interest, and development of infrastructure. The PNGRB Act borrows the concept of "restrictive trade practice" from the Monopolies and Restrictive Trade Practices Act, 1969.

Prior to 28 March 2002, marketing and pricing of petroleum products, including transportation fuels, namely motor spirit (MS) and high-speed diesel (HSD), was controlled by the government under a mechanism known as administered price mechanism (APM). The APM was dismantled by a notification[22] on 28 March 2002. A market-determined pricing mechanism for motor spirit/petrol and diesel was established. Even the gas sector had similar kind of pricing mechanism—which

[22]http://pib.nic.in/archive/releases98/lyr2002/rmar2002/28032002/r280320029.html accessed on 13 May 2016.

has been moving towards market driven pricing. Recent Hydrocarbon Exploration Licensing Policy offers greater pricing freedoms to contractors/producers of natural gas. In a competitive environment, competitors generally try to marginalize/kill competition. Therefore, the PNGRB has an important role to play by discouraging anti-competitive behaviour in petroleum and natural gas sector.

2.2 *Current Status*

The PNGRB has been set up as an independent sector-specific regulator to ensure fair trade and competition in the petroleum and natural gas sector. Since its inception until date, the Board focused mostly on the gas market development. However, the Board is gradually developing desired regulations to regulate mid-stream and downstream petroleum business. So far, the PNGRB notified 8 regulations on CGD network and 9 regulations related natural gas pipeline. Tables 2 and 3 present the regulations and subsequent amendments in detail. The regulator notified three important regulations related to petroleum product pipelines (Table 4). General understanding is that stable and transparent regulatory environment propels the growth of any sector. Therefore, the future natural gas market developments would depend on the role clarity and limitations of regulators.

The PNGRB notified the following regulations under other categories:

- Petroleum and Natural Gas Register, GSR 481(E): 07/06/2010
- Codes of Practices for Emergency Response and Disaster Management Plan (ERDMP), GSR 39(E): 18/01/2010
- Procedure for Development of Technical Standards and Specifications Including Safety Standards, GSR 323(E):14/05/2009
- Appointment of consultants, GSR 36(E):30/11/2007
- Conduct of business, receiving and investigation of complaints, GSR 744(E): 30/11/2007
- Meetings of the Board and levy of fees and other charges, GSR 731(E): 26/11/2007 and GSR 732(E):26/11/2007

The PNGRB has been actively pursuing the objective of ensuring greater natural gas penetration by authorizing number of CGD geographical areas across the country. So far, the PNGRB authorized 32, 17, and 10 CGD networks under the Regulations 5, 17, and 18 (1) respectively. A brief detail of the authorization, targets, marketing exclusivity, and performance guarantee is presented in Tables 5, 6, and 7. As of March 2016, PNGRB has authorized:

- 32 entities under Regulation 5;
- 17 entities under Regulation 17;
- 10 entities under Regulation 18 (1).

Table 2 PNGRB notified regulations on CGD network

No.	Title of regulation	Notified date	Amendments
1	Determining capacity of city or local natural gas distribution network	Infra/PNGRB/Capdet/ CGD/02/2015 19/05/2015	NONE
2	Integrity management system for city or local natural gas distribution networks	INFRA/IMP/CGD/1/2013 16/05/2013	NONE
3	Access code for city or local natural gas distribution networks	S-Admn./II/8/2010 29/03/2011	L-MISC/VI/I/2007 01/01/2015
4	Code of practice for quality of service for city or local natural gas distribution networks	GSR 720(E) 01/09/2010	NONE
5	Technical standards and specifications including safety standards for city or local natural gas distribution networks	GSR 612(E) 27/08/2008	L-MISC/VI/I/2007 01/01/2015 M(I)/T4S/CGD/1/2010 16/12/2014 GSR 750(E) 14/10/2009
6	Authorizing entities to lay, build, operate, or expand city or local natural gas distribution network	GSR196(E) 19/03/2008	PNGRB/CGD/Amendment/2015/2 11/12/2015 PNGRB/CGD/REGULATIONS/AMEND-2015 13/02/2015 L-MISC/VI/I/2007 01/01/2015 PNGRB/CGD/BID/4/2013-PRE BID. 07/04/2014 PNGRB/CGD/REGULATIONS/REVIEW-2011/2012-III 21/06/2013 GSR605(E) 19/07/2010 GSR 478(E) 07/06/2010

(continued)

Table 2 (continued)

No.	Title of regulation	Notified date	Amendments
			GSR 295(E) 30/04/2009 GSR 800(E) 19/11/2008
7	Exclusivity for city or local natural gas distribution network	GSR 198(E) 19/03/2008	L-MISC/VI/I/2007 01/01/2015 GSR604(E) 09/07/2010
8	Determination of network tariff for city or local natural gas distribution networks and compression charge for CNG	GSR 197(E) 19/03/2008	L-MISC/VI/I/2007 01/01/2015 M(C)/2012 21/03/2012 GSR 479(E) 07/06/2010 GSR 801(E) 19/11/2008

Source Compiled from PNGRB website

Table 3 PNGRB notified regulations on natural gas pipelines

No.	Title of regulation	Notified date	Amendments
1	Imbalance management services	PNGRB/M(C)/48 17/02/2014	NONE
2	Integrity management system for natural gas pipelines	Infra/IM/NGPL/1/2010 05/11/2012	NONE
3	Determining capacity of petroleum, petroleum products, and natural gas pipeline	GSR 476(E) 07/06/2010	L-MISC/VI/I/2007 01/01/2015
4	Technical standards and specifications including safety standards for natural gas pipelines regulations, 2009	GSR 808(E) 11/11/2009	L-MISC/VI/I/2007 01/01/2015 REG/T4S/NGPL/1/2012 13/03/2014 REG/T4S/NGPL/1/2012 03/09/2012
5	Guiding principles for declaring or authorizing natural gas pipeline as common carrier or contract carrier	GSR 273(E) 21/04/2009	L-MISC/VI/I/2007 01/01/2015
6	Determination of natural gas pipeline tariff	GSR 807(E) 20/11/2008	L-MISC/VI/I/2007 01/01/2015 PNGRB/M(C)/100 27/02/2014 PNGRB/M(C)/48 17/02/2014 PS/Secy./M(C)/2012 13/09/2012 PNGRB/M(C)/11/Final Tariff Filing 30/05/2012 GSR 986(E) 20/12/2010

(continued)

Table 3 (continued)

No.	Title of regulation	Notified date	Amendments
7	Affiliate code of conduct for entities engaged in marketing of natural gas and laying, building, operating, or expanding natural gas pipeline	GSR 540(E) 17/07/2008	L-MISC/VI/I/2007 01/01/2015 PNGRB/M(C)/42 19/02/2014
8	Access code for common carrier or contract carrier natural gas pipelines	GSR 541(E) 17/07/2008	L-MISC/VI/I/2007 01/01/2015 PNGRB/M(C)/48 17/02/2014 GSR 274(E) 21/04/2009
9	Authorizing entities to lay, build, operate, or expand natural gas pipelines	GSR 340(E) 06/05/2008	PNGRB/NGPL/REGULATIONS/AMEND-2015 22/05/2015 L-MISC/VI/I/2007 01/01/2015 PNGRB/REGULATIONS/AMEND-2014 08/08/2014 PNGRB/M(C)/48 17/02/2014 GSR 594(E) 09/07/2010 GSR 480(E) 07/06/2010 GSR 38(E) 18/01/2010 GSR 769(E) 20/10/2009 GSR 802(E) 19/11/2008

Source Compiled by the author from published sources including PNGRB website

Table 4 PNGRB notified regulations on petroleum product pipelines

No.	Title of regulation	Notified date	Amendments
1	(Guiding principles for declaring or authorizing petroleum and petroleum products pipelines as common carrier or contract carrier) regulations, 2012	M(C)/2009 05/07/2012	L-MISC/VI/I/2007 01/01/2015
2	Determination of petroleum and petroleum products pipeline transportation tariff	GSR987(E) 20/12/2010	L-MISC/VI/I/2007 01/01/2015 PNGRB/M(C)/62/2014 16/12/2014 PNGRB/M(C)/62/2012 13/12/2012
3	Authorizing entities to lay, build, operate, or expand petroleum and petroleum products pipelines	GSR722(E) 01/09/2010	L-MISC/VI/I/2007 01/01/2015

Source Compiled by the author from published sources including PNGRB website

Regulation 5: as per this regulation, the Board is empowered for selection of entity for expression of interest route and authorize any entity after due diligence as per the regulations. These regulations may be called the Petroleum and Natural Gas Regulatory Board (authorizing entities to lay, build, operate, or expand city or local.

Natural Gas Distribution Networks) Regulations, 2008.

As per the Regulation 10 (3), the grant of authorization to the entity shall not be renunciated by way of sale, assignment, transfer, or surrender to any person or entity during the period of five years from the date of its issue or till the achievement of Minimum Work Programme, whichever is earlier from the date of its issue. Initially, the restriction on renunciation was just for three years.

Regulation 17: as per this regulation, an entity authorized by the central government for laying, building, operating, or expanding CGD network before the appointed day needs to apply for authorization from PNGRB.

Regulation 18 (1): an entity laying, building, operating, or expanding a CGD network at any time before the appointed day but not duly authorized to do so by the Central government shall apply immediately for obtaining an authorization in the form as at Schedule I.

2.3 *Laws Pertaining Oil and Gas Industry*

2.3.1 Upstream Sector

The Directorate General of Hydrocarbons (DGH) regulates the upstream sector in India. Many experts in the industry including some of the officers in the DGH consider the DGH as a technical wing of Ministry of Petroleum and Natural Gas (MoPNG). The DGH develops policies relevant to exploration and production of

Table 5 PNGRB-authorized CGD networks under Regulation 5

Name of the CGD network	Area covered	State	Entity authorized	Authorised date of under: authorization	Date of authorization	Marketing exclusivity	Performance bank guarantee	Target at the end of marketing exclusivity		
						Months from the date of issue of performance guarantee	Million Rs.	Cumulative domestic PNG connections (,000)	CNG compression capacity (kg/day)	Steel pipeline (in. km)
1. Allahabad	Allahabad	UP	M/s IOCL-AGL	Regulation 5	8/5/2013	60	30	225		480
2. Amritsar	Amritsar	Punjab	GSPL	Regulation 5		60	22,120	24.44		953.28
3. Bengaluru rural and urban		Karnataka	Gail Gas Limited	Regulation 5	18/2/2015	60	51,999.92	131.15	NA	1582.2
4. Belgaum	Belgaum	Karnataka	MEIL	Regulation 5	14-Sep-15	60	176	49.19	NA	1800
5. Bhavnagar	Bhavnagar	Guajarat	M/s Gujarat Gas Company Ltd.	Regulation 5	5/3/2014	60	30	122.5	NA	2695.73
6. Chandigarh	Chandigarh	UT	M/s IOCL-AGL	Regulation 5	8-May-13	60	30	50	NA	1400
7. Daman	UT of Daman	UT	M/s IOCL-AGL	Regulation 5	1-Apr-15	60	2043.6	2.52	NA	33.15
8. Dewas	Dewas	MP	Gail Gas Limited	Regulation 5	1-June-09	60		40	NA	220
9. Dharwad	Dharwad	Karnataka	IOCL-Adani	Regulation 5	14-Sep-15	60	180	19.14	NA	1520
10. Dadra Nagar Haveli	Dadra Nagar Haveli	UT	GSPC	Regulation 5	1-Apr-15	60	2980	3.8	NA	320.4
11. East Godavari Dist.	East Godavari Dist.	Andhra Pradesh	CAH	Regulation 5	14-Aug-15	60	140	60.9	NA	1800
12. Ernakulam Dist.	Ernakulam Dist.	Kerala	IOCL-AGL	Regulation 5	14-Oct-15	60	42,480	40.7	NA	1099.44
13. Haridwar Ditst.	Haridwar Ditst.	Uttarakhand		Regulation 5	7/27/2015	60	200	16.9	NA	830
14. Jalandhar	Jalandhar	Panjab	M/s Jay Madhok Energy Pvt. Ltd.	Regulation 5	6-Sep-13	60	20	46.8	NA	7020

(continued)

Table 5 (continued)

Name of the CGD network	Area covered	State	Entity authorized	Authorised date of under: authorization	Date of authorization	Marketing exclusivity	Performance bank guarantee	Target at the end of marketing exclusivity		
						Months from the date of issue of performance guarantee	Million Rs.	Cumulative domestic PNG connections (,000)	CNG compression capacity (kg/day)	Steel pipeline (in. km)
15. Jamnagar	Jamnagar	Guajarat	GSPC Gas Company Limited	Regulation 5	17-Jan-14	60	30	92	NA	1049.04
16. Jhansi	Jhansi	UP	M/s Central UP Gas Ltd.	Regulation 5	26-Feb-14	60	15	0.62	NA	23.6
17. Kakinada	Kakinada	Andhra Pradesh	Bhagyanagar Gas Limited	Regulation 5	12-June-09	60	122.2	35	NA	92
18. Kota	Kota[a]	Rajasthan	Gail Gas Limited	Regulation 5	1-June-09	60	114.8	25	NA	343.6
19. Krishna District	Krishna District	Andhra Pradesh	MEIL	Regulation 5	14-Sep-15	60	150	50.8	NA	1800
20. Kutch	Kutch (East)	Guajarat	JMEL	Regulation 5	12-Mar-15	60	20		NA	
21. Kutch	Kutch (West)	Guajarat	GSPC Gas Company Limited	Regulation 5	19-Dec-14	60	20	41.5	NA	2768.3
22. Ludhiana	Ludhiana	Punjab	JMEL	Regulation 5	6-July-15	60	30	74.5	NA	2600
23. Mathura	Mathura	UP	JV of M/s DSM Infratech Pvt. Ltd. and M/s Saumya Mining Pvt. Ltd.	Regulation 5	11-June-09	60	97.6	57	NA	345
24. Meerut	Meerut	UP	Gail Gas Limited	Regulation 5	12-June-09	60	154	110	NA	643.18
25. Panipat	Panipat	Haryana	IOCL-AG	Regulation 5	1-Apr-15	60	4532	11.78	NA	650
26. Pune CGD	Pune Dist. CGD (excludes areas already notified)	Maharashtra	Mahesh Resources Pvt. Ltd.	Regulation 5	18-May-15	60	120	52.8	NA	1800

(continued)

Table 5 (continued)

Name of the CGD network	Area covered	State	Entity authorized	Authorised date of under: authorization	Date of authorization	Marketing exclusivity	Performance bank guarantee	Target at the erd of marketing exclusivity		
						Months from the date of issue of performance guarantee	Million Rs.	Cumulative domestic PNG connections (,000)	CNG compression capacity (kg/day)	Steel pipeline (in. km)
27. Raigar	Raigarh Dist. (exludes already notified areas)	Maharashtra	MGL	Regulation 5	1-Apr-15	60	1052	26.4		1800
28. Sonipat	Sonipat	Haryana	Gail Gas Limited	Regulation 5	12-June-15	60	1295	60		329
29. Thane	Thane Dist. (excludes earlier notified areas)	Maharashtra	GGCL	Regulation 5	1-Apr-2015	60	4120	51.45	NA	1800
30. Tumkur	Tumkur Dist.	Karnataka	MEIL	Regulation 5	14-Aug-2015	60	160	32	NA	1800
31. Udham Singh Nagar Dist.	Udham Singh Nagar Dist.	Uttarakhand	IOCL-AGL	Regulation 5	28-July-2015	60	192	15.4	NA	921
32. West Godavari Dist.	West Godavari Dist.	Andhra Pradesh	CAH	Regulation 5	14-Aug-15	60	140	54.57		1800

Note For Allahabad the targets are for 2017–18
[a]PNGRB approved transfer of authorization to JV of GAIL and RSPCL
Source Complied from the authorization grant documents available at PNGRB website

Table 6 Entities authorized by PNGRB under Regulation 17

Name of the CGD network	Area covered	State	Entity authorized	Date of authorization	Marketing exclusivity	Performance bank guarantee	Target at the end of marketing exclusivity		
					Months from the date of issue of performance guarantee	Million Rs.	Cumulative domestic PNG connections (,000)	CNG compression capacity (kg/day)	Steel pipeline (in. km)
1. Agartala	Agartala	Tripura	Tripura Natural Gas Limited	24/2/2015	NA	NA	NA	NA	NA
2. Agra	Agra	UP	Green Gas Limited	12/11/2009	60	60	22	1,080,000	388
3. Bareilly	Bareilly GA	UP	Central UP Gas Limited	22/4/2009	60	60	7	16,000	80
4. Delhi	Delhi CGD	Delhi and NCR	Indraprastha Gas Limited	9/1/2009	36	100	237	3,600,000	1100
5. Firozabad geographical area (Taj trapezium zone)	Firozabad, Fatepur sikri, Bharatpur, Govardhan, and Vrindavan geographical area	UP	GAIL Gas Ltd.	26/09/2011	36	30	24	92,260	1010
6. Gandhinagar, Sabarkanta, Mehsana	Ghandhinagar, Mehsana, Sabarkantha	Guajarat	Sabarmati Gas Limited	16/12/2009	60	100	99	280,800	1913
7. Gwalior	Gwalior	MP	Aavantika Gas Limited	4/1/2010	60	40	44	10,827	180
8. Hyderabad	Hyderabad	Telangana	Bhagyanagar Gas Limited	9/10/2009	60	100	266	849,836	1602
9. Indore	Indore and Ujjain	MP	Aavantika Gas Limited	31/8/2009	60	60	50	259,782	540

(continued)

Table 6 (continued)

Name of the CGD network	Area covered	State	Entity authorized	Date of authorization	Marketing exclusivity	Performance bank guarantee	Target at the end of marketing exclusivity		
					Months from the date of issue of performance guarantee	Million Rs.	Cumulative domestic PNG connections (,000)	CNG compression capacity (kg/day)	Steel pipeline (in. km)
10. Kanpur	Kanpur GA	UP	Central UP Gas Limited	22/4/2009	60	60	320	70,035	788.52
11. Kolkata	Kolkata Municipal Corporation, parts of 24 North Praganas, Howrah, Hooghly and Nadia	West Bengal	Greater Calcutta Gas Supply Corporation	2/2/2016	60		1417.96	790,200	7296
12. Lucknow	Lucknow	UP	Green Gas	15/3/2016	60		494.19	275,390	2543
13. Mumbai	Mumbai and Greater Mumbai	Maharashtra	Mahanagar Gas Limited	21/1/2009	36	100	406	1,795,000	1580
14. Mumbai CGD network (GA 2)	Thane city and adjoining contiguous areas including Mira Bhayender, Navi Mumbai, Thane city, Ambernath, Bhiwandi, Kalyan, Dombivily, Badlapur, Ulhasnagar, Panvel, Kharghar and Taloja	Maharashtra	Mahanagar Gas Limited	4/8/2009	60		225	574,000	1675

(continued)

Table 6 (continued)

Name of the CGD network	Area covered	State	Entity authorized	Date of authorization	Marketing exclusivity	Performance bank guarantee	Target at the end of marketing exclusivity		
					Months from the date of issue of performance guarantee	Million Rs.	Cumulative domestic PNG connections (,000)	CNG compression capacity (kg/day)	Steel pipeline (in. km)
15. Pune city including Pimpri Chichwad	Pune city including Pimpri Chiechwad and along with adjoining contiguous areas of Hinjewadi, Chakan and Talegaon GA	Maharashtra	Maharashtra Natural Gas Limited	1/6/2009	60	60	72,000	385,320	1313
16. Upper Assam	Upper Assam	Assam	Assam Gas Company Limited	6/2/2015	No exclusivity	NA	NA	NA	NA
17. Vijaywada GA	Vijaywada GA	Andhra Pradesh	Bhagyanagar Gas Limited	28/7/2007	60	40	97.267	227,000	635

Note For Delhi and Mumbai targets are for FY 2011–12. Kolkata and Lucknow: Milestone targets for the period of 2016–2021
Source Complied from the authorization grant documents available at PNGRB website

Table 7 PNGRB-authorized CGD networks under Regulation 18 (1)

Name of the CGD network	Area covered	State	Entity authorized	Authorised under	Date of authorization	Marketing exclusivity	Performance bank guarantee	Target at the end of marketing exclusivity		
						Months from the date of issue of performance gurantee	Million Rs.	Cumulative domestic PNG connections (,000)	CNG compression capacity (kg/day)	Steel pipeline (in. km)
1. Anand	Anand area including kanjari and Vadtal villages (Kheda)	Gujarat	Charotar Gas Sahakari Mandali Ltd.	Regulation 18 (1)	12/5/2015	No exclusivity	20	NA	NA	NA
2. Hazira	Hazira geographical area	Gujarat	GSPC Gas Company Limited	Regulation 18 (1)	17-July-14	60.48	5	5.48	1899	44
3. Khurja	Khurja geographical area	UP	Adani Gas Limited	Regulation 18 (1)	4-Dec-12		10	3.5	12,000	260
4. Morababad	Moradabad geographical area	UP	Siti Energy Limited	Regulation 18 (1)	30-Nov-12	60	20	20	29,000	300
5. Nadiad	Nadiad geographical area	Gujarat	GSPC Gas Company Limited	Regulation 18 (1)	1-Oct-13	60	30	58.58	72,300	336
6. Navsari	Navsari geographical area	Gujarat	GSPC Gas Company Limited	Regulation 18 (1)	1-Oct-13	60	30	67.19	64,140	345
7. Rajkot	Rajkot geographical area	Gujarat	GSPC Gas Company Limited	Regulation 18 (1)	2-July-13	60	30	166.63	196,160	1532

(continued)

Table 7 (continued)

Name of the CGD network	Area covered	State	Entity authorized	Authorised under	Date of authorization	Marketing exclusivity	Performance bank guarantee	Target at the end of marketing exclusivity		
						Months from the date of issue of performance gurantee	Million Rs.	Cumulative domestic PNG connections (,000)	CNG compression capacity (kg/day)	Steel pipeline (in. km)
8. Surat-Bharuch-Ankleshwar	Surat geographical area	Gujarat	GSPC Gas Company Limited	Regulation 18 (1)	8-Nov-12	36	53.3	405	475,000	4450
9. Surendranagar	Surendranagar geographical area	Gujarat	GSPC Gas Company Limited	Regulation 18 (1)	1-Oct-13	60	30	34	62,820	470
10. Valsad[a]	Valsad GA	Gujarat	GSPC Gas Company Limited	Regulation 18 (1)	20-Jan-15	60	30	115	86,760	802

Note
[a]Existing PNG domestic connections No.: (73503), steel pipeline (in. km): 507.84, compression capcity (kg/day): 166320. The target for Valsad is for the year 2020
Source Complied from the authorization grant documents available at PNGRB website

oil, natural gas, coal-bed methane (CBM), shale gas, and other unconventional sources of hydrocarbons. Unlike the PNGRB, DGH has limited scope, power, and authority.

Constitutional authority

The Constitution of India gives power to make laws related to petroleum and natural gas resources to the central and state governments. As per the Article 297 (1), "All lands, minerals and other things of value underlying the ocean within the territorial waters, or the continental shelf, or the exclusive economic zone, of India shall vest in the Union and be held for the purposes of the Union."[23]

As per the Territorial Waters, Continental Shelf, Exclusive Economic Zone and Other Maritime Zones Act, 1976, no exploration in the continental shelf and exclusive economic zone is permitted without central government's permission.[24]

Article 246, List-I (Union List)[25] entry 53 empowers the union of India to make laws and regulations for the development of oil fields and mineral oil resources, petroleum and petroleum products, and other liquids and substances declared by Parliament by law to be dangerously inflammable. Regulation of labour and safety in mines and oil fields comes under entry 55 of the same list. Natural gas being a petroleum product falls under the purview of the central government.

As per the Article 246, List-II (State list)[26] entry 25, "Gas and Gas Works" falls under the domain of States. Land, that is to say, rights in or over land, land tenures including the relation of landlord and tenant, and the collection of rents; transfer and alienation of agricultural land and land improvement comes under state list, entry 18.

As per the Oilfields (Regulation and Development) Act, 1948, the central government is empowered to make rules with regard to mining leases and development of mineral oil. The mineral oils include petroleum and natural gas. The mining lease defined exhaustively to cover all forms of exploring and exploiting mineral oils and all purposes connected thereto.

Petroleum and Natural Gas Rules, 1959, provides framework for grant of exploration licences and mining leases.

---

[23]http://lawmin.nic.in/olwing/coi/coi-english/coi-4March2016.pdf accessed on 9 May 2016.

[24]http://www.un.org/depts/los/LEGISLATIONANDTREATIES/PDFFILES/IND_1976_Act.pdf accessed on 9 May 2016.

[25]http://lawmin.nic.in/olwing/coi/coi-english/Const.Pock%202Pg.Rom8Fsss(35).pdf accessed on 9 may 2016.

[26]http://lawmin.nic.in/olwing/coi/coi-english/Const.Pock%202Pg.Rom8Fsss(35).pdf accessed on 9 may 2016.

Salient features of the Rules:

- Prohibition on prospecting and mining except under a licence or lease granted under the rules [Rule 4];
- Central government has the power to grant licences or leases in respect of any land vested with it or minerals underlying the ocean within the territorial waters or the continental shelf [Rule 5(i)];
- State government has power to grant licence or lease over lands vested with it [Rule 5(ii)]; and
- Person obtaining exploration licence obtains the exclusive right to a lease for producing (i.e. extracting) oil/gas over any part of area covered in licence.

2.3.2 Downstream

Land acquisition Act

The Land Acquisition (LA) Act, 1894, enacted during British rule was adopted by the Government of India in 1947. The LA Act has amended many times to improve its effectiveness. The LA Act deals with the acquisition of land for public purpose. The Act deals with the procedure and the conditions under which a land can be acquired and used.

The Constitution of India placed "Acquisition and requisitioning of property" as Entry 42 in the concurrent list. This meant that both the central and state governments could make laws governing land acquisition. However, in the case of a conflict between the central and state laws, the central legislation would prevail (Raghuram and Sunny 2015).

Right to Fair Compensation and Transparency in Land Acquisition, Resettlement and Rehabilitation (RFCTLARR) Act, 2013 repealed LA Act 1894. The RFCTLARR Act, 2013, was enacted[27] to ensure the land is acquired strictly for public welfare projects, and land owners are adequately compensated and rehabilitated.

The Land acquisition Act remains a very important Act for the oil and gas industry, especially downstream business. Of course, it is also equally important for the upstream business too. It is evident from the observation made by T K Ananth Kumar, Director (Finance), Oil India Limited, that the company is already facing some difficulties in land acquisition for their E&P activities (Cleantech 2013).

[27]http://dolr.nic.in/dolr/downloads/pdfs/Right%20to%20Fair%20Compensation%20and%20Transparency%20in%20Land%20Acquisition,%20Rehabilitation%20and%20Resettlement%20Act,%202013.pdf accessed on 9 May 2016.

The Petroleum Act, 1934

The Act deals with import, transport, storage, production, refining, and blending of petroleum. The Act is one of the oldest acts in the oil and gas sector. Earlier to this Act, the rules regarding the above-specified activities were separate for separate states. This Act brought about uniformity in this field.

As per the Petroleum Act, 1934, the central government may makes rules—(a) prescribing the conditions subject to which petroleum may be produced, refined, or blended; and (b) regulating the removal of petroleum from places where it is produced, refined, or blended and preventing the storage therein and removal there from, except as petroleum Class A[28] of any petroleum which has not satisfied the prescribed tests.

The Petroleum Minerals Pipelines (Acquisition of Right of users in Land) Act,[29] 1962

The Petroleum Minerals Pipelines (Acquisition of Right of users in Land) Act, 1962, was enacted to provide for the acquisition of Right of user (RoU) in land for laying pipelines for the transport of petroleum and any minerals.

On 20 November 2002, The Ministry of Petroleum and Natural Gas notified guidelines for laying petroleum product pipelines.[30] As per the above notification, the petroleum product pipelines would be categorized as follows:

i. Pipelines originating from refineries, whether coastal or inland, up to a distance of around 300 km from the refinery;
ii. Pipelines dedicated for supplying product to particular consumer, originating either from a refinery or from oil company's terminal; and
iii. Pipelines originating from refineries exceeding 300 km in length and pipelines originating from ports, other than those specified in (i) & (ii) above.

Right of user (RoU) in land for laying pipelines under the Petroleum Pipelines (Acquisition of Right of User in Land) Act 1962, for the pipelines falling under category specified in sub-clauses (i) & (ii) of clause 1, will be granted in favour of

[28]"Petroleum Class A" means petroleum having a flash-point below twenty-three degrees centigrade; "Petroleum Class B" means petroleum having a flash point of twenty-three degrees centigrade and above but below sixty-five degrees centigrade; "Petroleum Class C" means petroleum-having flash point of sixty-five degrees ("flash-point") of any petroleum means the lowest temperature at which it yields a vapour which will give a momentary flash when ignited, determined in accordance with the provisions of Chap. 2 of the Petroleum Act, 1934, and the rules made thereunder.

[29]http://www.petroleum.nic.in/docs/ref/Petroleum&MineralPipelines(P&MP)Act,1962andAmendmentsthereto.pdf accessed on 9 May 2016.

[30]http://www.petroleum.nic.in/docs/ref/Guidelinesforlayingproductpipelines.pdf accessed on 9 May 2016.

applicant company treating such pipelines as captive pipelines, i.e. for exclusive use by the proposer company.

Petroleum and Natural Gas Regulatory Board Act, 2006

This Act provide for the establishment of Petroleum and Natural Gas Regulatory Board (PNGRB) to protect the interests of consumers and entities engaged in specified activities relating to petroleum, petroleum products, and natural gas and to promote competitive markets and for matters connected therewith or incidental thereto.

3 PNGRB

As per the PNGRB Act 2006, the central government was empowered to appoint a Board called Petroleum and Natural Gas Regulatory Board and the first Board appointed on 1 October 2007 (PNGRB 2010). As per the PNGRB Act, 2006, the Board consists of a chairperson, a member (Legal), and three other members appointed by the central government.

The central government shall appoint the chairperson and other members of the Board from amongst persons of eminence in the fields of petroleum and natural gas industry, management, finance, law, administration, or consumer affairs: provided that no person shall be appointed as member (legal) unless he

(a) is qualified to be a Judge of a High Court; or
(b) has been a member of the Indian Legal Service and has held a post in Grade I of that service for at least three years.

3.1 Vision of PNGRB

PNGRB's vision is being developed to "create a vibrant energy market with rapid and orderly growth through facilitation of flow of investments into the basic infrastructure for efficient transportation and distribution of petroleum, petroleum products, and natural gas at minimum cost and high level of protection of consumer interests through fair trade practices and competition amongst the entities so as to ensure the enhanced competitiveness of Indian economy and customer satisfaction".

3.2 Functions of the Board

The Board has wide range of functions[31] that include the following:

- protection of consumers interest, fostering fair trade and competition among entities;
- registration of entities to market-notified petroleum products subject to the contractual obligation of the government; establish and operate liquefied natural gas terminals; establish storage facilities for petroleum products, including natural gas as per the regulations;
- authorise entities to (i) lay, build, operate, or expand a common carrier or contract carrier; (ii) lay, build, operate, or expand city or local natural gas distribution network;
- declare pipelines as common carrier or contract carrier;
- regulate, by regulations on common carrier or contract carrier, transportation rate for common carrier and contract carrier, access to city or local distribution network, technical specifications, and safety standards;
- other important functions include ensuring adequate supply of notified petroleum products, monitoring prices, and taking corrective actions;
- perform such other functions as may be entrusted to it by the central government to carry out the provisions of this Act.

3.3 Settlement of Disputes

The Act has an inbuilt provision to address if any disputes through a dispute settlement mechanism. A Bench comprising of the member (legal) and one or more members nominated by the chairperson of the PNGRB settles disputes between entities on matters relating to:

(a) refining, processing, storage, transportation, and distribution of petroleum, petroleum products, and natural gas by the entities;
(b) marketing and sale of petroleum, petroleum products, and natural gas including the quality of service and security of supply to the consumers by the entities; and
(c) registration or authorisation issued by the Board under Section 15 or Section 19.

The Bench has the same powers as a civil court. Disputes under the purview of the Consumer Protection Act, 1986, will not be taken up.

[31] http://www.pngrb.gov.in/Functions-of-Board.html accessed on 9 May 2016.

The Appellate Tribunal under the Electricity Act, 2003, serves as the Appellate Tribunal for this Act. The central government may appoint one or more technical members (petroleum and natural gas) on the Tribunal.

3.4 Offences and Punishments

The PNGRB Act, 2006, has many provisions for reducing/minimizing undesirable offences. Depending on the nature of offences, the Act recommends punishments or penalties ranging from financial to term imprisonment. Table 8 gives brief details about the nature of offences and recommended penalties or punishments.

Table 8 Provisions for punishment as the PNGRB Act, 2006

Sl. no.	Nature of offence	Punishment/penalties
1	Contravention of directions of the Board	• Fine which may extend to twenty-five crore rupees • Additional fine which may extend to ten lakh rupees every day during which the contravention continues
2	Wilful failure to comply with orders of Appellate Tribunal	• Fine which may extend to one crore rupees • Subsequent offence with fine which may extend to two crore rupees • Continuing contravention with additional fine which may extend to twenty lakh rupees for every day
3	Unauthorized activities	• Imprisonment which may extend to three years or with fine which may extend to twenty-five crore rupees or with both
4	Establishing or operating a liquefied natural gas terminal without registration	• Imprisonment for a term which may extend to three years or penalty of twenty-five crore rupees or with both
5	Laying, building, operating, or expanding a common carrier or contract carrier without authorisation	• An imprisonment for a term which may extend to three years or penalty of twenty-five crore rupees or with both
6	Wilful damages to common carrier or contract carrier	• Imprisonment which may extend to three years or with fine which may extend to twenty-five crore rupees or with both

Source PNGRB Act 2006

4 Important Regulations

4.1 Authorizing Entities to Lay, Build, Operate, or Expand City or Local

Natural Gas Distribution Networks Regulations, PNGRB (2008)

- These regulations shall apply to the following:

  I. An entity which is laying, building, operating, or expanding or which proposes to lay, build, operate, or expand a CGD network; or
  II. A CGD network shall be designed to operate at a pressure as specified in the relevant regulations for technical standards and specifications, including safety standards for maintaining the volumes of supply of natural gas on a sustained basis to meet the following requirements, namely:

    (a) customers having requirement of natural gas up to 50,000 SCMD shall be supplied through the CGD network;
    (b) customers having requirement of natural gas more than 50,000 SCMD and up to 100,000 SCMD shall be supplied:

      i. through the CGD network; or
      ii. through a pipeline not forming part of the CGD network;

    (c) customers having requirement of natural gas more than 100,000 SCMD shall be supplied through a pipeline not forming part of the CGD network.

As per the above regulation, an entity desirous of laying, building, operating, or expanding a CGD network shall submit an expression of interest to the Board in the form of an application at Schedule B along with an application fee as specified under the Petroleum and Natural Gas Regulatory Board (Levy of Fee and Other Charges) Regulations, 2007.

The Board may *suo-motu* invite bids from entities interested in laying, building, operating, or expanding a CGD network for any specified geographical area.

The Board may carry out a preliminary assessment of the expression of interest with respect to the following, namely:

(a) natural gas availability position;
(b) possible connectivity with an existing or proposed natural gas pipeline for supply of natural gas to the city gate of the proposed CGD network, including LNG supplies by tank trucks or tank wagons and CNG by cascades; and
(c) any other relevant issues as the Board may consider necessary.

Within fifteen days of receiving expression of interest (EOI), the Board may publish through open advertisement acceptance of EOI and commencement public consultation period of 30 days or reject the EOI. In case of rejection, the Board should inform the entity along with reasons for rejection.

Within a period of fifteen days after the end of the public consultation process period, the Board may invite the proposal for the development of CGD network and bids for the same publish through an open advertisement. The application-cum bids should be submitted within 60 days of advertisement.

The entities must submit bids in two parts:

(a) Part I (Technical bid) covering general particulars of the entity and technical details of the project including minimum eligibility criteria under sub-regulation (6);
(b) Part II (Financial bid) covering the financial details under Regulation 7.

The Board may extend the date of submission of bids up to a period of one month through an advertisement.

The Board may suo-motu form a view regarding the development of a CGD network in a specific city or a geographic area, and in such a case, the procedure as specified in Regulation 5 (except aspects relating to the expression of interest shall be replaced by project details as prepared by the Board) shall apply.

In the initial days, the Board evaluated bids using following criteria and assigned weightage:

(a) lowness of the present value of the overall unit network tariff for usage of the proposed CGD network was assigned 40% weightage
(b) lowness of the present value of the compression charge for CNG for dispensing in the CNG stations in the proposed CGD network over the economic life of the project was 10% weightage
(c) highness of the present value of the inch-kilometre of steel pipelines proposed to be laid in the CGD network during the period of exclusivity in terms of an exemption from the purview of the common carrier or contract carrier as specified in the Petroleum and Natural Gas Regulatory Board (exclusivity for city or local natural gas distribution Networks) Regulations, 2008 was 20% weightage
(d) highness of the present value of the number of domestic customers proposed to be connected by PNG by the entity during the period of exclusivity in terms of an exemption from the purview of common carrier or contract carrier as specified in the Petroleum and Natural Gas Regulatory Board (exclusivity for city or local Natural Gas Distribution Networks) Regulations, 2008 was 30% weightage

    The present value in the criteria as mentioned above shall be calculated by the entity using a discount rate equal to 14% (fourteen percent). The entity with high composite score declared as successful in the bidder and awarded the awarded the specific geographical areas subject to fulfilling other criteria as per PNGRB Act. In the case of equal composite present value score, the entity furnishing greater value of performance bond is awarded the bid.

In 2013, the PNGRB amended the bid evaluation criteria as follows:

(a) Lowness of the present value of the overall unit network tariff for usage of the proposed CGD network by all categories of customers of natural gas (including the supply of natural gas to online compressor for CNG) over the economic life of the CGD network project (twenty-five years from the date of authorization). "The entity is required to bid for the tariff charge for each year of the economic life of the project and the variation in unit network tariff between any two consecutive years should not be more than 10%. This shall have a weightage of 70%."[32]
(b) Lowness of the present value of the compression charge for CNG for dispensing in the CNG stations in the proposed CGD network over the economic life of the project. "The entity is required to bid for the compression charge for each year of the economic life of the project and the variation in compression charge for CNG between any two consecutive years should not be more than 10%. This shall have a weightage of 30%."[33]

Table 9 presents bid bond and performance bond requirements as per the population of a geographical area.

The entity authorized under Regulation 10 shall enter into a firm natural gas supply agreement for the proposed CGD network project with any entity owning natural gas in a transparent manner on the principle of "at an arm's length" for a period equal to or more than the exclusivity period for exemption from the purview of common carrier or contract carrier allowed under the Petroleum and Natural Gas Regulatory Board (exclusivity for city or local natural gas distribution networks) Regulations, 2008, within ninety days of the date of issue of the authorization.

- Provides enabling platform with
  - minimum technical qualification requirements
  - net worth and bid bond linkage with pipeline length (Table 10) ensures serious bidding

To ensure timely commissioning of the natural gas pipeline, winning bidder need to furnish "performance bond of an amount equal to 1% of the estimated project cost for the project under execution or 1% of the projects rounded off the nearest lakh rupees or Rs. 20 crores, whichever is less and it shall be applicable to all pipelines authorized under Regulation 4, 17, 18, or 19.

[32]Substituted vide PNGRB (Authorising Entities to Lay, Build, Operate or Expand City or Local Natural Gas Distribution Networks) Amendment Regulations, 2013. PNGRB/CGD/Regulations/Review-2011/2012-III [Regulation prior to Amendment] effective from 19/03/08 TO 20/06/13.

[33]Substituted vide PNGRB (Authorising Entities to Lay, Build, Operate or Expand City or Local Natural Gas Distribution Networks) Amendment Regulations, 2013. PNGRB/CGD/Regulations/Review-2011/2012-III [Regulations prior to the amendment] effective from 19/03/08 TO 20/06/13.

Table 9 Population as a criterion for bid and performance bond

Previous performance bond (Rs.) requirements	Current performance requirements	Population of the geographic area	Current bid bond (Rs.) requirements
100 (hundred) million or 5% of the estimated project capital cost, whichever is higher	• Four times of bid bond	• Five million	• 50 million and proportionately higher amount for population > five million
60 (sixty) million or 5% of the estimated project capital cost, whichever is higher	• Four times of bid bond	• >=one million < five million	• 30 million
40 (forty) million or 5% of the estimated project capital cost, whichever is higher	• Four times of bid bond	• >=Half a million < One million	• 20 million
30 (thirty) million or 5% of the estimated project capital cost, whichever is higher	• Four times of bid bond	• >=Quarter of a million < half a million	• 15 million
	• Four times of bid bond	• >=one-tenth of a million < half a million	• 10 million
	• Four times of bid bond	• <one-tenth of a million	• 5 million

http://www.pngrb.gov.in/OurRegulation/CGD-Network-GSR196.html accessed on 6 May 2016
Source PNGRB (2008) and subsequent amendments

Table 10 Bid bond linkage with length of a pipeline

Pipeline length including spur lines (in km)	Amount of bid bond (in millions of rupees)
>=2501	250
1751–2500	200
1001–1750	150
251–1000	80
<=250	20

Source http://www.pngrb.gov.in/OurRegulation/NGP-GSR340.html

Provided that in case the pipeline capacity is expanded for more than 50% of its authorized capacity, the performance bond shall be increased in proportion to the increase and shall be rounded of the nearest lakh rupees subject to a total ceiling of Rs. thirty crores".[34]

GAIL furnished performance bond of Rs. 200 million for Surat–Nagpur–Raipur–Paradip natural gas pipeline with capacity of 74.81 million standard cubic metre per day (MMSCMD).[35] Similarly, Indian Oil Corporation Limited furnished

[34]http://www.pngrb.gov.in/OurRegulation/NGP-GSR340.html accessed on 9 May 2016.

[35]http://www.pngrb.gov.in/pdf/ngp/Govt-Authorization/Grant_of%20Auth26april.pdf accessed on 9 May 2016.

Rs. 200 crores for Ennore–Tiruvallur–Bengaluru–Madurai–Tuticorin natural gas pipeline with capacity of 84.67 MMSCMD.

4.2 Determination of Natural Gas Pipeline Tariff Regulations,[36] *2008*

The Petroleum and Natural Gas Regulatory Board (Determination of Natural Gas Pipeline Tariff) Regulations (2008) shall apply to an entity:

I. Authorized by the central government under Regulation 17 of the Petroleum and Natural Gas Regulatory Board (authorizing entities to lay, build, operate, or expand natural gas pipelines) Regulations, 2008, for laying, building, operating, or expanding a natural gas pipeline before the appointed day;
II. Laying, building, operating, or expanding a natural gas pipeline before the appointed day and authorized by the Board for such activities under Regulation 18 of the Petroleum and Natural Gas Regulatory Board (authorizing entities to lay, build, operate, or expand natural gas pipelines) Regulations, 2008.
III. Laying, building, operating, or expanding a dedicated pipeline which is converted into a natural gas pipeline under the provisions of sub-regulation (1) or (2) of Regulation 19 of Petroleum and Natural Gas Regulatory Board (authorizing entities to lay, build, operate, or expand natural gas pipelines), Regulations, 2008.

Bidding criteria used for evaluating bids:

- Bidding criteria (25 years economic life):
  - lowness of PV of tariff bid for each year (70% weightage)
  - highness of PV of natural gas volumes (in MMSCMD) proposed to be transported (30% weightage)

  Tariff to be bid zone-wise (TZn >= TZn−1) with

  - 40% weight for TZ1
  - 20% weight for % increase over first zone
  - 10% weight for % increase over TZ2 (constant % thereon)

PV to be determined by using discount rate of 12%
entity with highest composite score is considered successful bidder

[36]http://www.pngrb.gov.in/OurRegulation/NGP-GSR807.html accessed on 9 May 2016.

- Provision for extra capacity in natural gas pipeline to be made available for use as common carrier by any third party on open access and non-discriminatory basis;
- PNGRB may modify or accept EOI depending upon gas availability or need to expand an existing pipeline and guided by objective (s) of:
  - promoting competition among entities;
  - avoiding infructuous investment;
  - maintaining or increasing supplies or for securing equitable distribution;
  - ensure adequate availability of natural gas throughout the country;
  - protection of customers' interest in terms of availability of natural gas at reasonable natural gas pipeline tariff;
  - incentivizing rapid development of natural gas pipeline infrastructure

4.3 Determination of Network Tariff for City or Local Natural Gas Distribution Networks and Compression Charge for CNG Regulations,[37] 2008

The Determination of Network Tariff for City or Local Natural Gas Networks and Compression Charge for CNG Regulations, 2008 shall apply to an entity:

(a) authorized by the central government for laying, building, operating, or expanding CGD network before the appointed day;
(b) laying, building, operating, or expanding CGD network before the appointed day and authorized by the Board for such activities under the Petroleum and Natural Gas Regulatory Board (authorizing entities to lay, build, operate, or expand city or local natural gas distribution networks) Regulations, 2008.

As per the regulations of the Board, the entity to which these regulations apply shall submit all technical, operating, financial, and cost data of the CGD network or CGD network project that may be required by the Board for the determination of the network tariff and compression charge for CNG.

The network tariff and compression charge for CNG in a CGD network shall be determined by considering a reasonable rate of return (14%) on normative level of capital employed plus a normative level of operating expenses in the CGD network.

[37]http://www.pngrb.gov.in/OurRegulation/pdf/Gazette-Regulation/English/GSR197(E)-E.pdf accessed on 9 May 2016.

The return on total capital employed shall be determined separately for the capital employed in:

i. the common infrastructure in the CGD network (i.e. consisting of the pipeline from the tap-off point in the natural gas pipeline up to the city gate station, if any, city gate station, city gate distribution network consisting of pipelines, district regulatory station, and distribution-related equipments and facilities, but excluding CNG compression- and dispensation-related equipments and facilities; and)
ii. online compressors and related facilities as specified

As per the above-mentioned regulations on network tariff and compression charge, the network tariff and compression charge for CNG as determined by the Board are subject to review after five financial years. The Board either may, on its own or on the entity's request, carry out review in between two tariff review periods as per the Schedule A [8 (d)] of the said regulations.

In the recent past, the regulator found to be overstepping its authority. In its judgement delivered on 1 June 2012, Hon'ble Delhi High Court had held that PNGRB is not empowered to fix or regulate maximum retail price at which gas is to be sold by IGL to its consumers. The High Court judgement had also held that the Board is not empowered to fix any component of network tariff or compression charge for any entity like IGL having its own distribution network, while quashing PNGRB order dated 9 April 2012 fixing maximum retail price and requiring IGL to disclose the network tariff and compression charges to its customers.[38]

4.4 *Exclusivity for City or Local Natural Gas Distribution Network Regulations,*[39] *2008*

Petroleum and Natural Gas Regulatory Board (Exclusivity for City or Local Natural Gas Distribution Network) Regulations (2008) shall apply to an entity which is laying, building, operating, or expanding or which proposes to lay, build, operate, or expand a CGD network under the Petroleum and Natural Gas Regulatory Board (authorizing entities to lay, build, operate or expand city or local natural gas distribution networks) Regulations, 2008.

[38]Statement by Mr. Narendra Kumar, Managing Director, IGL on the verdict of Hon'ble Supreme Court of India on SLP filed by PNGRB challenging decision of Delhi High Court quashing PNGRB's order on network tariff and compression charge, available at http://www.iglonline.net/Documents/MD_Statement_2015.pdf accessed on 29 December 2015.

[39]http://www.pngrb.gov.in/OurRegulation/CGD-Network-GSR198.html accessed on 9 May 2016.

As per the exclusivity regulations,

- in case an entity proposes to lay, build, or expand a CGD network on or after the appointed day, the entity can be granted exclusivity of 25 years (economic life of the project) for infrastructure from the date of authorization of the Board;
- in case an entity is laying, building, or expanding CGD network before the appointed day, where the entity has either an authorization from the central government before the appointed day or an authorization from the Board under the Petroleum and Natural Gas Regulatory Board (authorizing entities to lay, build, operate, or expand city or local natural gas distribution networks) Regulations, 2008, the economic life of the CGD Network project shall commence from the start-up date of the commencement of physical activities of laying, building, or expanding the CGD network.

After the expiry of infrastructure exclusivity, the Board may grant an extension of 10 years to the entity.

The Board may provide exclusivity to an entity proposing to lay, build, operate, or expand a CGD network from the purview of common carrier or contract carrier for a period of five years from the date of authorization.

4.5 Technical Standards and Specifications Including Safety Standards for City or Local Natural Gas Distribution Networks Regulations,[40] *2008*

The PNGRB (technical standards and specifications including safety standards for city or local natural gas distribution networks) Regulations, 2008, are applicable to all new and such aspects of already existing networks as design, fabrication, installation, testing at the time of construction and commissioning. However, if an entity has laid, built, constructed, or expanded the CGD infrastructure based on some other standards or is not meeting the standards specified in these regulations, the entity needs to carry out a detailed technical audit of its infrastructure through a Board authorized or approved third-party agency by the Board. The entity thereafter is required to submit the recommendations made by the third party along with its time-based mitigation plan and implementation schedule to the PNGRB for authorization within six months from the date of notification of these regulations (i.e. 27 August 2009).

[40]http://www.pngrb.gov.in/OurRegulation/CGD-Network-GSR612.html accessed on 9 May 2016.

4.6 Code of Practice for Quality of Service for City or Local Natural Gas Distribution Networks Regulations,[41] 2010

The PNGRB (code of practice for quality of service for city or local natural gas distribution networks) Regulations, 2010, are applicable to an entity authorized for laying, building, operating, or expanding city or local natural gas distribution network in accordance with the provisions specified in the PNGRB (authorizing entities to lay, build, operate, or expand city or local natural gas distribution networks) Regulations, 2008.

These regulations lay down code of practice for promoting reliable service to consumers and the public and obligations of consumers besides conforming to minimum levels of service to be provided by the entity authorized for laying, building, operating, or expanding CGD Networks.

These regulations define service norms for new PNG connections and obligations of consumers. Further, these regulations prescribe system and procedure of metering, billing. The regulations recognize the role of consumer convenience and critically of service failure.

For example, the Regulation 9 deals with quality of service pertaining to CNG station. The regulation suggests that filling time should be minimized. Further, in case any consumer is turned away either for want of gas or inadequate pressure or failure of equipment, the incident should be recorded and submitted to the PNGRB.

As prescribed in the Regulation 9 (2), the following basic amenities shall be provided at CNG stations:

(a) safe and potable drinking water;
(b) free air;
(c) clean and functional toilets;
(d) consumer complaint box;
(e) safety kits, fire extinguishers, and other safety related equipment;
(f) shelter for passengers off loaded before refilling of vehicles on a best endeavour basis.

The Regulation 10 deals with network interruptions, and Regulation 11 prescribes guidelines and procedures for complaint handling mechanism.

[41]http://www.pngrb.gov.in/OurRegulation/CGD-Network-GSR720.html accessed on 9 May 2016.

4.7 *Access Code for City or Local Natural Gas Distribution Networks Regulations,*[42] *2011*

Petroleum and Natural Gas Regulatory Board (access code for city or local natural gas distribution networks) Regulations, 2011, are applicable to an entity authorised to lay, build, operate, or expand a city or local natural gas distribution network in a geographic area under the PNGRB Act, 2006, immediately at the end of its exclusivity period, if any, allowed by the Board under the Petroleum and Natural Gas Regulatory Board (exclusivity for city or local natural gas distribution network) Regulations, 2008, from the purview of common carrier or contract carrier, and to any entity or shipper who wants access to entry point capacity, exit point capacity and delivery at CNG exit point capacity on such city or local natural gas distribution network for supply of natural gas to domestic, commercial, or industrial consumers:

Provided that

(a) the authorised entity shall allow access to shipper on non-discriminatory basis to a city or local natural gas distribution network in a geographic area subject to the threshold limits mentioned in Regulation 5 at the end of the exclusivity period, if any, allowed by the Board;
(b) the development of the infrastructure shall be governed by the Petroleum and Natural Gas Regulatory Board (authorizing entities to lay, build, operate, or expand city or local natural gas distribution networks) Regulations, 2008, and the shippers shall get their gas transported or delivered by the authorised entity:

    i. up to the suraksha hose connecting to the burner of the domestic connections;
    ii. up to the metering point in the case of commercial and industrial connections;
    iii. up to compressor exit point in case of CNG station.

These regulations lay procedures and guidelines for capacity declaration, capacity booking, obligations for shippers and entity, gas parameters, facilities at entry and exit points, the nomination, scheduling, metering, allocation, and billing; and gas accounting and reconciliation. Further, these regulations define following charges to be paid by the shippers to the authorized entity:

(a) network tariff which includes gas transportation in CGD network, odourisation, gas metering, gas reconciliation, and system use gas;
(b) compression charges;
(c) overrun charges, if applicable;

[42]http://www.pngrb.gov.in/OurRegulation/CGD-Network-S-Admn29-03-2011.html accessed on 9 May 2016.

(d) system imbalance charges, if applicable;
(e) off-spec gas charges, as agreed in access arrangement;
(f) applicable taxes;
(g) any other charges mutually agreed in the access arrangement such as

i. ship or pay;
ii. transport or pay;
iii. technical up gradation of system;
iv. R&D;
v. any other charges with the approval of the Board.

4.8 Integrity Management System for City or Local Natural Gas Distribution Networks Regulations,[43] 2013

The Petroleum and Natural Gas Regulatory Board (integrity management system for city or local natural gas distribution networks) Regulations, 2013, are applicable to all the entities laying, building, operating, or expanding city or local natural gas distribution networks. These regulations have been developed for designing and implementing an effective and efficient integrity management plan for city gas distribution networks.

The integrity management system (IMS) deals with:

(a) evaluating the risk associated with city gas distribution networks and effectively allocating resources for prevention, detection, and mitigation activities;
(b) improving the safety of city gas distribution networks so as to protect the personnel, property, public, and environment;
(c) bringing more streamlined and effective operations to minimize the probability of CGD network failure.

The regulations prescribe guidelines and process of developing, implementing, monitoring, and reviewing of IMS. Concurrence of IMS by the PNGRB is mandatory before approval by the governing board of an entity. With reference to IMS, these regulations fix responsibility and accountability of the entity.

[43]http://www.pngrb.gov.in/OurRegulation/CGD-Network-S-Admn12-07-2013.html accessed on 9 May 2016.

5 Challenges for the Regulator

Under the current circumstances, the PNGRB faces key challenges like protecting consumer interest and entity interest without diluting competition. Evolving natural gas market with some amount of price control by the government does complicate the matter. On top of that, unavailability of desired amount of domestic gas further raises serious concern over the growth and development of the sector.

Partially controlled pricing with domestic gas allocation policy poses significant challenges to the existing and prospective investors. Therefore, gas exploration and production considered less attractive proposition in India. Many intellectuals argue that current supply scenario could change with more market oriented gas pricing. The gas pricing should be market- and investment-oriented with some amount of focus on consumer too. Intellectuals are constantly arguing in favour of developing trading hub in India, which in turn would help developing trading platform for price discovery and gas market development in India.

This would help more investment find its way into the hydrocarbon exploration and infrastructure sector and ensure efficient usage of the infrastructure (PNGRB 2013). Under the Hydrocarbon Exploration & Licensing Policy regime, the producers have greater flexibility in terms of pricing and marketing of their gas.

Kelkar Committee (2014) recommends that the Petroleum and Natural Gas Regulatory Board (PNGRB) should be empowered to provide all policy guidance with reference to midstream infrastructure development across the country. Current ambiguity over its role and jurisdiction needs far more clarity. If desired, the government should amend the PNGRB Act to remove the ambiguity and bring much needed clarity. Learning experience from various litigations and intervention of judiciary has been quite helpful to improve the delivery of regulatory system. In an evolving regulatory environment, all stakeholders are learning the tricks of the trade. Gradually, the market is moving towards achieving transparency, competitiveness, and consumer welfare.

- Interchangeable use of common carrier or contract carrier in policy document and PNGRB Act, 2006—yet internationally gas markets have first been *contract carriage systems before graduating into mature common carriage or hybrid systems.*

Common Carriage:

- Shipper does not book capacity;
- Settlement of tariff on short-term nomination basis;
- Mechanism for adjustment of oversubscription of capacity;
- Advantage to shipper only; and
- No operator would like to build pipeline on common carrier basis.

Contract Carriage:

- Parties contract capacity on take-or-pay basis before putting-up pipeline;
- Capacity bookers can trade or sub-contract capacity while remaining responsible for payment of tariff;
- No automatic right of entry to third parties; and
- Once pipeline is fully depreciated, system moves on to a common carriage.

The regulator faces serious challenge of attracting talented, skilful, and experienced human resources. Till date, an ad hoc approach has been adopted by the regulator to recruit talent. New contracted recruits do not see any future with the regulator. The trained human resources exit after the contract period, and the regulator loses key-trained resources, who could have contributed immensely towards the development of PNGRB and the gas market.

5.1 Recent Developments

5.1.1 Government Moots Single Entity for Safety in Oil and Gas

There are multiple bodies like Directorate General of Mines Safety (DGMS), Oil Industry Safety Directorate (OISD), Petroleum and Explosives Safety Organisation (PESO), and PNGRB to develop guidelines/codes for operations of oil and gas sector. After several critical incidents related to safety aspects relevant to various value chains of the petroleum industry, the government is suggesting for the creation of an umbrella organisation to deal with safety standards of all oil and gas installations (ET 2015). In case this happens, the role of PNGRB with respect to regulating safety of pipelines may be reassessed.

5.1.2 Judicial Intervention on PNGRB's Authority

On 1 June 2012, the Delhi High Court ruled that the PNGRB is not empowered by law to set or regulate the maximum retail price at which companies such as Indraprastha Gas Ltd to sell gas to consumers or to fix any component of network tariff or compression charge for an entity that has its own distribution network (Rautray 2015). This means that the provisions of the PNGRB (determination of network tariff for city or local natural gas distribution networks and compression charge for CNG) Regulations, 2008, become irrelevant. PNGRB appealed in the Supreme Court against the decision of the Delhi High Court and The Supreme Court dismissed the Board's appeal. The judiciary intervention brought much needed relief to the marketing entities, which means they have more liberty to play with the price of natural gas.

6 Regulations and Investment

The government intends expanding CGD network to more than 200 cities within five years from the existing 59 cities. Each city would warrant an investment ranging between 65 million USD to 100 million USD (Chavda 2014). According to data released by the Department of Industrial Policy and Promotion (DIPP), the petroleum and natural gas sector attracted foreign direct investment (FDI) worth US $ 6640.13 million (DIIP 2016) between April 2000 and December 2015, which is 2.3% of the total inflows.

However, not much investment is coming for the city gas distribution. Possibly lack of foreign direct investment could be compensated by domestic investment by the government and private companies. The investors look for the government and regulator to facilitate holistic development of infrastructure projects and synchronization of the same. This would certainly enhance the utilization of infrastructure efficiently to discover new markets and customers.

7 Conclusion

Downstream petroleum regulatory developments have been noteworthy in India. Within a very short period, the regulator developed some critical regulations relevant to developing natural gas pipelines, developing CGD network, and ensuring safe operations. Largely the regulator achieved its objectives as defined in the PNGRB Act. During the short span of its existence, the regulator faced a fair amount of challenges to remain relevant and meaningful. On few occasions, the regulator acted beyond its regulatory authority leading to legal conflict with the entity. The judiciary system is showing the right direction to the regulator.

The downstream regulatory environment is passing through transformative stages. Many believe that the regulator overstepped on some occasions. Compared to other successful regulators, PNGRB is quite young, therefore we firmly believe that the regulator need time to mature and operate as an independent regulator. The government seems to have lesser amount of confidence on the regulator; therefore, the regulator is taking limited actions. The regulator needs to be more professional in terms of recruiting talents and retaining them for achieving better results.

Acknowledgments Research assistance is provided by Ms. Prachi Agarwal, MBA 2013–2015, Department of Management Studies, Rajiv Gandhi Institute of Petroleum Technology to prepare the first draft of this chapter is highly appreciated.

References

Baron PD (1989) Design of regulatory mechanisms and institutions. In: Richard S, Robert DW (eds) 2 Handbook of industrial organization 1349, 1349.

CCI (2012) Interface between competition act and the petroleum and natural gas regulating board act. http://cci.gov.in/images/media/ResearchReports/GAURAAN2012.pdf

Chavda K (2014) Oil and gas policy for the next government. Available at http://centreright.in/2014/03/oil-gas-policy-for-the-next-government/#.VpUFatIdDS4. Accessed on 12 Jan 2015

Cleantech (2013) Gas price hike will augment production. Available on http://chemtech-online.com/O&G/T_K_Ananth_Kumar_Sept13.html. Accessed on 12 Jan 2016

Dudley SE, Brito J (2012) Regulation: a primer, 2nd edn. Mercatus Center at George Mason University, USA

Deloitte (2013) IEC 2013 securing tomorrow's energy today: policy & regulations: resource allocation and pricing, pp 1–17

DIIP (2016) Statement on Sector-wise FDI Equity Inflows from April 2000 to December 2015. Available at http://dipp.nic.in/English/Publications/FDI_Statistics/2015/FDI_FactSheet_OctoberNovemberDecember2015.pdf

ET (2015) Government moots single entity for oil sector safety, installations http://economictimes.indiatimes.com/industry/energy/oil-gas/government-moots-single-entity-for-oil-sector-safety-installations/articleshow/48348904.cms. Accessed on 27 Jan 2016

Greenstone A, List JA, Syverson C (2012) The effects of environmental regulation on the competitiveness of U.S. manufacturing. Working Paper 18392, National Bureau of Economic Research, M.A.

Johan de Hertog JA (2010) Review of economic theories of regulation. Available at http://www.uu.nl/sites/default/files/rebo_use_dp_2010_10-18.pdf

Kelkar Committee (2014) Roadmap for reduction in import dependency in the hydrocarbon sector by 2013. Available at http://www.petroleum.nic.in/docs/FinalReportKelkarCommittee2014.pdf. Accessed on 6 May 2016

Orbach B (2012) What is regulation? volume: 30:1–10, Yale Journal on Regulation Online.

PNGRB (2008) Authorizing entities to lay, build, operate or expand natural gas pipelines. G.S.R.340 (E) 6th May 2008

PNGRB (2010) http://www.pngrb.gov.in/OurRegulation/pdf/Gazette-Regulation/English/GSR987E.pdf. Accessed on 2 Dec 2015

PNGRB (2013) Vision 20130. Natural gas infrastructure in India. Available at http://www.pngrb.gov.in/Hindi-Website/pdf/vision-NGPV-2030-06092013.pdf. Accessed 12 Jan 2016

Raghuram G, Sunny S (2015) Right to fair compensation and transparency in land acquisition, rehabilitation and resettlement ordinance 2014: A Process Perspective. W.P. No. 2015-07-03. IIMA, India

Rautray S (2015) SC dismisses PNGRB's plea on fixing price of natural gas. http://articles.economictimes.indiatimes.com/2015-07-02/news/64038691_1_compression-charge-network-tariff-pngrb. Accessed on 27 Jan 2016

Richard AP (1971) Theories of economic regulation. NBER Working Paper No. 41. May

Singh VV, Mitra S (n.d). Regulatory Management and Reform in India. Background Paper for OEC. http://www.oecd.org/gov/regulatory-policy/44925979.pdf. Accessed on 22 Dec 2015

Taxation Issues in Natural Gas Industry in India

Pankaj Gupta

Abstract The entire natural gas supply chain is plagued by plethora of taxes. These taxes are levied through separate statutes administered by different authorities from Central and State Government. Each statue levies tax at different stages in the value chain and therefore provides for separate valuation methodology. Most of these taxes are non-vatable having huge cascading effect. The customers mainly from power and fertilizer sector are also not able to take credit of most part of taxes paid on purchase of natural gas. This inefficiency is further compounded by administrative cost resulting from compliance of taxation statutes which are administered by different Government authorities. More often than not, divergent position taken on same issue by various wings of the taxation administrative authorities results in further complication. Consequent litigation takes anywhere between five to ten years for final position to emerge. Any adverse tax ruling results in huge unrecoverable tax burden as indirect taxes are largely pass-through and it becomes difficult to recover past tax dues from the customers after such a long lag. Therefore, a reform in the current tax regime is required for increasing the share of natural gas in the energy basket of India, being a cleaner fuel compared to other fossil fuels. But the expectation is being belied, as natural gas is being kept out of the proposed goods and services tax.

Keywords Natural gas · Taxation · Royalty · Service tax · VAT · CST · Income tax

1 Introduction

Taxation policy apart from revenue consideration should be simple, easy to comply, stable with long-term vision, and growth-oriented. It should be responsive to different sectors of economy complementing to overall national objective. The

P. Gupta (✉)
GAIL India Limited, New Delhi, India
e-mail: pangup@rediffmail.com

S.K. Kar and A. Gupta (eds.), *Natural Gas Markets in India*,
DOI 10.1007/978-981-10-3118-2_13

importance of the growth of Natural Gas sector in Indian economy, coupled with environmental obligations, have been duly discussed in the other sections of the book. The tax laws of India should therefore take into account the present impediments being faced by the natural gas industry and fiscal incentives to push it to a growth trajectory.

This chapter will focus on the present applicable taxes, complications peculiar to this industry, and suggestions for simplifying complication thereby reducing litigation. Further, the focus will be largely on the issues involved in taxation impact on marketing and distribution of natural gas and not on value-added products.

2 Taxation on Natural Gas Industry in India

Under the present scheme of the Constitution of India,[1] the subject matter of taxation is distributed between Union and State. As natural gas is distributed in various states through pipeline after exploration or import of goods, it is subject to multiple taxes throughout the distribution system. And as each state has rules and processes which are not uniform, it adds to the complexity and cost.

Taxes and duties levied presently from importation/exploration till delivery are as presented in Table 1.

At present, there is no excise duty on natural gas, and therefore, it has not been included above. Further, whether royalty will fall into strict definition of taxes or not may be an issue; however, as it is also a significant levy, the same has been included above and discussed below in brief.

- Royalty

The Central Government has the authority to enhance or reduce the rates of royalty in accordance to the Oilfields Regulation and Development Act, 1948 (ORDA). Entry 53 in the Union List of the Seventh Schedule to the Constitution of India which states "regulation and development of oil fields and mineral oil resources; petroleum and petroleum products; other liquids and substances declared by Parliament by law to be dangerously inflammable" provides such authority to the Central Government.

Rule 14 (1) of Petroleum and Natural Gas Rules 1959 framed under ORDA provides:

> (1) (a) Notwithstanding anything in any agreement, a lessee shall
>
> (i) where the lease has been granted by the Central Government, pay to that Government
> (ii) where the lease has been granted by the State Government, pay to that Government a royalty …

Royalty is therefore payable to the state for exploration onshore areas and to the center for offshore areas. Royalty is part of the cost, and it is either passed on or

[1] Article 246 of the Constitution of India read with Seventh Schedule of the Constitution.

Table 1 Existing taxes levied in the gas value chain in India

Nature of tax	Subject matter of tax	Right of tax
Royalty	Exploration	Onshore by the state, offshore by Union
Customs duty	Import of LNG/natural gas into India	Union
Service tax	Regasification of LNG	Union
Service tax	Transmission of natural gas	Union
Sales tax (VAT/CST)	Sale of natural gas	State
Entry tax	Import into the state	State
Income tax	Income	Union

included in the cost depending on the pricing order of the Government of India who as owner of natural resources has the power to determine price.

- Customs Duty

Customs duty arises on importation of any goods into India. At present, customs duty on LNG and natural gas is 5%[2] plus cess on CIF (carriage insurance freight) value. As there is no excise duty on LNG, no CVD (countervailing duty) is applicable. This duty is non-vatable and therefore adds into the cost of gas.

Annotation

One of the issues which have arisen in this aspect is the quantity and price for the purpose of levy of duty as there is difference in billed quantity and actual quantity received in Indian port. In the Ahmadabad bench of CESTAT order dated September 10, 2013 in the case of Reliance Industries Ltd, it was held that the value for the levy of duty is to be done on transaction value worked out as per contracts on the basis of ex-ship delivered quantity and unit price as declared in the final invoices. This decision was further followed in the case of GAIL India Ltd in similar case decided by the same bench in the following year.

The other issues relate to valuation of LNG in case of transaction between related parties as more often than not transaction happens between associated enterprises. In such a case, in terms of rules for valuation in case of import from associated enterprise, it is difficult to find comparables. Further, as more and more trade happens on spot basis, the difficulty increases. It is pertinent to point out that the price determined for the purpose of customs valuation wherein tendency is to increase the value for higher duty may not be readily acceptable to transfer pricing authorities for income tax purpose where lower valuation leads to higher tax collection. It is important that some kind of meeting grounds should be evolved between the two wings of the same Department of Government of India.

[2]Section 3 of the Customs Tariff Act read with heading 2711.11 of the First Schedule and Notification No. 12/2012 dated March 17, 2012.

- Service Tax

(i) *Regasification Services*

The regasification of LNG is subject to service tax. The present service tax rate is 14%[3] plus 1% on account of Swachh Bharat Cess[4] and Krishi Kalyan Cess.[5] When the person involved into regasification itself sells the goods, no service tax is applicable as it is a case of self-service and not a service from one person to another. If the service recipient is a reseller of the RLNG (regasified LNG), no credit of the service tax paid on the service will be available as no output service is provided by the reseller. This also gets added as a cost and passed onto customer. The Cenvat credit availability for the person involved in regasification on goods and services procured by him depends on the model followed by him as the output tax charged by him varies in each of the model.

Annotation

It may be noted that during the process of regasification, certain portion of LNG quantity will be lost which is inherent in the process. This loss in most cases might be predefined in the contract between the service provider and recipient. Revenue has sought to bring the value of such quantity of LNG as non-monetary consideration to be includible in the taxable value. This in my view is contrary to the provision and also abjectly ignores the peculiarity of the goods and services involved.

(ii) *Transmission Services*

Natural gas is transmitted through pipeline to customers located at different parts of the country which are connected by pipeline. The taxability of transmission charges will depend on the contractual relationship between the two persons. In case, the transporter is also reseller of the gas and the relationship is one of sale and purchase, the transmission value will be included in the value of gas which will be subject to sales tax. However, if the transporter merely provides transmission service to the shipper, then it will be subject to service tax.

Annotation

At present, Tariff on Transmission charges is regulated and determined by Petroleum and Natural Gas Regulatory Board (PNGRB). However, the gas prices are either Government-determined or market-determined. The gas consumers have entered either gas transmission contracts or gas sale contracts wherein delivery of gas till doorstep is made by the seller. But as the transmission is regulated, it is separately mentioned in the contract and in most cases in the sales invoice. Ambiguity may arise regarding leviability of service tax on transmission portion in

---

[3]Section 66B of the Finance Act 1994.

[4]Chapter VI of the Finance Act 2015.

[5]Chapter VI of the Finance Act 2016.

cases of gas sale contract. Though in my view, the same may not be legally tenable. Further, in the proposed GST Amendment Bill, it is only sale of natural gas that has been kept out of GST ambit, but the service of transmission will be exigible to GST and one cannot entirely rule out litigation in this front.

- Sales Tax

Before touching on the issue of sales tax (VAT or CST), it is important that factual matrix of marketing of natural gas in India and relevant provisions of laws are understood. The sources of natural gas which are at present transported in various parts of country are located in Gujarat and Andhra Pradesh. Other domestically produced gases are normally consumed within the state and therefore do not pose any major problem so far as taxability is concerned. The imported LNG at present is mainly at **Dahej and Hazira** terminals in Gujarat and Dabhol terminal in Maharashtra. The LNG imported at Kochi Terminal can presently be consumed in the state of Kerala though in future it will flow to Karnataka as well. However, in future, other terminals in different states are proposed. The pipeline connects all the sources and transports gas to customers located inside these states and to states of Madhya Pradesh, Uttar Pradesh, Delhi, Haryana, Punjab, Rajasthan, and Himachal Pradesh in north and Goa and Karnataka in south. En route, these pipelines extraction plants are situated to manufacture value-added products.

The complexity in taxing power arises mainly when there is movement of goods from one state to another and is primarily dependent on whether the movement of goods falls within the ambit of Section 3 of Central Sales Tax Act. In case the interstate transaction falls within the ambit of Section 3, then the transaction will be subject to CST to be charged by the originating state even if the gas is ultimately consumed in the destination state. In such a case, because of constitutional embargo, no tax on sales can be charged by the destination state. Alternatively, when the transaction is out of the ambit of Section 3, then the destination state can levy local sales tax, i.e., VAT. It is pertinent to note that states through provisions of input credit restriction or levy of entry tax have tried to overcome the revenue deficiency.

The VAT rates across states vary with the highest tax rate of 26% in the state of Uttar Pradesh. It is also important to point out that in some states credit is not allowed at all or in some states credit is denied when Natural Gas is used as fuel. As one of the main gas consumers is power and it is not a taxable good from local sales tax perspective, the entire tax goes into the cost.

Annotation

Major controversy in this area revolves whether the transaction is interstate sale or not, which revolves around the interpretation of Section 3 of the CST Act.

Section 3 of the CST Act provides the following:

> **3. When is a sale or purchase of goods said to take place in the course of inter-State trade or commerce** - A sale or purchase of goods shall be deemed to take place in the course of inter-State trade or commerce if the sale or purchase -

(a) occasions the movement of goods from one State to another; or
(b) is effected by a transfer of documents of title to the goods during their movement from one State to another.

Explanation 1 - Where goods are delivered to a carrier or other bailee for transmission, the movement of the goods shall, for the purposes of clause (b), be deemed to commence at the time of such delivery and terminate at the time when delivery is taken from such carrier or bailee.

Explanation 2 - Where the movement of goods commences and terminates in the same State it shall not be deemed to be a movement of goods from one State to another by reason merely of the fact that in the course of such movement the goods pass through the territory of any other State.

The above section has been a matter of numerous litigation, and important legal principle emerging out of the decision is that it is immaterial in which state the property passes but where there is direct nexus between the sale and movement of goods, then it falls within the ambit of Section 3.

In the case of RIL versus State of UP 2012 (194) ECR 293 (Allahabad), the question before the court was whether the gas sold by petitioner, where the property in goods is transferred in the state of Andhra Pradesh to customers located in the UP, would be an interstate sale or not. The customers in turn have transmission agreement with the pipeline companies for getting the gas transmitted to their plants in UP. As explained above, the gases so transmitted are commingled with gas from other sources and belonging to various persons. As per the transmission agreement so entered, the gas may have been processed en route by the transporter, and the quantity and quality of gas may also have undergone changes till it is redelivered by the transporter to the plant of the customer. Accordingly, it was argued by the revenue that it does not constitute a interstate sale, and therefore, the state of Uttar Pradesh has rightly assessed local sales tax (UP VAT) on the transaction. It was argued on behalf of the state that transportation of gas in common pipeline belongs to different buyers; hence, it becomes unascertained goods, and as such, sale shall be deemed to take place in the state of UP; hence, it cannot be an instance of interstate sale. It was further argued that the gas while moving in common pipeline is in commingled form; hence, it is not known as to which portion of gas belongs to whom, and thus, it becomes ascertained goods only at the delivery point where appropriation takes place. The court rejecting the argument held it as an interstate sale and said that in case the above argument is accepted, then the seller or buyer of natural gas, who does not possess his own pipeline, shall be prevented to transport his gas and everyone will have to install his own pipeline, which shall not be feasible or practical. Transportation of natural gas cannot be compared with transportation of tangible goods. Mixture of natural gas of the common quality during the course of transportation shall not affect the right of the buyer. Every buyer or shipper may draw its natural gas from open access gas pipeline with due measurement at exit point. The court also observed that it is necessary to consider the quality of gas and international practice as well as scientific views with regard to measurement and transport of gas from one place to another. Accordingly, the court took into consideration the special characteristic of natural gas and requirements for trading of gas including regulatory requirement of common carrier which necessitates comingling of gas.

The view of the Allahabad High Court was consented by Ahmadabad High Court in the judgement dated May 8, 2015 in the case of British Gas Exploration and Production India Ltd versus State of Gujarat in the Special Civil application no: 5159 of 2010, where the major contention was the situs of sale whether it was offshore or onshore in which case the state of Gujarat would have had the right to levy local sales tax. The court while concluding that the sale concluded outside the state of Gujarat has made the following important observation:

> 35. The question as regards the stage when appropriation of the goods has taken place has to be considered in the light of the peculiar nature of the goods, viz. Natural Gas, which cannot be transported in the manner in which ordinary goods are transported …
>
> Reference was made to Complete Auto Transit v. Brady, (1977) 430 US 274, wherein it was held that as a general rule, where fungible goods belonging to different persons are so intermingled as to be undistinguishable, whether by consent of the owners or by someone's wrongful act, the owners become tenants in common of the mass. The co-mingling of a fungible commodity does not affect ownership unless the parties intend to transfer title. The way the system works, each shipper is simply entitled to a volume of gas thermally equivalent to that which is placed into storage regardless of where it was placed when stored or from where it is taken when removed from storage.
>
> The court also made reference to the Canada Federal Court of Appeal decision in The Ministry of Public Safety and Emergency Preparedness (Canada) v. Tenaska Marketing Canada, 2007 FCA 223, wherein the court held thus:-
>
> "Due to its unique physical properties, large volumes of natural gas can only be transported in a continuous stream. Once delivered into a pipeline for transportation, it becomes commingled with other natural gas. Individual molecules are not separately identifiable, and cannot be accurately tracked or traced. As a result, natural gas is sold and purchased on a "quality and quantity basis", and treated as a fungible goods, with title taken on a quality and quantity basis.
>
> Accordingly, and at the ultimate point of delivery, what the purchaser would actually receive is the same general volume and quality of natural gas (less any fuel consumed in transport), and having the same effective heat content that was delivered at the upstream point.
>
> Pipeline transportation contracts generally provide for the commingling of the natural gas delivered to the pipeline with other natural gas, and require specified quality standards for natural gas being delivered to or by the pipeline.
>
> Given the fundamental properties of natural gas (i.e., it is a fungible commodity, commingled with the contents of the pipeline on delivery, and therefore not separately identifiable once delivered), all of Tenaska's shipments of the Canadian natural gas via the TCPL/GLGC Pipeline were commingled with like natural gas and lost their separate identities once delivered to that pipeline. Furthermore, and when transported through the U. S, the Canadian natural gas would likely have been further commingled with U.S. produced natural gas, being delivered to the GLPS at various points in the U.S.
>
> In the end, the change of ownership of the Canadian natural gas during its transit through the U.S. and the commingling of Canadian and American gas on the American part of the pipeline are not relevant to determine the purpose of the delivery from one place in

Canada to another place in Canada. These facts form an intrinsic part of the economic reality of transporting natural gas via pipeline through the United States. This economic reality was very much in the mind of the drafter of the legislation and found its way in the terms used by Parliament.

47. From the submissions made on behalf of the petitioners as well as the averments made in the memorandum of petition and the documents annexed therewith as well as from the above referred decision of the Allahabad High Court, it is apparent that due to its unique physical properties, large volumes of Natural Gas can only be transported in a continuous stream. Once delivered into a pipeline for transportation, it becomes co-mingled with other natural gas. Individual molecules are not separately identifiable and cannot be accurately tracked or traced. As a result, natural gas is sold and purchased on a "quality and quantity basis", and treated as a fungible good, with title taken on a quality and quantity basis. … However, merely because the gas which is delivered at the Offshore T-Junction is co-mingled with other gases, does not detract from the fact that prior thereto, the natural gas sold by the seller to the buyer, was ascertained at the off-shore processing facility and appropriated to the contract of sale and delivered at the Delivery Point. The subsequent sweetening of the gas post appropriation, does not change the situs of the sale.

Taxation on Swapping

It is important to point out that apart from the issue deliberated above, it is in the nature of natural gas business that swapping of natural gas between buyers and sellers located in different states is a necessity. However, as per Section 3 of the CST Act, physical movement of natural gas is sine qua non. Therefore, swapping of natural gas results in multi taxes, thereby increasing the cost. However, appreciating the difficulty, Ministry of Finance, Government of India, F.No: 34011/04/2015/SO-(ST) dated July 21, 2015 has issued the following clarification with regard to Section 3 of the CST Act:

Where the gas sold or purchased is transported through a common transport/distribution system (such as common carrier pipeline), where the gas may be commingled and fungible with the gas of other parties, so long as an equivalent quantity of the gas introduced into the system in one state is taken out of the system in other state (as evidenced through commercial documentation) such contractual movement of gas will be considered to be physical movement of gas from one state to another.

In absence of any express provision in the Act, it is highly unlikely that, if challenged, court may not uphold the clarification looking into the substantive change the clarification provides.

Appreciating the above legal lacuna, the Finance Act 2016 has tried to mitigate the legal deficiency by inserting the following explanation to Section 3 of the CST Act:

Explanation 3 - Where the gas sold or purchased and transported through a common carrier pipeline or any other common transport or distribution system becomes co-mingled and fungible with other gas in the pipeline or system and such gas is introduced into the pipeline or system in one State and is taken out from the pipeline in another State, such sale or purchase of gas shall be deemed to be a movement of goods from one State to another

- Entry Tax

Apart from VAT, some states are charging entry tax. Entry taxes are levied on goods brought into the state for sale, use, or consumption. In some states, entry tax is not leviable/exempted when local VAT is paid on resell of gas; however, no credit is allowed when it is consumed in the process of manufacturing. Apart from the constitutionality of entry tax levied by states on goods is pending before the constitutional bench, disputes relating to valuation of goods specially when the goods are brought into the state in its own pipeline are pending.

All the above taxes levied by different departments have huge cascading effect leading to taxes on taxes. Further, the credit availability is restricted not on the traded product, but even the credit of taxes on the development of infrastructure is also limited. This greatly hampers the development of taxes.

- Income Tax

In the field of income tax, under Section 35AD, accelerated depreciation incentive is available for laying and operating a cross-country natural gas pipeline network for distribution, including storage facilities being an integral part of such network. The pipeline must have been approved by the Petroleum and Natural Gas Regulatory Board and has made not less than such proportion of its total pipeline capacity as specified by regulations made by the Petroleum and Natural Gas Regulatory Board available for use on common carrier basis by any person other than the assessee or an associated person. In such case, deduction in respect of the whole of any expenditure of capital nature incurred, wholly and exclusively, for the purposes of the pipeline business is carried on by him during the year in which such expenditure is incurred by him.

Provided that the expenditure incurred, wholly and exclusively, for the said business, shall be allowed as deduction during the year in which he commences operations of the business, if—

(a) the expenditure is incurred prior to the commencement of its operations and
(b) the amount is capitalized in the books of account of the assessee on the date of commencement of its operations.

The said expenditure does not include acquisition of any land or goodwill or financial instrument. Therefore, the said capital expenditure is not only limited to plant and machinery but also includes other items such as furniture and other equipment. The only limitation is that it should be wholly and exclusively for the said business. One of the major expenditures that are incurred for laying a pipeline is acquisition of ROU (Right to Use). As the restriction is only on acquisition of land, in my view, the whole expenditure on ROU can be claimed under Section 35AD.

However, once a pipeline for which deduction is claimed has commenced operation, then any expenditure incurred which is capital in nature can be claimed as deduction even if the said item may not have been capitalized in the books of accounts. It may be noted that there is ring fencing, and in case of unabsorbed capital expenditure, the same has to be carried forward and cannot be set off against any other business, including from business of pipeline which does not fulfill conditions specified above. The income generated should be arising out of the said business. Further, it has been provided to other specified business like facility of cold chain, housing, fertilizer etc., the same dispensation has not been extended to Natural Gas pipeline sector.

3 Way Forward Under Current Regime

The entire indirect taxation regime is going to go a mammoth change with the introduction of GST. However, natural gas has been clubbed with other petroleum products and as per the Constitution (One Hundred and first Amendment) Act 2016, present taxation regime will be continued to be applicable on Natural Gas for the time being.

But before we discuss the arguments in support for and benefit of including natural gas within the GST framework, let us discuss the steps that Government should take for pushing the growth of gas industry in the present regime. The above taxes form a huge portion of tax mainly because of cascading effect and noncredit eligibility availability in case of gas. It may, in some case, be high as one-third of total cost to the consumer.

Customs duty of 5%[6] has been waived for the power sector subject to certain conditions for supply to power-generating companies for public distribution. Hence, the exemption is not even allowed for generation of power by companies for their own consumption or for commercial purpose. And it is not available for other sectors such as fertilizers and CGD business. Crude, however, attracts nil customs duty, and therefore, natural gas being a cleaner fuel makes a better case.

The VAT rate on natural gas varies in different states which is as high as 26%[7] in the state of Uttar Pradesh for a consumer manufacturing non-taxable goods like power. No credit is available for these taxes. This high rate of taxes is a big impediment in the growth of the industry and also creates a huge arbitrage between different business models of CST sale or local sale. As explained earlier, there is huge dispute in respect of the taxability of cross-country movement, in particular, the genesis of the dispute is result of peculiar character of the good and its distribution mechanism. Further, swapping of gas, which is also a requirement of the industry, is facing hurdle because of tax cost arising out of present tax provision.

The threefold suggestion for removing the distortion is as follows:

- Under Section 14 of the CST Act, goods of special importance are declared (DECLARED GOODS), which then prohibit any state of keeping VAT rate above 5%. Coal and crude oil are included in the list. It has been long argued

[6]Notification no: 12/2012 dated March 17, 2012.

[7]Entry 8 (c) of Schedule IV to Uttar Pradesh VAT Act 2008 read with Section 3A of the UP VAT Act and notification no: 307 dated February 19, 2012.

that natural gas should be included in the list and well appreciated even at the highest echelon of Government of India.

- Amendment in sale of goods acts to provide sale of a portion of goods out of bulk (unascertained goods) drawing reference from amendment in English Sale of Goods Act, 1979 where Section 20A and Section 20B were added in 1995 which enables transfer of property and thereby sale of a portion of an identified bulk before ascertainment of specific portion of the bulk. This will not only make high sea sale practically possible but also remove unnecessary litigation.
- The taxation regime should appreciate the peculiarity of trade arising out of the characteristic of good and legitimate business requirement and therefore should issue suitable notification/clarification such that unnecessary litigation should be avoided.

In the Income Tax Act, as explained earlier, one of the major incentives allowed to natural gas industry is for the pipeline business. Under Section 35AD of the Act, weighted deduction is allowed to certain sectors covered under the said section; however, an important sector which has a potential to fuel other sectors of the economy has been left behind. Otherwise, investment allowance under Section 32AC should be extended to this sector more so when Government is trying to develop national gas grid.

GST

Discussion on taxation aspect will be incomplete without touching upon the proposed biggest and ambitious tax reform in post-independence India. It is ambitious because it will create a Pan-India market by allowing free flow of goods and services and it will require the Union and 28 states to work in tandem to make it a success story. Under the GST, the excise, service tax (presently levied by Union), and sales tax and entry tax (presently levied by state) merge into a common GST.

The proposed Constitutional Amendment Bill[8] excludes natural gas (clubbing it with other petroleum goods) and excluded it from GST regime. The consideration for such an exclusion is obviously revenue consideration of state. Though natural gas may not contribute any significant taxes to the state coffer, but in any case, both economic and environmental consideration should be weighed in. Even the Task Force of 13th Finance Commission[9] in its first discussion paper for GST issued in 2009 gave the following recommendation:

> Natural gas like petroleum products, is derived from the same source. However, unlike petroleum products, natural gas does not generate negative externalities. Therefore the tax regime for natural gas should be distinctively different from the regime applicable to petroleum products. Accordingly, natural gas should be subjected to only GST (both central and state) with all benefits of input credit as in case of other normal goods. We recommend accordingly.

---

[8]One Hundred and Twenty Second Constitution Amendment Bill.

[9]Para 2.33 of Report of Task Force on GST of 13th Finance Commission dated December 15, 2009.

Even, the committee headed by chief economic advisor has recommended for inclusion of petroleum goods into the fold of GST.[10]

> Considering Natural Gas as an industrial input for the industry and also environment friendly, Natural Gas deserves the separate treatment than petroleum products. Further, presently excise duty is levied on CNG as deemed manufacture. Further some states are taxing PNG for domestic use as separate good than Natural Gas for the purpose of levy of VAT. It is not clear treatment to be given to these goods post implementation of GST. In case of a company present in entire value chain manufacturing goods (which are inside GST) where Natural Gas is the raw material, administration of tax regime will be very difficult and costly affair. It may also have huge cost implication in case tax credit between GST and Non GST goods is not allowed.

[10]Para 6.5 of Report on the Revenue Neutral Rate and Structures of the Rate for the Goods and Services Tax dated December 4, 2015.

Part V
Other Important Areas

Human Resource Challenges for Gas Sector in India

Ayush Gupta, Sanjay Kumar Kar, Amit Ray and D.V. Shastry

Abstract The authors highlight how natural gas (NG) industry across the world is facing a severe crunch of talented, skilled, and experienced manpower. The industry is losing experience by way of superannuation of experienced employees, and losing talent to other industries as oil and gas sector is no longer a preferred sector among younger generation. The chapter discusses critical issues and emerging challenges related to human resource requirements for NG sector in India. Further, the authors throw some lights on human resource demand, skilled manpower availability, academic and professional institutions offering specialized courses/training programs and skill sets required to succeed in the sector. The authors discuss about the ways to address prominent HR challenges and meet existing gap in the NG value chain in India.

Keywords Human resource · Natural gas · Skill · Gap

1 Introduction

Natural Gas (NG) industry across the world is facing a severe crunch of talented, skilled, and experienced manpower. The industry is losing experience by way of superannuation of old employees, and also losing talent to other industries, as oil

A. Gupta (✉) · A. Ray · D.V. Shastry
GAIL India Limited, New Delhi, India
e-mail: ayushgupta@gail.co.in

A. Ray
e-mail: amitray@gail.co.in

D.V. Shastry
e-mail: dvshastry@gail.co.in

S.K. Kar
Department of Management Studies, Rajiv Gandhi Institute of Petroleum Technology, Jais, India
e-mail: skar@rgipt.ac.in

S.K. Kar and A. Gupta (eds.), *Natural Gas Markets in India*,
DOI 10.1007/978-981-10-3118-2_14

and gas sector is no longer a preferred sector among younger generation. Over and above this, the job requires a high skill level to ensure the safety and integrity of plants and assets. In particular, across the entire NG value chain starting from upstream activities of exploration and production, to transmission and distribution to the downstream activities of petrochemical production, the entire sector is reeling under the shortage of manpower that has the requisite skills and experience to carry forward the much needed growth in this sector. With signing up of international trade agreements and expansion of NG-based infrastructure including cross-country and city gas pipeline network, LNG terminals, gas-based petrochemical plants, and setting up of cross-border supply networks, what is needed is the development of a competitive workforce to meet the requirement of skilled manpower in the NG sector.

2 The Natural Gas Industry in India

NG industry is still in the nascent stage of development in the country. A fairly young sector, NG industry, has made significant contributions, especially in the development of gas-based power and fertilizer sector in the country. With limited availability of domestic gas, there is a lot of dependency on gas imports. The infrastructure across the entire value chain of NG right from exploration and production in the upstream to the industrial and domestic connections in the downstream is expanding at a quick pace. India is aggressively scouting for new sources of gas within and outside the country. Still, there is a huge gap between the demand and availability of NG in the country. India today has several players in the NG transmission and distribution business. It has over 16,000 km of NG pipelines and out of that GAIL has about 11,000 km gas pipelines infrastructure with a capacity of about 210 MMSCMD.[1] GAIL controls about 69% of the NG transmission pipeline network in the country. GAIL has about 4300 employees (3400 executives and 900 non-executives) on its rolls. Out of this workforce, about 2000 employees (1500 executives and 500 non executives) are engaged in the projects/pipelines division. These are regular employees of GAIL in technical disciplines only and do not include the contract manpower. This does not include the support services of HR, marketing, and finance, etc. The pipeline and project regular employment statistics of GAIL suggests that approximately 182 technical persons are employed per 1000 km of NG pipelines.

The NG business in India was more of a monopoly business till the late 1990s, that is, when the Government of India decided to open up the NG sector for other government and private players. The administrative price mechanism (APM) under which the price of NG was being regulated was also decided to be dismantled and prices to be brought in parity with market-driven prices in a phased manner. With a

[1]http://www.gailonline.com/final_site/naturalgas_transmission.html accessed on March 12, 2016.

significant growth in demand for NG, another significant event in the development of NG sector in India was the import of NG in the form of liquefied natural gas (LNG). Regasified LNG (RLNG) started flowing into the pipelines from April 2004. With APM prices still in place, it was a challenging task for the business organizations to secure contracts for marketing of RLNG which was priced significantly higher as compared to the APM gas. In an yet another significant development, other government and private players were also allowed by the government not only to market gas available with them but also to develop their own transmission and distribution networks.

In order to boost the gas pipeline infrastructure development and also to provide a level playing field to all the players in the industry, Government of India has set up a Petroleum and Natural Gas Regulatory Board (PNGRB). It has started functioning in 2006 and has the mandate to formulate and govern policies and regulations for development and operations of oil and gas pipeline infrastructure in the country. Under the new regime, all the interested organizations are now required to participate in the bidding process to get authorization for development and setting up of the pipeline and city gas distribution networks in any of the new areas or region.

Apart from these business challenges for the NG industry due to changes in the external environment, the industry is also facing a serious challenge related to attracting, recruiting, retaining, and developing the requisite human resources required to match with the growing pace of the NG sector in the country.

3 HR Challenges Faced by Natural Gas Industry in India

One of the key challenges for the NG industry is building up the capacity and capability of the most important resource, i.e., human capital. Knowledge and skills of their human capital are going to be the key differentiators among companies across the globe.

The challenge before the industry today is not just to stop its talented and trained human resource from moving away and to effectively train and develop them to meet the business requirements but also to provide an enabling environment under which they are not only completely engaged but also contribute more effectively towards the growth of the organization. The HR Report 2015 published by International Gas Union (IGU) on talent development has identified the challenges (Fig. 1) that are being faced by the NG industry across the globe.

3.1 Attracting Youth

Companies in the Middle East and African regions tend to have the youngest workforce while globally bigger companies tend to employ an older workforce.

Fig. 1 HR challenges faced natural industry across the globe

Fresh graduates represent only a minor percentage of total new hires made annually by gas companies.

The hardest skills to find in the gas industry globally are technical skills and project management skills. The biggest overall skill shortage is felt in the Middle East and African regions. Companies here report the lowest satisfaction with the number and quality of available science, technology, engineering, and mathematics (STEM) students. However, vast efforts are being committed to filling this gap. Companies in the Middle East and Africa hire on average more fresh graduates and invest more in training them than companies in any other region of the world. Despite these facts, more time is required to alleviate the skill shortages in this region.

European companies maintain the closest ties with Academia and sponsor a higher percentage of students. This strategy seems to be paying off, as companies in Europe report on an average the highest level of satisfaction regarding both the numbers and the quality of STEM students available.

It appears that larger gas companies are the ones attracting the highest quality of fresh graduates, perhaps on account of their well-known brands. These companies seem also to have more development programs in place and invest more time in grooming their employees before these can become independent and later on, leaders in their fields. The extra training is bearing fruit as large gas companies are inclined to promote their senior managers mostly from within the company.

Low compensation package is reported to be the main issue keeping graduates and expats away from the gas industry. This is especially true for large companies and companies in the Middle East where it is a big reason for people leaving the company as well.

Gas companies prefer to contact potential employees directly, seldom do they turn to external recruitment agents.

3.2 Attracting Expats

For 90% of gas companies globally, expats represent less than 10% of their total hires. Low compensation and legal issues represent the biggest barriers to international recruitment.

3.3 Attracting the Female

Women account for a reduced percentage in the gas industry's workforce and tend to be deployed mostly in support/non-technical functions having executive positions largely restricted to them. The gas industry continues to be a male-dominated industry where 3 out of 4 employees are men while over half the gas companies globally employ over 95% male managers.

The gas industry is intensifying its efforts to become more female friendly and about three quarters of gas companies globally already have in place one or more special programs dedicated to support their women employees.

3.4 Developing Talent

There are more career ladders in place for technical professionals than for commercial professionals. Furthermore, gas companies use more individual development programs than they do career ladders.

For European companies, classroom training has overtaken on-the-job training as the method of professional development in which both junior as well as experienced technical staff invests most of its time.

American companies use on average significantly less career development programs and individual development programs than companies in any other region in the world. This issue is felt by the youth, and 'insufficient career development opportunities' are given as a refusal reason more often in the Americas than anywhere else in the world. Perhaps related to the lower investment in training and development, companies in the Americas tend to promote fewer senior managers from within the company.

The size of gas companies seems to have an inverse correlation with the number of senior managers hired externally. That is, the smaller the size, the more likely the company is to hire its senior managers from outside the company.

3.5 *Retaining Talent*

Bonus schemes are the most widely spread long-term incentive tool used by gas companies in order to motivate employees. Gas companies in the Americas and Middle East and African regions tend to have lower attrition rates, and they also tend to have in place more long-term incentives available to their employees.

While compensation often discourages young professionals and expats to join a gas company, it represents much less of a reason for gas professionals to leave their companies.

The highest percentage of gas professionals leaving their companies usually tend to head into retirement. Among those who will continue working, most will be doing so with another energy company.

4 Challenges for the Gas Industry in India

The challenges for the gas industry in India are no different. An earlier study carried out by M/s EY in 2009 on HR Challenges in the Indian Oil and Gas Sector[2] had identified five key challenges presented in Fig. 2.

Another study[3] carried out by M/s Petrofed in the year 2006 highlights similar issues and challenges before the gas industry in India. An analysis by Booz & Company suggests that by the end of the decade, the demand for key technical positions, especially in the upstream activities, will exceed supply by 30%. As a consequence of these conditions, the wage bill is rising, as is the cost per employee. It is becoming more and more difficult to have the right people in the right jobs and in the right numbers at almost every level of India's hydrocarbon companies, including executive management, technical specialists, engineers, and operators.

Even after 07 years of the study conducted by M/s EY, the situation on the ground has not improved significantly. According to the PNG Statistics 2014–15 published by MOP&NG, the overall manpower employed in the Indian petroleum industry has rather gone down from 139,956 in 2008 to 134,019 in 2015, which is about 4.2% negative growth. Although these statistics include the manpower engaged in oil sector also, but the overall dwindling numbers are a clear indication of one of the biggest the human resource challenge being faced by the gas industry in India. While we take a little deeper insight into some of these issues under the current scenario of the Indian NG sector, it would be appropriate to regroup and analyze these factors under the following heads:

[2]HR Challenges in Indian Oil and Gas Sector. http://www.ey.com/Publication/vwLUAssets/HR_challenges/$FILE/HR_challenges_in_the_Indian_oil_and_gas_sector.pdf accessed on March 5, 2016.

[3]Workforce Sustainability and Talent Management in Indian Oil and Gas Upstream Industry http://petrofed.winwinhosting.net/upload/manpowerstudy2006.pdf accessed on March 12, 2016.

Fig. 2 Challenges for natural gas industry in India

(1) Manpower required or succession planning on account of aging workforce and retirements,
(2) Talent retention-managing the attrition,
(3) Managing the Gen-Y,
(4) Increase in manpower requirement on account of aging as well as expansion of existing pipelines and plants,
(5) Talent development: requirement of specific knowledge, skills, and the experience, and
(6) Industry expectations from the Academia.

4.1 Succession Planning

Like in other parts of the world, the manpower engaged in NG sector in India is growing older. Apart from the natural superannuation from the service, many employees are seeking premature retirements. With every superannuation, the organization loses over 35–40 years of cumulative work experience. When an employee at a senior-level decision-making position leaves the organization, it becomes critical for the organization to plan and develop a suitable successor to take over the vacant role. Considering that the major development of gas industry in India started between 1980 and 90, the industry is about to experience a large-scale retirement of the first generation of gas professionals in the country. According to the analysis carried out by Booz & Company, an alarming 68% of the employees are over 40 years and nearly one-third of them are expected to retire by 2020.

A well-defined and structured succession planning is thus required to ensure proper transfer of knowledge and skills from one generation to the other. A PwC report[4]—'Succession planning: What is the cost of doing it poorly…or not at all'—suggests a five-step approach to succession planning: (i) that such planning is integrated into the organization's long-term business strategy; (ii) that the planning is owned by the senior management team; (iii) that key talent is continually assessed; (iv) that succession planning is linked to other talent management processes and practices; and (v) that companies monitor and evaluate the performance of their succession planning efforts.

4.2 Talent Retention

Some of the conventional interventions used by organizations across the globe for retaining employees and drawing upon the best out of them are; coaching/mentoring/counseling, performance appraisals, performance-based incentives/promotions, perks/fringe benefits, job enlargement and job enrichment, etc.

But these are not just sufficient for retaining the key employees. What is required is to look beyond these conventional tools of human resource development to a more systematic and integrated approach toward creating an atmosphere for continuous learning and career advancement within the organization. The strength of an organization lies in its roots, i.e., its human capital. Therefore, the HR or training division of an organization has a more strategic role to play not just in attracting and retaining the talent but also in motivating and developing them and most importantly in bringing the transformation by creating an enabling climate within the organization.

4.3 Managing the Gen-Y

Owing to the complex and volatile nature of the business and the requirement of specialized skilled set for ever expanding NG infrastructure has led to the significant influx of Gen-Years/millennials in the workforce in the last 4–5 years in the gas industry.

Managing the expectations of millennials/Gen-Y in organizations has emerged as the biggest challenge for HR in recent times. Not satisfied with their work, young executives are leaving the company abruptly, thereby disrupting the continuity of major long-term deals. It takes three to four years to nurture and develop the niche talent and skill set required for important assignments in oil and gas sector. Further,

[4]Succession planning: What is the cost of doing it poorly…or not at all? https://www.pwc.com/gx/en/oil-gas-energy/publications/pdfs/energy-company-succession-planning.pdf accessed on March 5, 2016.

middle-level executives in plants are finding it difficult to get work done from the young executives and manage their expectations.

According to Deloitte's third annual Millennials Survey[5] conducted in 2014 covering nearly 7800 millennials from 28 countries across Western Europe, North America, Latin America, BRICS, and Asia-Pacific about business, government, and innovation, the key observations include:

- 50% of millennials surveyed want to work for a business with ethical practices.
- Millennials believe that the biggest barrier to innovation is management's attitude.
- Millennials believe that the success of a business should be measured in terms of more than just its financial performance, with a focus on improving society among the most important things it should seek to achieve.
- These young people in their 20s would like new jobs and new assignments every 12–24 months and will not wait for 3–5 years for a promotion.
- Millennials prefer organizations that have open, transparent, and inclusive leadership styles.
- They thrive on fairness and performance-based appraisals; not tenure and seniority.
- Millennials would prefer to have access to a number of peers and other leaders, rather than be limited to working with one leader or manager.
- Millennials thrive on innovation and change.

Millennials want to view their boss or manager as their 'friend,' rather than viewing them more as a peer, coach, or mentor.

5 Additional Manpower Requirement

Over the past few years, there has been a significant growth in the NG industry in the country. The total length of cross-country pipeline network has reached about 16,000 km and about 11,000 km under various stages of execution. City gas distribution networks have been set up in 57 geographical areas (GAs) authorized by PNGRB and the plan is to expand the network to over 50 GAs by 2019.[6] The PNGRB plans for authorizing 240 GAs by 2022. As the NG industry is going through a growth stage, the demand for additional manpower is bound to increase across the entire gas value chain (Fig. 3). The NG value chain is a very complex chain. The value chain for domestic gas production and consumption is less complex compared to import-based gas value chain. LNG import-based gas chain is

[5]The Millennial Survey 2014 http://www2.deloitte.com/al/en/pages/about-deloitte/articles/2014-millennial-survey-positive-impact.html accessed on March 5, 2016.

[6]Vision 2030: Natural Gas Infrastructure in India http://www.pngrb.gov.in/Hindi-Website/pdf/vision-NGPV-2030-06092013.pdf accessed on March 7, 2016.

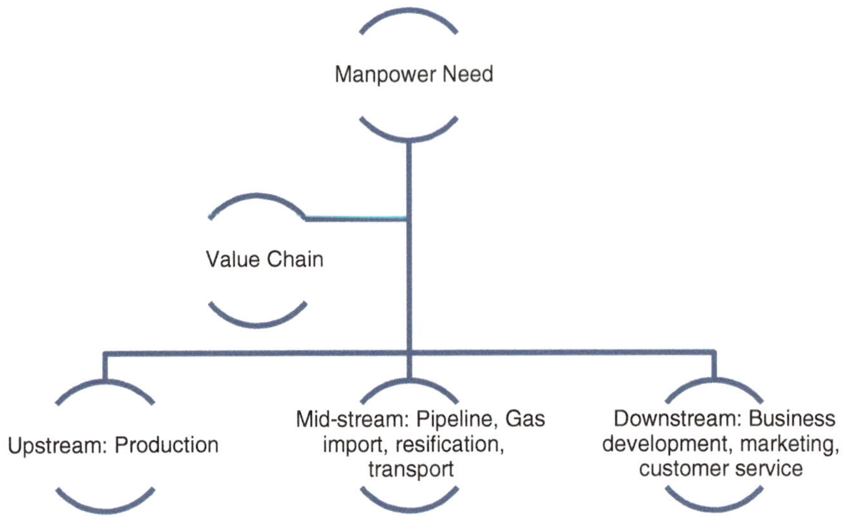

Fig. 3 Gas value chain

the most complex with greater scope for direct as well indirect employment opportunities. In case of LNG import-based gas value chain, employment opportunities would be in the following areas:

- Shipping,
- LNG Sourcing and contracting,
- Project evaluation financing,
- Consulting,
- Engineering, procurement, and construction of LNG receiving and regasification terminal,
- Pipeline construction and maintenance,
- R-LNG sales and marketing,
- CGD bidding, development, and marketing, and
- Operations and maintenance of CGD network.

In 2014–15, Petronet LNG deployed 461 employees for managing Corporate Office, Dahej and Kochi plants, with 71% of the employees having technical background and 29% non-technical background.[7] The company had attrition rate of only 4.5%. A rough estimate suggests that to run an LNG terminal with 5-MMTPA capacity requires about 120–125 employees. Table 1 presents estimated additional manpower requirements in the gas value chain in India by 2019. Core areas in the gas value chain which offers direct employment opportunities are gas sourcing, contracting, marketing, operations, business development, etc. Indirect employment

[7]Corporate profile, available at http://www.petronetlng.com/PDF/USD-Corporate-Presentation-Oct-2015.pdf accessed on March 8, 2016.

Table 1 Estimated additional manpower requirements by 2019

	Direct regular employment (No.)	Indirect employment (No.)
Pipelines	2000	5000–8000
CGD	2000	6000–8000
LNG	400	1000–1500
Services and consulting		250–500
Total	4400	12,250–18,000

Source Based on discussions with industry experts

could be in various areas such as consulting, legal services, finance and taxation, projects, maintenance, and metering.

City gas distribution offers better opportunity than LNG. Today, a CGD company like Indraprastha Gas Limited (IGL) deploys 550 regular employees and 7000–8000 contractual employees for managing its CGD business in Delhi and NCR. It is estimated that by 2019, about 2000 direct regular employment and 6000–8000 indirect and contractual employment opportunities would be created by the CGD industry across the country.

Apart from setting up a new gas-based petrochemical plant of 2.8-MMTPA capacity at Assam, the capacity of another mega petrochemical plant at Pata has been doubled from 4.1 to 8.1 MMTPA. Further, the Government has given a challenging target of providing PNG connections to over 01 crore households in the country in the next 5 years. India today has four LNG regasification terminals and 06 more under proposal/construction stage to facilitate import of gas. Developments are also shaping up in the areas of coal bed methane and shale gas. In addition to the acquisition of manpower in the gas industry, the aging infrastructure is also building up the pressure on the organizations with the requirement of additional manpower with the right set of skills and experience of working on NG pipelines/plants. Apart from the manpower required to fill the gaps on account of superannuation/retirements of employees, the organizations will face a tough task of fulfilling the manpower requirements for successful take off these wide spread and ambitious growth and expansion activities in the NG sector in India. According to the analysis carried out by Booz & Company, there will be gap of about 30% between the demand and supply of technical manpower in the hydrocarbon sector by the end of 2020. With relatively lower compensation and postings in remote and difficult locations, it is becoming increasingly difficult to attract the fresh pass-outs to take up a career in the gas industry. A still greater challenge is with respect to the number of institutes offering specialized domain-specific courses in the oil and gas sector. For such a large-scale growth and expansion, the industry is going to face a shortage of qualified, skilled, and trained workforce.

6 Supply Scenario

As of now, there are only a few institutes such as Rajiv Gandhi institute of Petroleum Technology (RGIPT), ISM Dhanbad, Pandit Deen Dayal Petroleum University (PDPU), University of Petroleum and Energy Studies (UPES), Great Lakes Institute etc. which are offering specialized degree courses in oil and gas domain. In spite of the students coming out with specialized domain specific degrees from these institutes, the students are not getting enough opportunities to work in the area of their specialization. As a result, the students passing out from these institutes are also forced to seek employment in other sectors.

Due to relative newness of these industry-specific sectoral degree programs, these institutions are yet to receive preference from the recruiters. Still the recruiters from the petroleum industry are comfortable with graduates with traditional branch such as mechanical, electrical, and chemical with generic master of business administration/postgraduate diploma in management form established institutes. Unlike the oil sector, the gas sector seems to be less attractive for prospective candidates. On campuses, the CGD companies remain as the less preferred companies to work. Interactions with prospective graduates suggest that reasons behind less attraction for CGD entities are:

- CGD companies are offer less attractive salary,
- limited career growth, and
- relatively smaller cities with lesser facilities.

7 Requirement of Specific Knowledge, Skills, and the Experience

Although some of the previous studies have tried to estimate the projected number of manpower required for employment in the NG sector in India, however, on account of various other factors including the prevailing uncertainty in the gas sector, it is very difficult to give any projections at this stage. What can be said with certainty is that with the large-scale retirements that are expected to take place within the next 4–5 years in the gas industry, the industry will need not only adequate numbers to fill this gap but more importantly the people with the specific domain knowledge, skills, and the experience of working in the gas industry. Also the continuous upgradation of the technology as well as emergence of new technologies like shale gas development calls for continuous upgradation and acquisition of required knowledge, skills, and technical expertise.

8 Actions to Be Taken

The challenge in meeting the requirement of the NG sector is not just in fulfilling the requisite numbers but also in understanding the current and futuristic skills required and developing the workforce in those skills. This will require all the stakeholders in the process, right from the academic institutions, industry including suppliers and contractors, government, training organizations, quality and certification bodies, and regulatory bodies, to come together and work upon in creating the required talent pipeline.

The way forward to overcome these challenges and to develop a workforce that is not only dynamic, flexible, and productive but also possesses skills for mobility and growth may include the following recommendations:

(1) Effective Industry—Academia Interface

The scope of existing interface and association between the industry and the academic institutions needs to be strengthened and extended beyond the conventional vocational trainings and industry sponsored projects to a more structured and regular interaction between the two, to support:

- Development of strong faculty by deputing employees from industry as faculty,
- Development of faculty by providing industry orientation and field-related research projects,
- Joint review of course curriculum and introduction of new courses to meet industry expectations,
- Industry exposure programs and sponsored projects for students, and
- Involvement of retired industry professionals as faculty in academic institutes.

The above-mentioned areas have been spoken about on many occasions. However, the progress in these areas is not noteworthy. Institutions like RGIPT encourage and invite industry experts to teach full courses and engage with students, faculty, and the institute. Deputing employees from industry to academic institutions has not happened so far. Remuneration structures of academic institutions are not comparable and compatible with the industry structure. Therefore, academic institutions would find very difficult to match the industry on remuneration structure. The deputation option would remain as challenging as before. However, the faculty deputation to industry is viable.

The idea of industry-academia joint research is very exciting and interesting. Industry and academia should create Joint Research Fund to promote joint research culture and encourage such research programs.

For the time being, the industry exposure programs are largely limited to summer internship, outcomes of such internship programs have limited use to the industry or academics. The industry exposure programs should be strengthened further to generate meaningful and productive outcomes.

Involvement of superannuated experienced industry experts in academics has been a practice in many of the academic institutions, especially sector specific institutions. Possibly more intense engagement of superannuated experienced resources from the industry would be helpful in the knowledge transfer process. The academic institutions and the budding young graduates are going to benefit from such progressive and meaningful engagements. The experienced and professional experts from the industry would be helpful to share their field experience, which would help the graduates to learn and inculcate the industry practices from the college days. In a sense, these graduates would be ready for any real-life future challenges to be encountered in the corporate world.

(2) Talent Attraction, Retention, and Development Programs

Hiring good people is tough, and retaining them is tougher but developing them in alignment with the organizational goals and objectives is the real challenge. This would call for multipronged initiatives like:

- Strengthening the gas sector branding in academic institutes through special industry-specific campaigns, screening of films, distribution of videos and literature, institution of scholarships, industry sponsored projects & higher education opportunities, etc.
- Offering an attractive compensation package, facilities, and a challenging work environment, reward, and recognition schemes focused at talent retention.
- Competency mapping, skill gaps identification, short- and long-duration training and development program, focused induction orientation programs, effective job rotation, mentoring, higher education incentive programs, etc.

(3) International Exposure Programs

There is a need to have focused international exposure programs to help the students as well as working professionals to assimilate the global best practices in gas sector and understand the importance of cross-cultural issues in managing business, trade, and industry. This will help the individuals to have a better understanding of the global business requirements and motivate and prepare them for effectively taking up the related assignments as and when required.

(4) Common Qualification and Certification Programs

In the wake of the expanding network and the safety hazards associated with it, it is essential to introduce a system of common minimum qualification and certification for the workmen engaged in design, construction, and operations and maintenance functions which directly affect the safety and integrity of the infrastructural facilities. Right from the upstream to downstream, the industry today is highly dependent on outsourced manpower across the entire gas value chain. Out of these, quite a big number of outsourced personnel do not have any professional qualification, certification, or even formal training of working in the NG sector. In order to take care of this situation, critical tasks and operations expected to be performed by these workmen needs to be identified. Based on these tasks, specific programs need to be developed for training, qualification, and subsequent skill certification of individual competence to satisfactorily carry out the particular task.

9 Conclusion

Globally, the oil and gas companies are experiencing a lot of 'change' and this change can be best managed if and only if the organizations can 'change or transform its human resource' to keep them aligned to organization's objectives and changing business requirements. The recent fall in crude oil prices has shaken the entire oil and gas industry across the globe and its impact can easily be seen on the NG industry in India also.

The prevailing uncertainty and volatility in the global market has forced the organizations to revisit their business strategies and shift their future outlook from 10–15 years to a much closer outlook of 3–5 years. This will compel the organizations to revisit their HR strategies as well make it more aligned to the changing business environment. Fresh manpower requirement may just be limited to fill the existing gaps for a next few years and the focus will be more to enhance the productivity and efficiency of the existing human resources. More than the capacity, it will be the intellect and capability. Thus, it is the people who will remain at the core of any organization, and as long as the roots of the organization, i.e., its human capital, are strong, no storm howsoever strong can bring down the organization.

Negotiating and successfully navigating this uncertain and volatile environment requires a highly efficient, vigilant, innovative, flexible, and agile workforce. In the nut shell, a talented human resource enriched with intellectual capital, i.e., professionals with sound and updated knowledge, skills, and a positive attitude, will hold the key to future direction and growth of NG sector in the country.

Health, Safety, Environment: Issues and Challenges for Gas Value Chain in India

Jitin Saxena

Abstract With growing demand of energy in country, activities in sectors are increasing due to which managing health, safety and environment (HSE) is becoming major and critical concern for gas companies. Any negligence in upkeep/handling of highly inflammable gas products can be a reason of accident taking place which possess high risk to life, property and environment. Over the period, accident rates in sector have gradually reduced which shows that companies are becoming more vigilant and proactive on safety issues. Still there are number of safety-related issues and challenges across gas value chain in India which needs to be addressed by the companies in support of government. This chapter describes various safety issues and challenges emerging these days in sector and highlights the needs for developing safety culture and well-established HSE management system to address those issues.

Keywords Health · Safety · Environment · Oil and gas

1 Introduction

Gas sector has a significant role in the nation's economic development. The sector is of strategic importance and plays a pivotal role in addressing India's energy concerns, thereby influencing decisions on all other spheres of economy. Due to reduction in domestic gas production and lower off-take of gas, there is a predominant decline recorded in sales of Natural Gas during 2015 (PPAC 2015). However, due to "Gas Pooling Policy" introduced by Govt. of India, demand for natural gas in India has increased significantly in the power and fertilizer sectors. Growing demand of natural gas has also driven by the increasing use of natural gas in the city gas distribution (CGD) sector due to rapidly increasing urbanization and also due to rising concerns to reduce pollution by using clean fuel in industrial

J. Saxena (✉)
GAIL India Limited, Noida, India
e-mail: jitin.saxena@gail.co.in

S.K. Kar and A. Gupta (eds.), *Natural Gas Markets in India*,
DOI 10.1007/978-981-10-3118-2_15

sector. Govt. of India is encouraging investments to address the significant unmet demand for gas. As per Snapshot of India's Oil and Gas data (PPAC 2015), India had a natural gas transmission network off 16,065 km, with an aggregate transmission capacity of 350 million standard cubic metres per day (MMSCMD) and another 11,302 km of natural gas pipeline projects with an aggregate transmission capacity of 468 MMSCMD are under execution or construction. In view of the increasing activities in exploration, LNG import, processing and transmission of natural gas, the safety concerns have become a major and critical factor in the gas industry.

Since gas is highly inflammable, the impact of any accident taking place can be very serious as such incidents will affect not only the accident site, but also the surroundings. They possess serious threat for lives, damage to material and environment. Hence, it is imperative that operating companies should not only follow the applicable standards of health, safety and environment (HSE) protection, but also be responsible to maintain even high-level safety in their operations so as to avoid any accident condition.

2 Accidents—Safety Performance Indicators

Fatal accident rate (FAR) and lost time injury rate (LTIR) indicate the measurement of safety performance of any organization. High FAR or LTIR impedes the progress of safety performance. FAR is defined as the number of fatalities in company per 100 million hours worked. FAR is based on 1000 employees working their entire lifetime. Assuming that an employee works for an average of total 50 years, the FAR is based on 10^8 working hours. LTIR is defined as the number of lost time injuries, i.e. fatalities and lost workday cases, multiplied by 1 million divided by the number of man-hours worked in the reporting period. A lost time injury is a work injury or disease where the injured person has at least 1 complete day or shift off work.

Generally, companies include safety statistics in their annual reports. Their management always has strong desire to show their safety performance indicators look good. But, the more is the pressure to keep these figures look good, the more is the possibility of personal damages and losses taking place reason being that falsely projected improvements in safety performance can be a major reason for over confidence in employees and workers which leads to lack in compliance levels and overlooking of safety issues.

2.1 Global Scenarios on Accident in Gas Sector

In spite of difference in cultural, historical background, geographical conditions and administrative setup in different countries across the world, the safety practices and procedures followed by companies are almost same for construction, operation and maintenance. Moreover, governments as well as companies are taking initiatives to gather data of these accidents taking place in industry globally, which is indeed extremely useful for improvement in safety systems of respective companies by timely identification and taking suitable actions for major causes of accidents taking place in the similar setups.

In USA, during 2003–2008, 648 oil and gas extraction workers were fatally injured on the job, resulting in an occupational fatality rate of 29.1 deaths per 100,000 workers—eight times higher than the rate for all US workers (NORA 2011). By the time, this report was published in 2011 by National Occupational Research Agenda (NORA) for Occupational Safety and Health Research and Practice in the US Oil and Gas Extraction Industry, another big accident of explosion and subsequent fire on Deepwater Horizon semi-submersible Mobile Offshore Drilling Unit (MODU) was reported on 20 April 2010. Total 11 workers were killed and 16 others were injured in the accident. The explosion caused the Deepwater Horizon to burn and sink, resulting in a massive offshore oil spill in the Gulf of Mexico affected around 200 miles. It was considered one of the largest accidental marine oil spills in the world.

Safety performance indicators of major companies globally are collected each year by the International Association of Oil and Gas Producers (OGP). Safety statistics for 2013 were contributed by 50 major companies in E&P and is covering operations in 110 countries worldwide. The statistics collected by OGP does not give a completed picture for the entire world's upstream sector; however, these figures may be taken as indicative to understand the trend.

The 2013 OGP Safety Performance Indicators shows that the FAR (Fig. 1) for reporting companies has decreased by 11% compared with 2012 (OGP 2013). However, the number of days lost per lost workday case overall has increased by 6% in 2013 compared with the results of 2012.

Eighty fatalities occurred in 43 separate incidents were reported in 2013, which were 8 fewer than those were reported in 2012 and 15 more than reported in 2011.

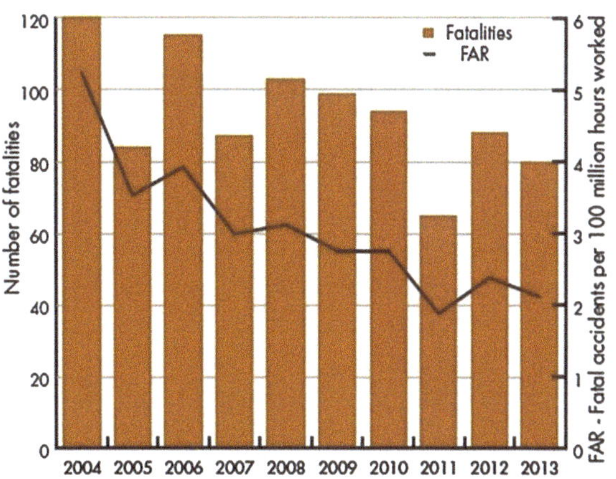

Fig. 1 Numbers of fatalities and fatal accident rate (2004–2013). *Source* OGP Safety Performance Indicators—2013 data, report July 2014

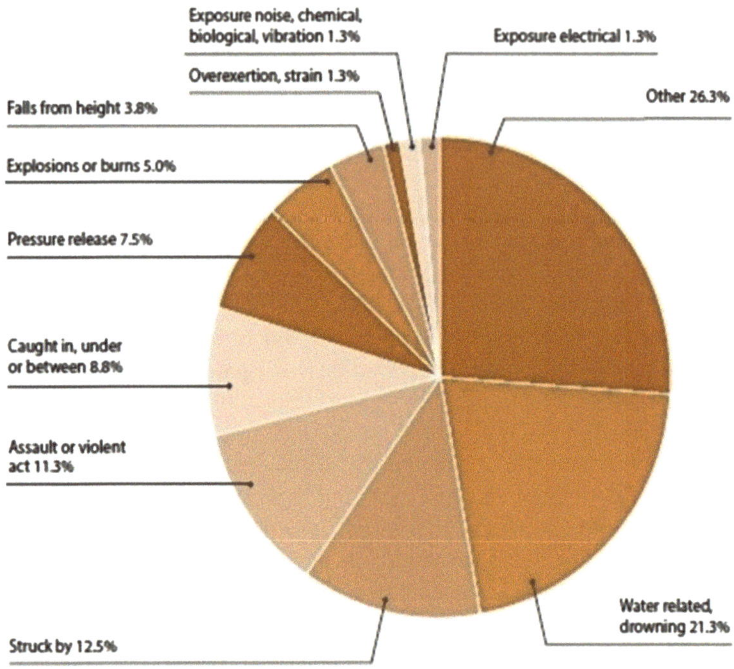

Fig. 2 Fatalities by reporting Category 2013 (% fatalities). *Source* OGP Safety Performance Indicators—2013 data, report July 2014

The largest portion of these fatalities reported in 2013 were categorized as Others (26%) (2% in 2012)—this mainly accounted for helicopter crash incidents reported at Peru and UK. Water-related drowning is another category in which 21% fatalities were reported in 2013 (1% in 2012) and fatalities occurred as result of individuals being struck by falling or moving objects accounted for 13% (Fig. 2) compared to 16% in 2012. Terror attacks are another category that has emerged as major reason for fatalities; 11% fatalities (Fig. 3) during 2013 (3% in 2012) were reported in a single incident of terror attack at Algeria.

There were 1627 reported lost workday cases resulting in off from work for at least one day, which equates to an average of 31 injuries resulting in at least one day off from work every week of the year. Although absolute numbers of lost workday cases have reduced (1699 in 2012), the time of absence from work due to the accidents has increased by 6% compared to 2012. Figure 4 presents severity of lost working days trend during 2004–13.

In gas transmission business, seventeen gas transmission system operators in Europe are collecting incident data on more than 143,000 km of gas pipelines every year. The total exposure, which expresses the length of a pipeline and its period of operation, is 3.98 million km year. As per the 9th report of EGIG (European Gas Incident Data Group) (EGIG 2015), incidents caused by external interference, corrosion, ground movements and construction defects/material failures (Fig. 5) are

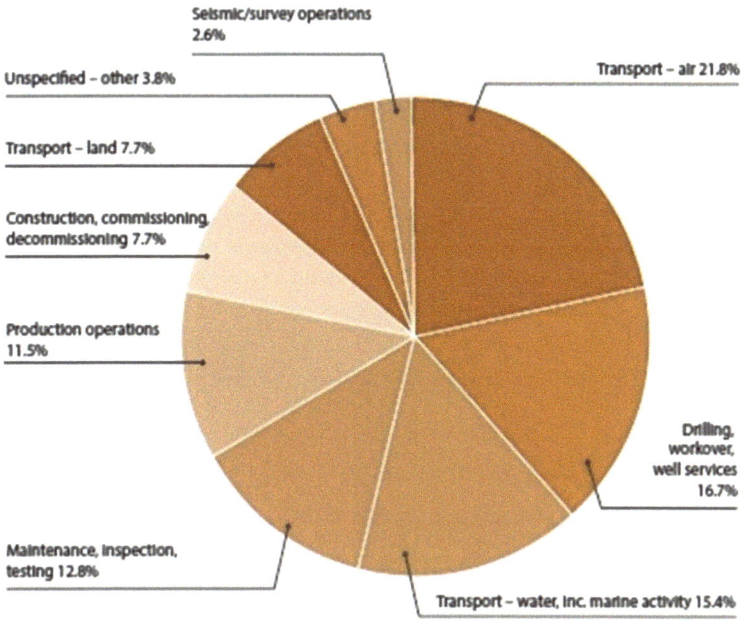

Fig. 3 Fatalities by Activity 2013 (% fatalities). *Source* OGP Safety Performance Indicators—2013 data, report July 2014

Fig. 4 Severity of lost workday cases (average days lost per LWDC). *Source* OGP Safety Performance Indicators—2013 data, report July 2014

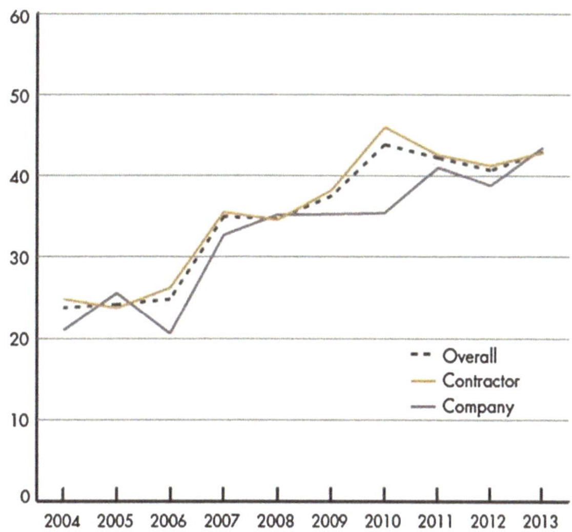

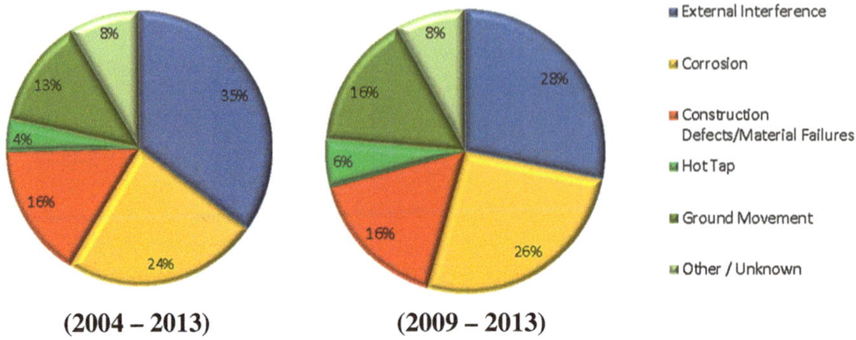

Fig. 5 Distribution of incidents in gas pipeline. *Source* European Gas Pipeline Incident Data Group

characterized by potentially severe consequences which emphasizes their importance to pipeline companies.

EGIG study reveals that external/third-party interferences have major contribution of 35% of accidents in transmission business reported during period of 10 years (2004–2013), while corrosion failures and construction defects/material failures are other major challenges for safety of transmission pipelines contributing 24 and 16%, respectively, for the causes for accidents recorded in last 10 years.

2.2 *Indian Scenarios on Accident in Gas Sector*

India witnessed the worst industrial disaster on midnight of 3 December 1984, when a pesticide plant in Bhopal accidently released around 40 metric tons of methyl isocyanate into atmosphere creating a catastrophic condition in entire Bhopal with estimated 2000 fatalities, 100,000 injuries and a huge damage to agriculture. Naming a few of major incidents of hydrocarbon disasters in India after the Bhopal Gas Disaster a fire in an oil well in Andhra Pradesh in 2003; an explosion in the Hindustan Petroleum Corporation Limited Refinery (HPCL), Vishakhapatnam, in 1997; ONGC Bombay High ship collision with platform and riser fire in 2005, an explosion in Indian Petrochemicals Corporation Limited (IPCL) Nagothane, Maharashtra, in 2008, Major fire at Indian Oil depot at Jaipur in 2009, and gas pipeline accidents Kakinada, Andhra Pradesh (GAIL), in 2013, has made operating companies extra vigilant and more responsible for adopting stringent safety measures in handling these products. It has also attracted the attention of government and regulatory bodies not to lax operating companies maintaining the high level of safety in their operations.

Bottom line for focusing the efforts on managing HSE by gas operating companies is incident and accident data. These data are often considered as a single defining measure of the industry's state of safety and the authenticity of these data is hardly challenged. Indeed, such data are often the initiator for expensive and

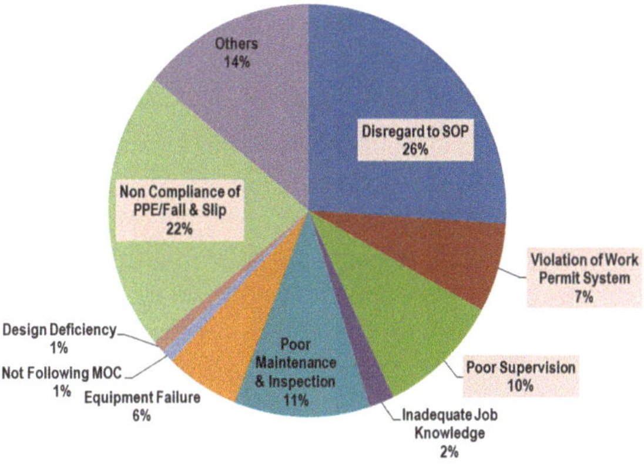

Fig. 6 Analysis of major incidents in refining and GPUs (2004–2015). *Source* Oil Industry Safety Directorate

elaborate safety interventions for improvements by the companies involved. Interestingly, Oil Industry Safety Directorate (OISD) illustrated in Analysis of Major Incidents in Oil and Gas Industry (2004–2015) revealed that disregarding the Standard operating procedure (SOPs), Non-compliance of Personnel Protective Equipment (PPEs), Poor Maintenance of Equipment and Poor Supervision (Fig. 6) are the major reasons of accident taking place at refining and gas processing plants.

OISD studies revealed that 94 nos. incident were reported during the period of 2004–2015 in pipeline transportation sector. Thirty-nine out of these were the fire incidents and 55 incidents were the non-fire incidents. 29 and 27 incidents were attributed due to Pilferage/miscreant activity/others and equipment failure, respectively, during the Root Cause Analysis of Major Incident Reported. Poor supervision and non-compliance of personal protective equipment/fall and slip were reasons identified for 22 incidents. Disregard of standard operating procedures and violations of work permit system were analysed as root causes for 7 incidents. Poor maintenance and inspection were identified as reason for another 7 incidents during 2004–2013. Figure 7 presents percentage share of major accidents in cross-country pipeline network in the country during 2004–15.

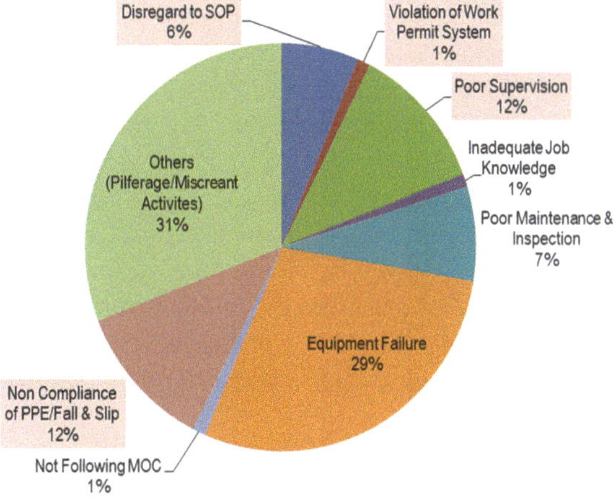

Fig. 7 Analysis of major incidents in cross-country pipeline (2004–2015). *Source* Oil Industry Safety Directorate

3 Factors Resulting into Accidents

Over the last decade, the overall frequency of accidents happening in gas industry has gradually reduced demonstrating increasing integration of safety with systems, i.e. proper design and construction, adequate maintenance and safe operation. This improving trend shows that the regulatory bodies and companies are becoming more vigilant and proactive on safety issues. Further, advancements in information technology have now made it possible to get quicker information about the effectiveness of safety measures and systems and also helping in making emergency response more effective. Still there are number of common causal factors related to accidents of high potential of fatality and damages taking place which includes inadequate/noncompliance of standard operating/work procedures, inadequate supervision, inadequate knowledge/competency level, inattention/lack of awareness of hazards leading to improper decision-making, design deficiencies, accident excavations/external interference, corrosion and construction defects/material failures etc.

A few of such safety issues and challenges gas sector is facing are elaborated below.

3.1 Human Factor

Deeply analysing the reasons for high potential of fatality and damages taking place such as inadequate/non-compliance of standard operating/work procedures, inadequate supervision inadequate knowledge/competency level and inattention/lack of awareness of hazards leading to improper decision-making, it became evident that numbers of factors converge to a common element of human interference or negligence by any person involved in act or condition responsible for the accidents. In fact, many factors are embedded in this main factor of human error. Both employer and workers recognize the seriousness of losing precious lives because of accidents and want to prevent the accidents resulting human tragedies and financial loss to the companies. There are certain gaps which are the root causes for most of the human-related factors resulting into accident situations. One such gap is temporary workers engaged on contract basis lacking awareness on hazards, safety procedures or perhaps less committed to comply with them. More work pressure on these workers may be due to reason that their compensation is linked with quantity of job they have to perform. A plumber engaged through a contract agency working for city gas distribution (CGD) company is getting his compensation based on number of domestic PNG connections he is making. Being a natural tendency to earn more he may overlook certain safety procedures or may execute his job in hurry to complete them. Sometimes they have been assigned with more dangerous jobs to perform with a perception or reality that accident happening with contract workers may have less direct implications on company in comparison with accident of regular employee or contract workers may be more so acquainted with work in poor working conditions and environment. Many such incidents have been reported so far in which such reasons were cited as leading causes for accidents taking place. Other side of this, it is not always true that contract or outsourced workers have been provided with inadequate resources and information, sometime workers from contracting companies supporting the operating companies by providing engineering solutions, operating or maintaining specialized equipment or by construction of new faculties carry difference in safety practices, procedures and also safety behaviours at times with them. In such case, those workers shoulder much of responsibilities with associated risk. These differences may have high implications towards factors contributing towards accidents taking place.

This aspect has created a serious need for organizations to align their policies and practices with intension to reduce chances of human errors. They should not merely develop safe procedures and practices and educate the workers at all levels, but also should encourage and motivate the workers with various reward or incentive schemes so as to create a safety culture within organization.

3.2 *Third-Party Interventions*

Gas transmission business presents unique features of long length of pipelines, high value of product and often inaccessible or remote locations where these pipelines are passing through. Unawareness of third parties or other utility providers about the presence of underground gas pipeline and lack of coordination of state agencies and gas companies is one of the major concerns for accidents taking place due to the third-party interferences in gas pipelines. In the OISD studies for analysing of major accidents reported during 2004–2015, maximum number of 29 incidents out of 94 incidents was attributed due to pilferage/miscreant activity/others. Third-party interventions are becoming a big challenge especially for pipeline operators.

In an incident reported on 16 January 2015 around 09:45 h (PNGRB—Analysis of incidents 2015), a gas leakage and fire incident took place due to drilling hole in 12″ natural gas pipeline of a major gas company by contractor in National Capital of Delhi. The contractor was doing soil investigation survey by drilling at affected location without any intimation and permission from company. After puncturing of pipeline gas leakage started which travelled through casing pipe and storm water drain. Fire spread around 20–30 m away from punctured pipeline due to possible naked fire in the surrounding areas. Hole size of around 92–103 mm diameter on 18″ casing and 08 mm × 47 mm size cut in curve form on 12″ carrier pipeline found made by the contactor. Due to fire, three cars parked near the site were burnt completely, and one three-wheeler passing through area also got burnt partially. Auto driver suffered from burn injuries and was discharged after first aid from hospital. Major causes for the accident were identified as mentioned below:

(a) A contactor was working in the affected area without informing the entity, through joint inspection for construction of underpass was already carried out with company and agency performing the job.
(b) Work carried out in the absence of technical supervision of company and agency performing the job.
(c) Ignorance of pipeline warning boards by the contractor.

In another incident reported, on 13 April 2015 around 22:30 h (PNGRB Analysis of incidents 2015), MDPE pipeline of gas company of diameter 63 mm and 4 bar pressure was damaged while HDD activity was being undertaken by third party resulting in leakage of natural gas. The leaking gas caught fire while rectification of the damaged pipeline was in progress to two workers who were rectifying the damage sustained minor burn injuries. Natural gas supply was stopped by using squeezers and isolation valves. Fire was extinguished and injured personnel were taken to the nearest hospital. In this case, if the third party had started HDD activity with due information to gas company the accident would not have taken place. In CGD, where gas pipelines are of smaller diameter and running close to utilities services/lines such as water/sewer mains, telecommunication and electricity, there is further high potential of third-party incidents taking place. As per the EGIG (European Gas Incident Data Group) recent report (EGIG 2015), failure frequencies (Fig. 8) are quiet high in small diameter pipelines in comparison with large diameter pipelines.

Pipeline systems are primarily maintained by round-the-clock monitoring of pipeline parameters such as flow, pressure and temperature with the help of integration of SCADA systems and instrumentation devices. In case of any unsafe state, such systems immediately detect and indicate to the operator for taking action. This certainly helps in making early actions to mitigate the losses and handling emergency but does not help to an extent for avoiding the accident situations taking place.

Pipeline patrolling through line walk and by helicopters has a proven method for timely identifying the third-party interferences. Most companies have increased the frequency of such patrolling activities against what is prescribed in national and international standards. Pipeline failures by third-party activities can also be

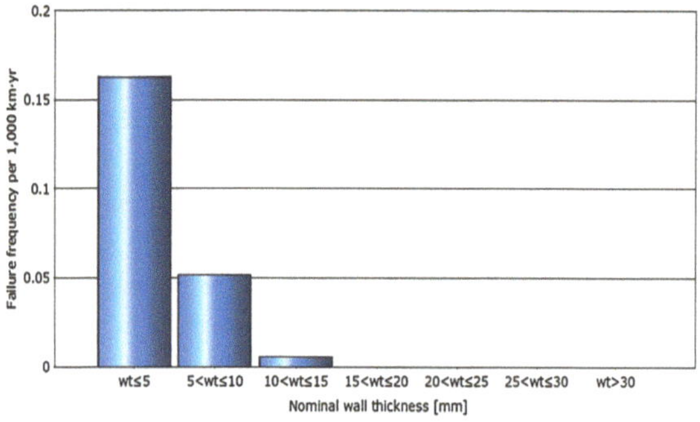

Fig. 8 Failure frequencies due to external interference in gas pipeline. *Source* European Gas Pipeline Incident Data Group—9th Report 2015

reduced by ensuring that the members of public, landowners and utilities services are always aware of the pipeline nearby and by proving adequate protection to the pipeline in areas where the risks appear high. This can be done with the help of media by conducting awareness campaigns explaining the needs to obey and follow the safety instructions and emergency procedures. In particular in cities, pipeline should run in dedicated pipeline corridors providing a protected zone so that the likelihood of failure by external third parties can be reduced. Moreover, there is a serious need emerging out needing for a nodal agency to coordinate permission to work in common ROU. This nodal agency can be put into place by joint efforts of central level regulatory bodies such as PNGRB and NDMA and concerned state agencies.

Increasing third-party interference in pipeline transportation urges also for a need of development of new technologies for monitoring and early detection of third-party damages that can help in strengthening the emergency procedures.

3.3 Operation of Assets Beyond 25 Years

Most of the existing infrastructure of gas industry is ageing and has been exposed to a severe environment and heavy usage. More than 30% of cross-country pipelines in our country are operating for more than 25 years. The ageing of these pipelines in public spaces pose safety risk to nearby population. Most of pipeline and installations are now operating beyond their design life. Integrity of these ageing pipeline and installations is facing challenge due to the presence of highly corrosive environment and operating conditions such as high temperature and pressure. To maintain integrity, much care is necessary to ensure that corrosion and other such factors have not been affecting the structural strength of pipelines or installations.

Older pipelines were built when standard requirements for design, construction, testing and operating were inferior in comparison with those of today. There is a significant improvement in standard requirement for design, construction, testing and operating because of technological advancements and availability of more historical data to analyse the need for upgrading those standards. Failures due to corrosion in old pipelines are also one of the major concerns. Different parameters considered for corrosion failures are year of construction of pipeline, type of coating and wall thickness. Corrosion is not directly depending on thickness of wall of pipeline, but pipelines with less thickness are more like to fail. In fact, requirement of cathodic protection for pipeline was not there in codes before the 1950s. The welding procedure was less advance and materials that are used now days to prevent corrosion were not available. Older pipelines with tar coatings have higher failure frequency in comparison with modern coatings such as polyethylene coating. Further, the impact of leaks from these pipelines on environment was also not given much consideration in early days.

Exhaustive risk assessment studies for such older pipeline and plants should be carried out and thrust should be given for minimizing the risk of accidents taking place in older pipelines and reducing the consequences of accidents. Integrity management of older pipeline and plants should be given due importance and must be addressed at every stage of life cycle, including application of inherent safe design, operation within design limits, regular preventive maintenance and inspection of safety critical equipment and managing engineering changes. It certainly requires investment in design reviews and increased maintenance activities. Integration of the old assets with new facilities may possess treats for their safe and continuous operation. Thus, due considerations must be given on before assimilation of old and new facilities. The additional investment and increasing maintenance requirements to secure asset integrity mean the industry faces another challenge to ensure sufficient and competent professionals to meet these challenges.

3.4 Increasing Urbanization Nearby Passing Pipelines and Plants

Density of population expressed as the number of persons per unit area helps in getting a better understanding of the spatial distribution of population in relation to land. Rapid increase in urbanization has been observed in many parts of the country. There has been a steady increase of more than 200 persons/km^2 over the last 50 years as the density of population increased from 117 persons/km^2 in 1951 to 382 persons/km^2 in 2011 (NCP). This trend of population growth and the location of this growth imply the need to manage the increasing number of people near gas transmission pipelines. Increase in population density has increased the risk for old pipelines due to high potential of losses in case accidents taking place in and around gas pipelines. Pipelines codes are using “Location class” to prescribe design pressure, type of construction, pipe thickness and operating parameters for construction of pipelines. Location with high density of populations is prescribed with higher design factor which implies to more strengthen in type of construction and thickness of pipeline. Old pipelines were laid as per the design requirements applicability at the time of lying. However, the substantial growth in population density near these pipelines making their design becomes obsolete but there is no option available with gas companies except to operate them with the increased risk. Say in the National Capital Territory of Delhi present population density is 11,320 persons/km which has grown drastically in last decade. The gas pipelines designed and constructed in National Capital Territory of Delhi around 20–30 years back are now categories in high risk level due to continuously increasing population density around these pipelines and thus requires additional protection.

Increase of risk due to high consequences of accident taking place in the pipeline and plants where the population density has grown up become further manifold due to increasing possibility of accidents taking place due to ageing of these pipelines and plants.

3.5 *Shortage of Skilled/Technical Manpower*

Due to growing needs of energy in country, there is a high demand for oil and gas and is further expected to see the increase in years to come. While the sector is one of the world's higher wage payers, it has been suffering serious shortages of manpower that can affect its ability to operate and achieve the needed grow rate. The sector is facing this challenge in terms of high attrition rate, ageing workforce, inability to attract new talent and unavailability of skilled workforce for both core and outsourced operations. This is in turn creating a serious concern for maintaining safe operations for which sufficient, well-trained and capable manpower is must. There are several other reasons why this particular sector is facing this challenge and one of them is that major working levels are outsourced where contractors and subcontractors intend to gain maximum profits and pay only minimum wages to the workers. This discourages the contract workers to entry in this sector. Due to high level of safety compliance, working conditions and environment in gas companies are also generally above average, although most gas installations are remotely located with odd working timings, demotivating the capable people to choose their carrier for coming into this sector.

Hydrocarbon Sector Skill Council (HSSC) was established by Ministry of Petroleum and Natural Gas in 2014 to address this concern for shortage of manpower in the sector. HSSC in its meeting held on 17 December 2014 has projected and proposed certification-based skill development programmes for around 19.5 lakh (Fig. 9) skilled workers in various categories in hydrocarbon sector in next 10 years. Looking into the huge requirement of skilled manpower in hydrocarbon sector it is required that private agencies should come forward and supplement the steps taken by the government.

S No.	Category	Estimated No. in 10 years	Areas covered
1.	Retirements of permanent staff (PSUs alone)	25,000	Oil & Gas Companies
2.	Additional requirement for Core Operations (PSUs & Private)	50,000	E&P, Refining, Pipelines, Sales & Distribution, Pipelines
3.	Contract Workers (approx. 3.5 times permanent staff)	6,12,500	Project/Construction/ Annual Rate Contracts
4.	Workers engaged through Service Providers (approx. 6.5 times permanent staff)	11,37,500	Retail Outlets, Service Station, Rigs, etc.
5.	Reskilling of the existing personnel	1,25,000	All sub-sectors
	TOTAL	**19,50,000**	

Fig. 9 Projection for certification-based skill development programmes in hydrocarbon sector in next 10 years. *Source* Hydrocarbons Sector Skill Council

4 Leadership and Commitment—Managing Safety Culture

"*How can we improve our safety performance to the next level*"—This is an obvious question for almost all companies. It is generally difficult for senior-level executives of companies with even a well-established safety management system in place to spend their much time directly dealing with HSE issues. Usually HSE professionals are conducting safety meetings, inspections and arranging audits. Is this can be considered enough for management to demonstrate their commitment towards managing HSE? Certainly not, role of management towards safety in much beyond providing sufficient competent workers to operate safety, to provide and maintain safe equipment and operations and to ensure settlement of environment concerns emerging out of operations. A proactive management is focused for establishing a safety culture which is much above complying the requirements and procedures.

There are two known ways indicating safety performance of an organization—*reactive way*, which are outcomes of breaches in systems and procedures indicted by accident and incidents taking place and *active way*, which show how successfully HSE plans are implemented through compliance with systems and procedures.

Safety performance can be managed only by keeping the focus on reactive indicators, i.e. taking remedial actions on accidents, incidents and unsafe accidents; however, if safety performance has to be brought to next level, there must be a clear demonstration of commitment by management. They have to make it clear with their statements and by actions that they are totally committed for safety and expect those below them to place a high priority on safety. *Performance you demonstrate is the performance you expect.*

A general theory is that numbers of risk and at-risk behaviour are a function of the safety culture of any organization. Safety culture is generally developed and established by management by virtue of setting examples of safe behaviour in their action and decision-making. Safety culture makes workers empowered and educated towards need to maintain safety and transformed them proactively towards managing safety performance. Even when a new executive enters the organization, he or she gets influenced from fellow colleagues on how to perform certain jobs and how to get the jobs done from subordinates. What level of safety compliance (Fig. 10) is acceptable and what type of safety violations will have no significant impacts. This develops a culture in them which generally sustains during their entire period of their association with the organization. However, it also indicates on a serious need of a tailored initial induction training programmes covering the orientation about the hazards, associated risks and the control mechanism in place of the new employees.

Realizing the need to address the safety in behaviour of workers both regular employees and those working on contract basis, major companies have put in place a number of extensive safety interventions for behavioural-based safety (BBS). Behaviour-based programmes need to have a top-to-bottom approach. Programmes

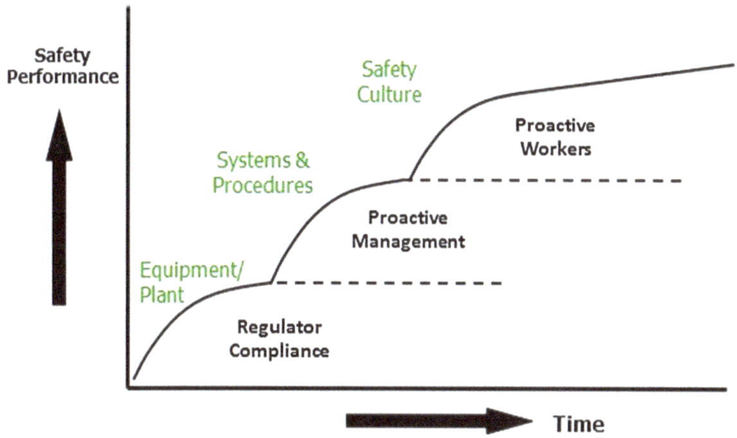

Fig. 10 Managing safety performance

can be most effective when used in systematic manner in combination with engineering solutions. These programmes are geared towards educating the workforce the consequences of their actions when dealing with hazardous equipment and operations. This creates awareness amongst workers towards hazards specific to the actions they are going to do or the operations in which they are involved. A need for a tripartite national commitment and national action in fostering a preventive approach and a safety culture was also pointed out during the 91st session, 2003 of International Labor Organization (ILO) on Global Strategy on occupational strategy on occupational safety and health (ILO 2003).

5 Need to Establish Health Safety and Environment Management System

Understanding the importance of HSE issues, organizations have established HSE management system customized to their requirements, although the principle approach of such HSE management systems is mainly to achieve operating excellence and avoid losses of all kinds. Generally, HSE management systems are focused on providing management commitment, implementing risk–control mechanism, a structured performance measurement programme and a systematic auditing in place. All activities of organization are covered in HSE management system describing detailed processes, procedures and other programmes by which organization should deliver simultaneously improving business performance and safety conditions.

Many guiding frameworks are globally available helping organizations in establishing HSE management system. A systematic approach provided on occupational health and safety management by International Labour Organization

(ILO) contains the main elements of policy, organizing, planning and implementation, evaluation and action for improvement. Another similar in structure and internationally recognized accredited standard for occupational health and safety management is OHSAS 18001. Exception is that OHSAS is only awarded to organizations those follow a successful audit by an awarding institution.

In the Indian context, Oil Industry Safety Directorate has developed framework on Safety Management System in Petroleum Industry (OISD GDN—206 2001) which provides a useful guideline for establishing a structured HSE management system for managing safety and intends to cover all aspects of plant and facilities to control loss to personnel, equipment, material and environment. OISD guidelines specify the requirement of following 19 elements for Safety Management System in hydrocarbon industry:

(a) Safety Organization
(b) Employees Participation
(c) Process Safety Information
(d) Process Hazard Analysis
(e) Operating Procedures
(f) Training
(g) Contractors
(h) Pre-startup Safety Review
(i) Mechanical Integrity
(j) Work Permit
(k) Management of Change
(l) Incident Investigation and Analysis
(m) Emergency Planning and Response
(n) Compliance Audit
(o) Occupational Health
(p) Off-the-Job Safety
(q) Customers and Products
(r) Road Transportation
(s) Trade Secrets

HSE management system of any organization is a true mirror which reflects the well-being and image of the company. Success of any HSE management system depends on the way it is implemented within organization. HSE programmes and interventions must be aligned with normal business planning processes for better acceptance at all levels. HSE requirements must be agreed within employees and built into the appropriate groups and personnel objectives. Active participation of employees at all levels in the HSE programmes will certainly help in fair and effective implementation of HSE Management system.

Safety monitoring and auditing of HSE management system plays a vital role in sustaining and developing the systems and procedures. Organizations that are successful at managing HSE have periodic comprehensive internal and external audits. The main purpose of such audits is to determine that how well the HSE requirements and procedure are understood, the degree of compliance with or deviations from the HSE management system and any step necessary to achieve further improvement.

6 Conclusion

With growing HSE issues in sector, it is imperative for the gas companies to have well-developed systems and practices in place. They should keep the employees motivated to not only follow the strict compliance of safety norms, but also develop and sustain the culture promoting safety. The safety mechanism and approach adapted by the companies to maintain safety should be regularly monitored and supervised by the government and regulators.

Despite the safety being of paramount concern, establishing and implementing of safety measures are yet to be addressed at large portion of gas industries and gas consumers. Since there exists the difference in processes, size, culture and other characteristics in different organizations, their need of HSE management system is different. Approach of establishing HSE management system must provide a structured mechanism to manage occupational health and safety risks and to define strategies for implementing control actions, resulting operating excellence and avoiding all kinds of losses.

An appropriate technique for safety performance measurement is also necessary for improvement of HSE management system. The absence of the undesired events is by itself a sufficient indicator of good HSE management system. The performance against any laid down criteria in HSE management system, e.g. lost time accidents, number of near misses, safety index frequency rate, followed by a trend analysis from one year or more often provide some safety performance indicator.

References

EGIG 2015—9th report of the European Gas Pipeline Incident Data Group (1970–2013), February 2015

Hydrocarbons Sector Skill Council Presentation—3rd NSQC meeting 17th December 2014. Available on website of National Skill Development Agency. http://www.nsda.gov.in

ILO 2003—Global strategy on occupational strategy on occupational safety and health: Conclusions adopted by the international labour conference at its 91st Session, 2003

NCP—National Commission of Population, Ministry of Health and Family Welfare. http://populationcommission.nic.in/content/625_1_index.aspx

NORA 2011—National Occupational Research Agenda (NORA) for Occupational Safety and Health Research and Practice in the U.S. Oil and Gas Extraction Industry, October 18, 2011

OGP (2013) Safety Performance Indicators-2013 Data, report July 2014 by International Association of Oil and Gas producers

OISD—Analysis of major incidents of oil and gas industry 2004–2015 by Oil Industry Safety Directorate
OISD GDN—206 "Safety management system in Petroleum Industry" 1st edition, September, 2001
PNGRB—Analysis of incidents 2015
PPAC MoPNG 2015—Industry performance review report of Petroleum Planning and Analysis Cell (PPAC), Ministry of Petroleum and Natural Gas: December, 2015 and Snapshot of India's Oil and Gas data, December, 2015

Reviewing Europe's Transition from a Gas Market to Renewables

Gerard Kreeft

Abstract Not only this chapter focuses on the European gas market, but it is also an overview/summary of trends and innovations that have started in Europe and helped shape renewable energy—trends that can indeed influence various parts of the Global Energy Market. Much of this has been done within the confines of the Global Energy Village, an annual storage event conducted by EnergyWise. An event brings together the best minds so that new trends and innovations can be explained and compared and thereby scenarios developed. Key issues include Europe's diminished gas production and its increased dependency on gas imports, in particular from Russia. While gas imports will probably stay long term at their current level, long-term this can change. There is a question whether Gazprom can finance its Yamal's assets to ensure a continued supply of gas to Europe. And Europeans are increasingly turning to renewable fuels. How this will play out remains to be seen. Europe's energy politics is in the short-to-medium term a muddle. The rise of shale gas in the USA has given a sharp rise to coal imports to Europe. CO_2 emission trading prices are too low; hence, they are irrelevant. Continuous subsidies for green energy continue to distort the marketplace. And in Germany 'brown coal' continues to be popular as an industrial fuel. Also an analysis is given of the oil and gas industry. The basic tenant of a hydrocarbon base, known in the industry as Reserve Replacement Ratio (RRR), is critically examined. Instead, it is argued that the industry should introduce an 'Energy Unit', based on renewables, and be CO_2 neutral. This can be the basis of a new Energy Road Map.

1 Introduction

Europe's gas dependency will no doubt increase in the foreseeable future. Yet Europe's domestic production will continue to decrease, thereby increasing gas imports from abroad, in particular from Gazprom. How this will play out is

G. Kreeft (✉)
EnergyWise, Arnhem, The Netherlands
e-mail: g.kreeft@energywise.nl

S.K. Kar and A. Gupta (eds.), *Natural Gas Markets in India*,
DOI 10.1007/978-981-10-3118-2_16

dependent on Gazprom's long-term ability to ensure that its gas assets in Yamal can continue to be the reliable sources of gas for Europe. It is argued that it is questionable whether Gazprom has the necessary capital to upgrade these fields: continuing development of these fields is seen as essential for Europe's long-term gas requirements.

Europe's acceptance of renewables has indirectly been stimulated by the growing uncertainty of gas imports from Russia. Renewables, indeed, are becoming an important part of the energy mix in Europe. Witness the Power-2-Gas developments in the transportation and chemical sectors; Carbon, Capture and Storage projects, and redefining how energy companies are viewed.

2 Europe's Domestic Production: A Diminishing Asset

Europe's gas production continues to diminish and now accounts for less than 50% of what is consumed on an annual basis: over the last decade, Norway, the Netherlands, and the UK have provided the bulk of natural gas production (Fig. 1).

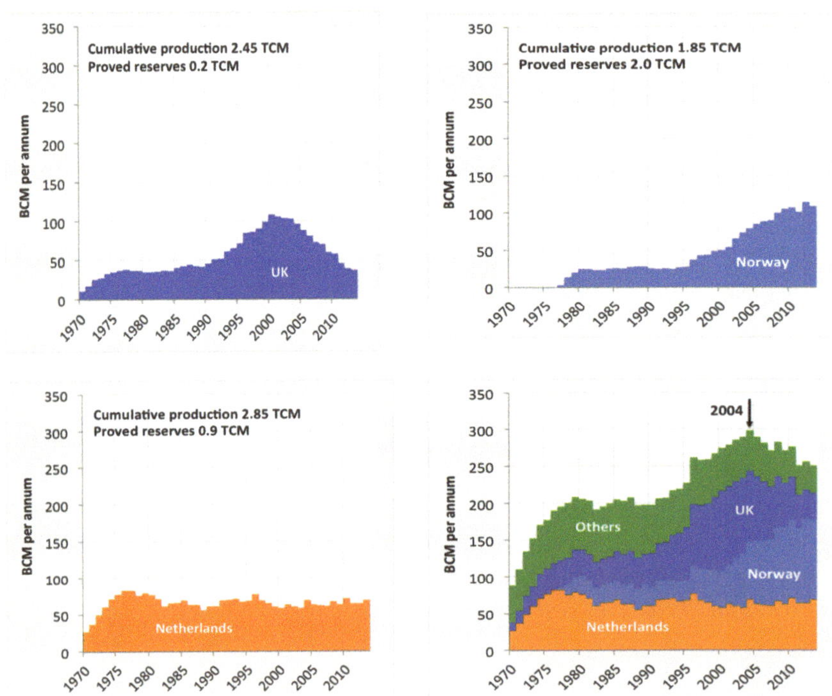

Fig. 1 European gas production and consumption. *Source* Euan Mearns (2015), Energy Matters Presentation GEV 2015 www.euanmearns.com

Fig. 2 Breakdown of EU consumption and supplies. *Source* Euan Mearns (2015), Energy Matters, Presentation GEV 2015 Fig. 10

Consumption			BCM
	EU	438.1	
	Norway	4.4	
	Switzerland	3.6	
	Turkey	45.6	
Total consumption			491.7
Indigenous Production			
	EU	146.8	
	Norway	108.7	
			255.5
Pipeline imports			
	Russia	162.4	
	Algeria	24.8	
	Libya	5.2	
			192.4
LNG imports			
	Qatar	23.4	
	Algeria	13.5	
	Nigeria	6.9	
	Trinidad	2.2	
	Other	3.3	
			49.3
LNG net exports			
			5.6
Total supply			491.6

Gas imports from (Russia = FSU) will continue to play a predominant role, as well as gas imports from North Africa and LNG imports (Fig. 2).

2.1 *Gazprom's Dilemma and Gas Future*

In Western Europe, gas imports from Russia are approx. 25–30% in Western Europe and in some cases 100% in Central–Eastern Europe. According to Gazprom, European gas consumption in 2013 reached 541 BCM, of which 161.5 BCM or 30% was supplied by Gazprom; Russia's market share in Europe increased significantly to 26% due to a growth in exports (+23 BCM or 16%) and a stagnating European demand (Clingendael 2014).

While Gazprom's market share has increased, it is important to note that total European gas consumption has decreased. For the long term, it could be a sign that Gazprom exports have reached their apex not only because of the European economic slump, but also because of the growing importance of renewables in Europe (see below).

What is important to understand is that the Ukraine with its extensive network of pipelines and storages is the main transit route for Russian gas to Europe. In 2013, approximately half of the Gazprom's exports to Europe crossed the Ukraine. Yet the Ukraine remains a major headache for Gazprom. Ukraine, an important transit country for Russian gas to Europe, has also proven to be a major obstacle for guaranteeing Russia's gas to Europe. The on-again off-again gas negotiations between Russia and the Ukraine in the period 2013 and 2014 did not enhance Gazprom's reputation in Europe. This is also a major reason why Gazprom constructed Nord stream and other major pipeline systems, to enhance its ability to delivery gas to Europe.

According to Gazprom, Gazprom's gas supply to Europe brings in more than a half of the Group's revenues (www.gazprom.com). Quite simply Gazprom requires investment capital from Europe to ensure that it can maintain and modernize its gas networks. European gas consumers bankroll Gazprom's investment programme; Russia's two-tier gas structure means that for its domestic market Gazprom is subsidizing the Russian consumers. Not an exercise in energy efficiency.

Russia has staked its gas future out with the development of the Yamal Gas Project which consists of:

> 11 gas and 15 oil, gas and condensate fields with approximately 16 tcm of explored and preliminary estimated gas reserves and nearly 22 tcm of in-place and forecast gas reserves;
>
> Reserves of condensate are estimated at 230.7 mln t and those of oil – 291.8 mln t.

In terms of gas reserves, the Bovanenkovskoye field is the most significant one in the Yamal Peninsula (4.9 tcm). The initial gas reserves of the fields (Kharasaveyskoye, Kruzenshternskoye and Yuzhno-Tambeyskoye) amount to about 3.3 tcm.

Three production zones have been created—Bovanenkovo, Tambey and Southern:

- The Bovanenkovo production zone includes three basic fields: Bovanenkovskoye, Kharasaveyskoye and Kruzenshternskoye (the licences are held by Gazprom Group). The total production is projected to reach up to 220 bcm of gas and up to 4 mln t of condensate per annum.
- The Tambey production zone involves six fields: Severo-Tambeyskoye, Zapadno-Tambeyskoye, Tasiyskoye, Malyginskoye (the licences are held by Gazprom Group), Yuzhno-Tambeyskoye and Syadorskoye. The total production is projected to reach up to 65 bcm of gas and up to 2.8 mln t of condensate per annum.
- The Southern production zone involves nine fields: Novoportovskoye (the licence is held by Gazprom Group), Nurminskoye, Malo-Yamalskoye, Rostovtsevskoye, Arkticheskoye, Sredne-Yamalskoye, Khambateyskoye, Neytinskoye and Kamennomysskoye (located onshore). The total production is projected to reach up to 30 bcm of gas and up to 7 mln t of oil per annum.

Gazprom estimates that the maximum annual gas production on Yamal is comparable to the volume of Gazprom's current gas supplies to the domestic market and exceeds twofold the volume of gas exported to the countries beyond FSU.

Much attention is now being given to the Yamal LNG Project, which is part of Tambey South. The Yamal LNG Project is based on the estuary of the Ob River which is ice-bound nine months of the year. The project ensures production and marketing of the Russian Arctic's vast natural gas reserves and has involved the construction of a major new maritime route for transporting liquefied natural gas to Europe and Asia.

The project is one of the largest industrial undertakings in the Arctic. It will eventually involve the drilling of more than 200 wells, the construction of 3 LNG trains, each with a capacity of 5.5 million tons per year, and a vast gas terminal, and the commissioning (a world first) of 16 icebreaker tankers, each able to transport 170,000 m^3. Yamal LNG is a joint-venture company owned by Novatek (**60%**), Total (**20%**) and CNPC (**20%**).

In an assessment of Gazprom's needed investment capital, Yulia Grama, Department of Diplomacy, National Chengchi University (Taiwan) estimates that until 2030 Gazprom will require some US$560–590 billion. Of this US$150–170 billion is required for the Yamal Peninsula, the source of most of the Russia's gas exports to Europe (Grama 2012).

Three major fields—Yamburg, Urengoy and Medvezhye—make up about 45% of Gazprom's gas reserves. These fields are mature and have been in operation since the late 1960s and early 1970s. Not only do they need upgrading, but there is also a strong need to ensure that there is a strong reserve base for the next 25–30 years.

The Shtokman Gas Field, which received much attention earlier in the decade, and which has Statoil and Total as international partners, was heralded as a new gas source. But for all intents and purposes, the project has been put on hold.

Other major pipeline projects, have in the past months, become relics of the past. Take for example the Nabucco pipeline project.

The Nabucco pipeline project, which was to have transported gas from the Caspian Sea to Europe in order to bypass Russia, has been cancelled. The pipeline, sponsored by the European Union (EU), had already been reduced last summer in length from the original 3900 to 1300 km. The eastern section, which was to have run from Azerbaijan across Georgia and Turkey to the Bulgarian border, was abandoned. Instead, the Trans Anatolian Pipeline (TANAP), funded by Azerbaijan and Turkey, is due to come into operation in 2018.

Nabucco-West, which was to have carried gas from Turkey to Austria, through Bulgaria, Romania and Hungary, was the only remaining part of the original project. At the end of June 2014, it was announced that this project would also be dropped.

2.1.1 Going Eastward—Turkmenistan and China

In the past, Gazprom could rely on Turkmenistan to ensure that gas molecules from that country provided a gas buffer for Europe—not only it was a buffer, but also it was cheap to buy. Gas prices in Turkmenistan were at the domestic Russian price. Gazprom added the necessary surcharges and moved the gas onwards to Europe. In 2008, Gazprom raised gas prices to the European level; Turkmenistan took note of this and looked eastwards to China (Online Asian Times, December 17 2009, www.atimes.com).

Since 2006, China has invested heavily on developing gas imports from Turkmenistan, opening its Central Asia Gas Pipeline (CAGP) across Uzbekistan and Kazakhstan at the end of 2009. CNPC (China National Petroleum Corporation) has announced that cumulative gas imports into China through the Central Asian Pipeline have now reached 100 BCM. The pipeline connects Central Asia and China's gas pipeline network.

The 1000-km D line starts from Turkmenistan and would pass through Uzbekistan, Tajikistan, and Kyrgyzstan to reach China. Once completed, it will have an annual capacity of 30 bcm.

Natural gas imports from Central Asia are now supplied to 25 provinces, municipalities, autonomous regions and Hong Kong (Fig. 3).

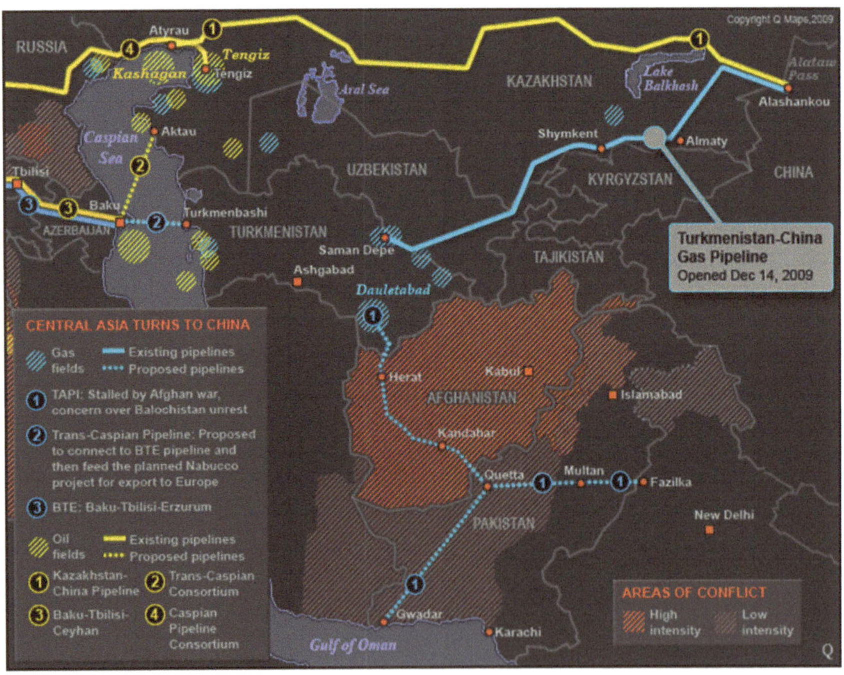

Fig. 3 Central Asian Gas Pipeline (CAGP). *Source* Online Asian Times, December 17 2009, www.atimes.com

2.1.2 China–Russia: A New Alliance?

Gazprom finalized a deal with the Chinese National Petroleum Corporation (CNPC) in May 2014. Under the first phase of the new 30-year contract, Russia will supply China 38 BCM, per year of natural gas starting in 2018. Future phases could increase this volume to as much as 60 BCM per year. The contract links the natural gas price to international crude oil prices and operates as a take-or-pay scheme: the buyer, CNPC, must pay for the contracted natural gas even if it decides not to receive it.

New natural gas production in Russia will mainly come from fields in eastern Siberia, which currently lack export infrastructure. The planned Siberia pipeline will export gas south to China and east to a liquefied natural gas (LNG) plant on Russia's east coast.

While the gas deal was heralded in the press, little has been said how this infrastructure will be funded. While it is anticipated that the Chinese will finance it, Gazprom, no doubt, will have to provide substantial discounts on the delivery of its gas.

Gazprom's contact with the Chinese is also a move to create a greater diversification of gas buyers. How this will affect Europe is still an open question. Nonetheless, Europe will continue to be an important market for Russia, and it is certainly given that the Chinese will not pay the higher gas prices that the Europeans have paid.

2.2 *Underground Gas Storage (UGS)*

UGS plays a pivotal role in Europe's gas industry. Working gas volume (WGV), representing the amount of gas stored at any point in time, is between 17 and 20% of gas demand. Total WGV globally is 358.8 bcm more or less evenly divided among North America (38%), CIS (32%) and Europe (28%).

Based on these figures, it would appear that there are 693 UGS facilities on a global basis and 183 planned and potential developments (see Fig. 4); Europe operates 148 facilities and has plans to develop 100 new facilities (see Fig. 5). Where this optimism comes from is hard to understand, given the current lack of upgrading and building of new UGS facilities (Fig. 6).

The working committee drew up a series of low-average-high storage demand scenarios for up to the period 2030. Based on an average storage scenario for 2030, it is estimated that 500 bcm will be required—up from the current 359 bcm (Fig. 7).

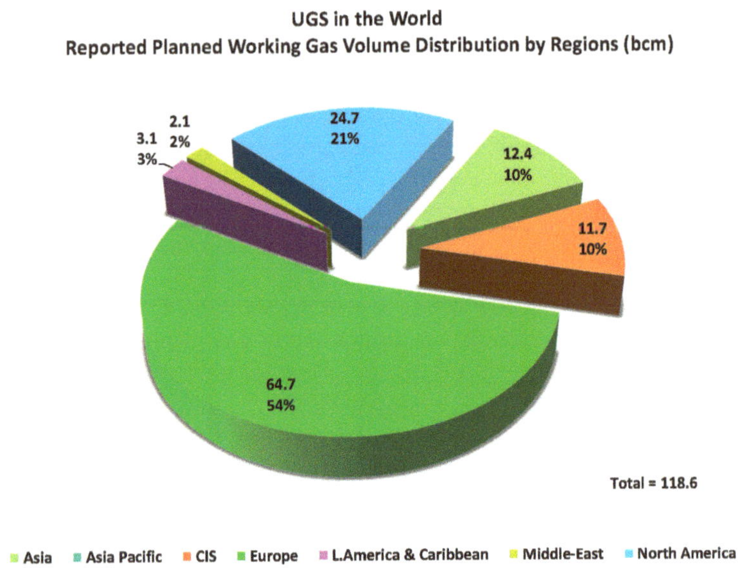

Fig. 4 Working Gas Volume by Region (BCM). *Source* IGU 2009–2012, June 2012, Working Committee: Underground Gas Storage, Fig. 7

World	No. of UGS Facilities	Working Gas Volume (bcm)
UGS in operation	693	358.8
Planned developments in existing UGS	32	16.1
Planned new (Greenfield) UGS	120	102.5
Potential	31	22.0
Planned and potential developments	183	140.6
Total - UGS in operation and planned	844	499.4

Fig. 5 Summary of Global UGS in operation, planned and potential UGS facilities. *Source* IGU 2009–2012, June 2012, Working Committee: Underground Gas Storage, Table 2

Europe	No. of UGS Facilities	Working Gas Volume (bcm)
UGS in operation	148	98.7
Planned developments in existing UGS	26	10.6
Planned new (Greenfield) UGS	58	54.1
Potential	16	8.1
Planned and potential developments	100	72.7
Total - UGS in operation and planned	222	171.4

Fig. 6 Summary of European UGS in operation, planned and potential UGS facilities. *Source* IGU 2009–2012, June 2012, Working Committee: Underground Gas Storage, Table 3

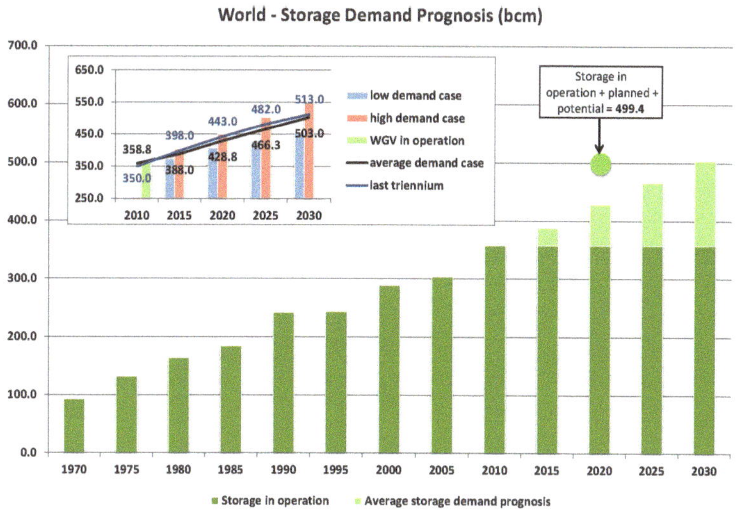

Fig. 7 European Storage Demand Prognosis (bcm). *Source* IGU 2009–2012, June 2012, Working Committee: Underground Gas Storage, Fig. 15

3 The European Dilemma: Examining the Issues Towards a Sustainable Energy Mix

In December 2014, the IEA published its Energy Report concerning the direction the EU should follow. Key recommendations included:

A new commitment to the internal energy market across the EU, with an interconnected energy network and competitive retail markets to ensure:

- In electricity, the market integration of variable renewable generation with strong co-ordination of electricity system operation; generation adequacy; and demand-side response, balancing and intra-day markets across interconnected systems.
- In gas, access to and efficient use of gas storage and liquefied natural gas terminals and unconventional gas sources.
- Timely adoption of market-based and governance rules for an integrated 2030 Climate and Energy Framework with priority to energy efficiency, a strong EU ETS, and support to all low-carbon technologies, by integrating technology, RD&D and innovation foresight.
- Enhanced EU-wide co-operation on upgrades, safety upgrades, and extensions of the lifetimes of existing nuclear power plants to ensure highest safety standards and regulatory stability needed for the investment decisions in those countries that opt for nuclear energy (EU 2014).

3.1 Europe's Contradictions

Increased coal demand, thus displacing natural gas, running gas-fired power plants at a loss, or mothballing them, and subsidies on renewable fuels have sent shockwaves through the European Power Sector. Ernest & Young in their review of the European power and utility sector calculate that in the period 2010–2012 some €30 billion was written off the balance sheets of 16 companies surveyed (EY 2013).

RWE and EON, two of the Germany's largest utility companies, best represent the dilemma that the continent is facing. Since the Fukushima I nuclear disaster, the Germany Government introduced a nuclear moratorium which would phase out all nuclear plants by 2022. This was most painful to RWE. Not only was their business model focused on nuclear energy, but also the production of electricity was also dependent on lignite coal. In 2014, 4.8% of the electricity it generated came from renewables, less than half compared to EON (18 June 2015, Bloomberg.com).

In the meantime, EON has taken the step to divest itself from 'old energy sources', a company to be called Uniper, and a new EON focused on renewables—A Good Bank and Bad Bank development.

The European Energy Market is thus confronted with the following paradoxes:

- Coal demand and coal imports are rising because of the cheaper price, displacing natural gas as a fuel of choice;
- Switching from gas-to-coal means that CO_2 emissions have not decreased;

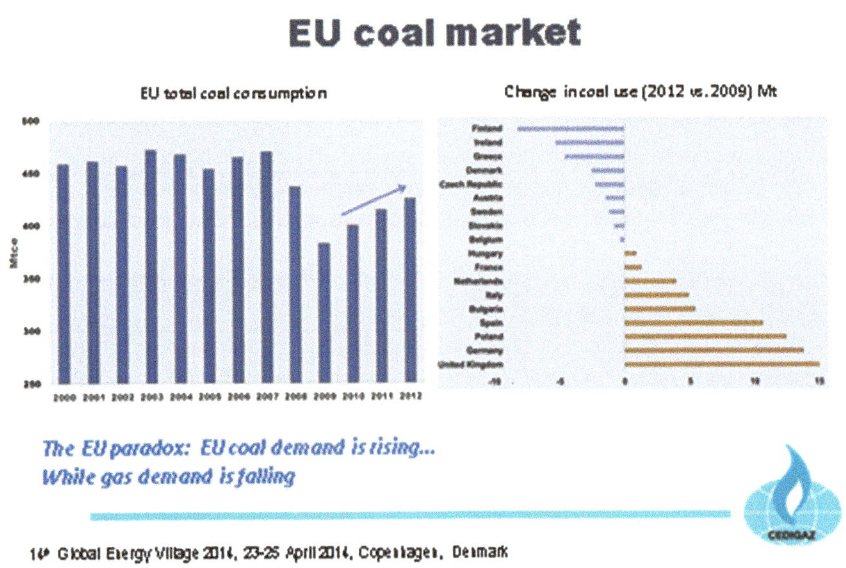

Fig. 8 EU coal market. *Source* Sylvie Cornot (2014), SeniorEconomist, Cedigaz, Presentation'at Global Energy Village 2014, entitled 'How do we move from the Cheap Coal Era to One Where Green Energy Storage starts to make economic sense?'

- Running gas-fired power plants is a loss-making business;
- Closing or mothballing of gas-fired power plants is ongoing;
- Subsidies for renewable energy are not economically feasible.

In the short-to-medium term, Europe's road to a sustainable future looks rather bleak. Coal consumption is rising and gas demand is decreasing (Fig. 8).

Because of the US shale gas revolution, cheap coal has become an attractive fuel in Europe (Fig. 9).

Because of the increase in gas imports and the collapse of CO_2 prices, coal has become an attractive fuel (Fig. 10).

3.2 *The Case of Germany*

No where are contrasts between the demand for coal and use of renewables more startling than in Germany, which is also symbolic for much of the European Union.

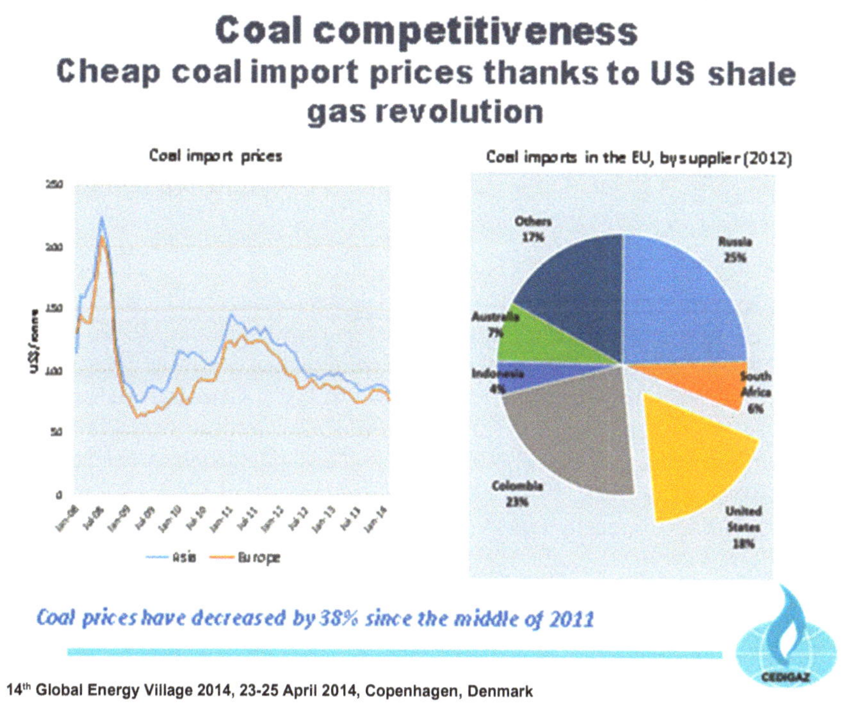

Fig. 9 Coal competitiveness: cheap coal import prices thanks to US shale gas revolution. *Source* Sylvie Cornot (2014), SeniorEconomist, Cedigaz, Presentation'at Global Energy Village 2014, entitled 'How do we move from the Cheap Coal Era to One Where Green Energy Storage starts to make economic sense?'

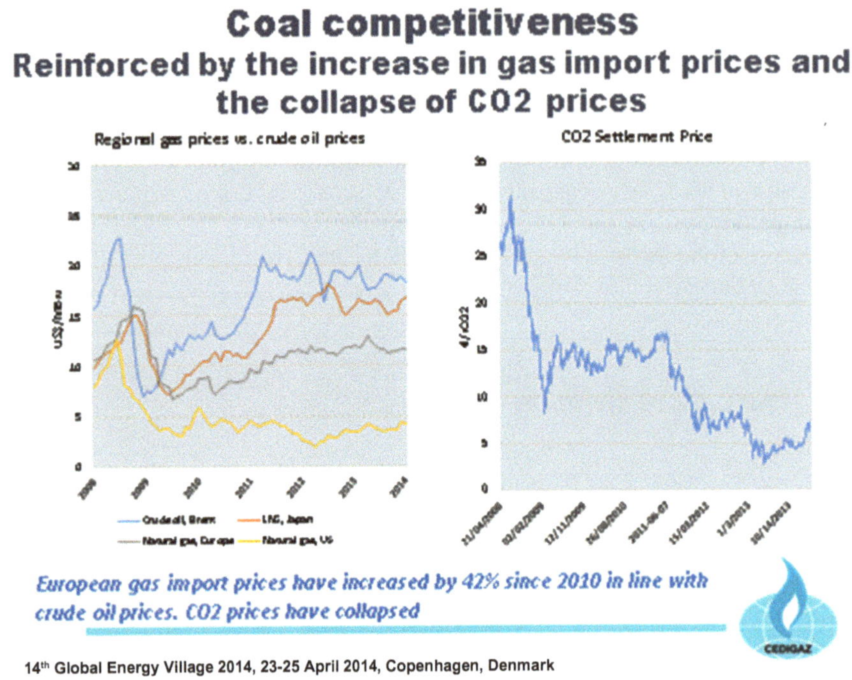

Fig. 10 Coal competitiveness: reinforced by the increase in gas import prices and the collapse of CO_2 prices. *Source* Sylvie Cornot (2014), SeniorEconomist, Cedigaz, Presentation'at Global Energy Village 2014, entitled 'How do we move from the Cheap Coal Era to One Where Green Energy Storage starts to make economic sense?'

Coal is Germany's most abundant indigenous energy resource, and it accounted for about 24% of Germany's total primary energy consumption in 2012, a slight increase over the previous few years. Coal consumption increased after Japan's Fukushima reactor accident occurred in March 2011, and Germany used coal as a substitute for nuclear power in electricity generation. Germany was the world's eighth largest producer of coal in 2012. Nearly all coal production serves the power and industrial sectors.

Deutsche Bank Research has developed two scenarios for the Germany Energy Market (Fig. 11):

Domestic lignite is virtually unexportable because of its low energy value. The use of this relatively high emission but price competitive fuel has been part of Germany's industrial fabric (Fig. 12).

Deutsche Bank Research anticipates that a more rapid increase in coal prices is possible—should US exports be lower than was anticipated? A more ambitious environmental agenda would trigger a huge surge in the cost of emission allowances. This would hit coal harder than gas, and gas could benefit.

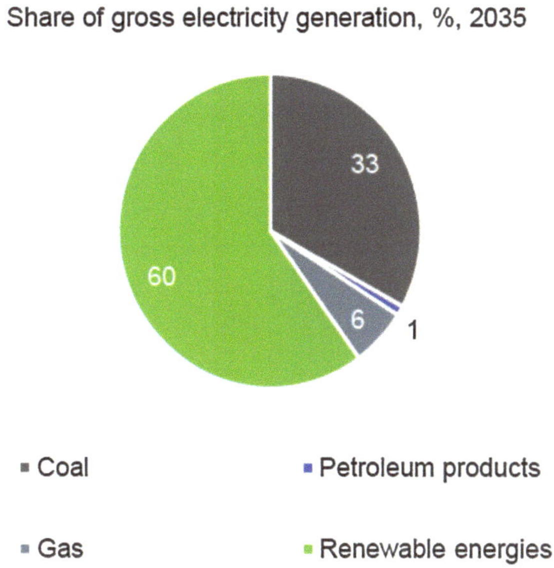

Fig. 11 Scenario 1, coal remains important for generating electricity 2035. *Source* Auer Josef and Vasilios Antolitis, 'The changing energy mix in Germany', June 26, 2014, Deutsche Bank Research

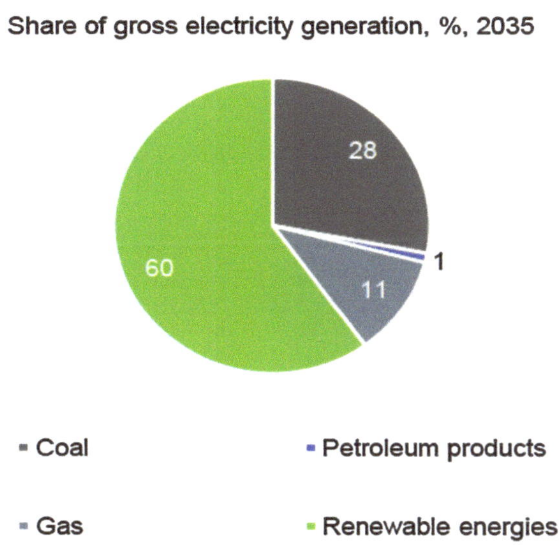

Fig. 12 Scenario 2, gas cements its position in the electricity generation market. *Source* Auer Josef and Vasilios Antolitis, 'The changing energy mix in Germany', June 26, 2014, Deutsche Bank Research

4 Putting Renewables in a European Perspective

According to BP's Annual Statistical Review (2014), renewable energy in power generation grew by an above average of 16.37% in 2013. Europe/Eurasia once again contributed the largest growth increment, even though growth in Europe's

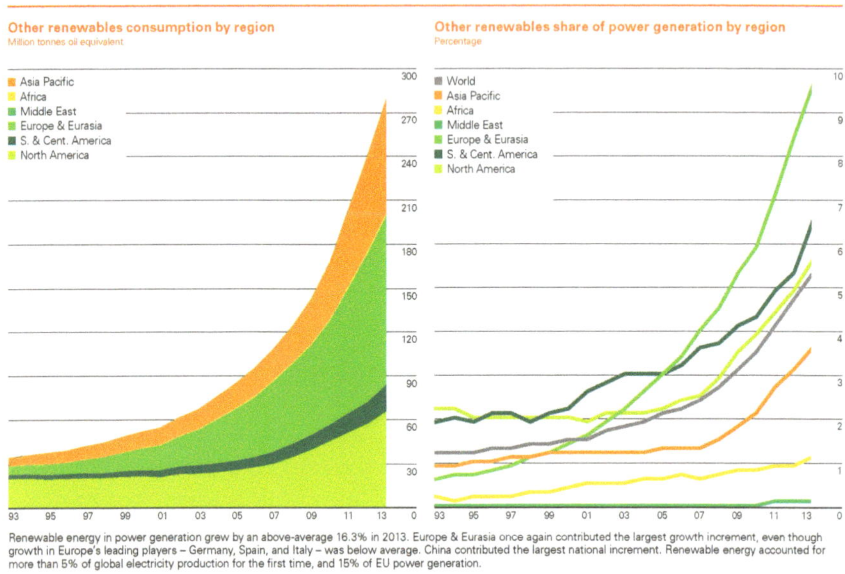

Fig. 13 Renewable energy consumption/share of power by region. *Source* BP Statistical Annex 2014, p. 38

leading players—Germany, Spain and Italy—was below average. China contributed the largest national increment. Renewable energy accounted for more than 5% of global production for the first time and 15% of EU power generation (Fig. 13).

Yet the impact of renewables on the energy mix is very visible (BP 2014):

- The share of fossil fuels in the energy mix fell to 77% in 2013, a record low, while renewables reached a record high of 6.6%.
- The largest increments in renewables was in wind (7 Mtoe, 16%), followed by biomass (3 Mtoe, 9%), then solar (3 Mtoe, 17%) and finally geothermal (0.1 Mtoe, 5%).
- The EU's carbon intensity of primary energy declined by 1.3% to 2.3 tonnes of CO_2 per tonne of oil equivalent in 2013 as carbon emission dropped by 1.9% to 86.2 Mt.
- The consumption of all three fossil fuels declined, with oil undergoing the largest decline (13.6 Mtoe, 1.9%) followed by coal (8.1 Mtoe, 2.5%) and then natural gas (5.4 Mtoe, 1.1%).

4.1 Carbon Capture Storage (CCS)

Carbon capture and geological storage (CCS) is a technique for trapping carbon dioxide emitted from large point sources such as power plants, compressing it and transporting it to a suitable storage site where it is injected into the ground. This technology has significant potential to help mitigate climate change both in Europe and internationally, particularly in countries with large reserves of fossil fuels and a fast-increasing energy demand.

Before carbon dioxide gas from power plants and other point sources can be stored, it must be captured and stripped of most associated substances. This is not a new technology, as CO_2 is routinely separated and captured as a by-product from industrial processes. Captured CO_2 then needs to be stored (in compressed form) and transported to the place of sequestration.

CO_2 can be stored in geological formations including oil and gas reservoirs, unmineable coal seams, and deep saline reservoirs. It is generally accepted that appropriately selected and managed geological reservoirs are very likely to retain over 99% of the sequestered CO_2 for longer than 100 years and 'likely' to retain 99% of it for longer than 1000 years.

Key Issues

Ensuring environmental integrity to ensure that the CO_2 captured and stored remains isolated from the atmosphere and that it does not present a risk to human health or ecosystems;

- high costs, in particular the capture component;
- a clear regulatory framework.

Nonetheless, the oil and gas industry has already initiated a number of CCs projects.

Statoil is involved in four large-scale CCS projects (Statoil):

- The Sleipner field in the Norwegian sector of the North Sea
- Snøhvit LNG production of gas from the Norwegian sector of the Barents Sea
- The In Salah field in Algeria
- Technology Centre Mongstad (TCM)

Shell is involved in the following CCS projects

Gorgon Australia: At the Gorgon gas fields off the coast of Western Australia, natural gas will travel through undersea pipelines to a liquefied natural gas plant on nearby Barrow Island. Once injection operations are at full capacity in 2015, 3–4 million tonnes a year of naturally occurring CO_2 produced with the natural gas will be captured and injected into a deep sandstone formation around 2.5 km beneath the island. Chevron is leading the Gorgon project, with Shell and ExxonMobil as partners. Gorgon is the world's largest CCS project.

Quest, Canada: The Quest project will capture, transport and store over a million tonnes of CO_2 a year from Shell's Scotford Upgrader in Alberta, Canada, starting in 2015. It is the first commercial-scale CCS project in the oil sands industry. Shell is leading this project on behalf of the other Athabasca Oil Sands Project joint-venture owners, Chevron and Marathon.

Scotford Upgrader, Canada: where a planned CCS project would capture CO_2.

4.2 Power-2-Gas (P2G)

Power-to-Gas (P2G) basically converts electrical power to a gas fuel by means of electrolysis: splitting water into hydrogen and oxygen; there are basically two methods:

- Hydrogen is injected into the gas grid or used in the transport or industrial sector
- Hydrogen is combined with CO_2 and converted to methane and fed into the natural gas grid.

The attractiveness of P-2-G lies in its potential ability to store excess wind and solar energy so that it can be used at a later stage. Before switching to natural gas, the German gas networks used 'Towngas' which consisted of 50–60% hydrogen. The storage capacity of the German natural gas network, including UGS, is more than 200,000 GW h which is enough for several months of energy requirement. By comparison, the capacity of all German pumped storage power plants amounts to only about 40 GW h. The transport of energy through a gas network is done with much less loss (<0.1%) than in a power network (8%). The storage costs per kilowatt-hour are estimated at €0.10 for hydrogen and €0.15 for methane (Power-to-Gas Wikipedia).

The ChemCoast project study is perhaps an early example of the feasibility of hydrogen as a potential fuel, especially for the transport and chemical sectors. Clients include ChemCoast and various industry partners (chemical, energy and process technology), regions' governments (Lower Saxony, Schleswig-Holstein, and City of Hamburg with Transport Authority of the City of Hamburg).

Hydrogen infrastructure timeline:

2015	Pilot project
2020	Extend existing H2-pipeline (PPP financed) to broaden regional H2-market, plan and develop first H2-cavern for large-scale H2-storage and connect to H2-pipeline, complemented by H2-trailer transport, utilize cost advantage of cheap by-product H2 in early transition phase
2025	Extend production portfolio of wind-H2, H2-production costs decreasing, H2-marketplace growing further, add H2-pipeline capacity, networking of regions, relevance of transport sector growing (Bünger 2014)

4.2.1 Falkenhagen

EON power-to-gas pilot unit injects more than two million kilowatt-hours of hydrogen into natural gas pipeline system in the first year of operation. According to EON, its P2G pilot unit in Falkenhagen in eastern Germany has delivered a positive performance in its first year of operation, injecting more than two million kilowatt-hours of hydrogen into the gas transmission system.

The Falkenhagen unit uses renewable source electricity to power electrolysis equipment that transforms water into hydrogen, which is injected into the natural gas transmission system. With a capacity of two megawatts, it can produce 360 m^3 of hydrogen per hour. This stored energy becomes part of the natural gas supply and can be used for space heating, industrial applications, in areas such as mobility, and power generation. EON is currently building a second P2G pilot unit in Reitbrook, a suburb of Hamburg. The purpose of this unit, which will enter service in 2015, is to optimize the transformation process by means of more compact and efficient electrolysis equipment.

4.2.2 Thüga Power-to-Gas Energy Storage Project

Thüga reported that the system has been declared 'technically able to participate in the secondary power market'. This means that the electrolyser is technically capable of offering grid-balancing services, which increases the economic viability of the technology.

According to ITM Power, which is implementing this project, 'The system has now undergone and passed its first annual reassessment by TUEV Hessen. The electrolyser system has met every aspect of its specification including efficiency, remote control functionality and response time. The data for the plant has been shared with the partners under the project partnering agreement. The 300 kW plant is the first and only rapid response PEM electrolyser plant to be injecting hydrogen into the German gas network. The first injection was on 26th Nov 2013' (ITM Power 2015).

5 Profiling the Energy Company of the Twenty-First Century: The Quandary They Face

While renewable energy is gaining legitimacy throughout the energy sector, there is still a gaping structural problem: the inability to see that oil companies are not sustainable by their very nature. The primary purpose of oil and gas companies is to produce hydrocarbons—nothing more and nothing less.

Current oil producers are caught up in a quandary: IOCs must seek to replace their reserves, which at best have received mixed reviews. In the trade, this is known as the Reserve Replacement Ratio (RRR) and is used by financial analysts to benchmark the international oil and gas companies. A modest 100% RRR is required if oil companies are to maintain their reserve equilibrium.

Even the goal of 100% RRR is hard to achieve. Oscillating RRR values of between 49 and 120% exist. Regardless of which company is examined, and regardless of their CapEx budget, the major publicly traded oil and gas companies based in Europe and North America all have oscillating RRR values of between 49 and 120%.

The global benchmark for establishing reserves is currently the 'Petroleum Resources Management System'. This document reflects the consensus of the Society of Petroleum Engineers (SPE), American Association of Petroleum Geologists (AAPG), World Petroleum Council (WPC) and the Society of Petroleum Evaluation Engineers (SPEE). It was published in 2007 and is generally referred to as the SPE-PRMS (SPE 2007). In short, this document defines the various categories of hydrocarbon resources, including unconventional resources such as shale gas. In 2011, additional 'Guidelines for Application of the Petroleum Resources Management System' was released, which further details the calculation procedures for the SPE-PRMS (SPE 2011). The SPE-PRMS by now has become the worldwide standard for reserves classification and reporting and has been adopted by many government and stock exchange regulators.

For companies listed at a US stock exchange, the rules set by the US Securities and Exchange Commission (SEC) apply. In the past, the SEC permitted oil and gas companies to disclose only proved reserves, and the calculation rules were rather different from those used the SPE, resulting in often significant lower proved reserves (sometimes even 30% less) as when the SPE rules would have been applied. In 2010, the SEC changed their guidelines significantly; the rules for proved reserves became much more aligned with those of the PRMS, but not completely. For example, the SEC uses the average price over the past year, while the PRMS requires a reasonable forecast of prices and economic conditions. As shown during 2014/15, this can still make a remarkable difference, while during the first quarter of 2015 the oil price had dropped to around 50 \$/bbl—the SEC required the use of the 2014 average of 95 \$/bbl to be used for the 31 December 2014

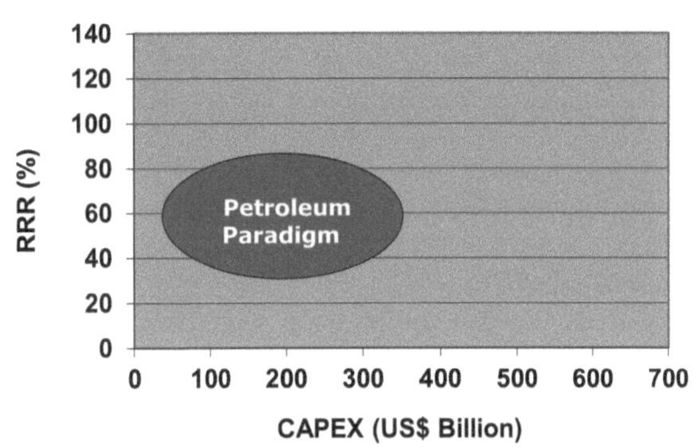

Fig. 14 Petroleum paradigm. *Source* Gerard Kreeft (2015) EnergyWise Presentation: GEV 2015

reserves estimates. Contrarily, in the past the SEC now also allows disclosure of probable and possible reserves, to better reflect the uncertainty that is inherent in any reserves estimation. For the calculation of the probable and possible reserves, the SEC has ruled that the definitions from the PRMS apply.

While the rules of engagement regarding RRR continue to evolve, there is no denying that the oil majors are struggling with varying success on this front. For lack of a better term, this could be called the 'Petroleum Paradigm' (Fig. 14).

The underlying assumption is based on a hydrocarbon reserve base. Is this realistic? And if the RRR is virtually assured of staying around the industry norm of 100% is it not an illusion to think that this will change? And the fact that the industry will continue to stumble under the 100% norm only helps generate a public perception that energy supplies are indeed dwindling and threatening shortages for future generations.

The industry's track record is not encouraging. An examination of the capital investment programmes of the sector indicates a $US 700 billion investment programme for 2014. Much of this is devoted not only to the upstream sector, but also to the midstream and downstream sectors. The lion's share is devoted to search or develop hydrocarbons (Schlumberger 2014).

But why is it that RRR is based on hydrocarbons? Well, basically that is because the oil and gas industry have established and followed this tradition, including SEC regulations, and how the financial community reports on it. Yet to change this tradition, is to disrupt industry consensus. But given that it was the oil and gas industry that set up this tradition why shouldn't the industry also point out new trends, so that a new consensus can emerge? A consensus that is in keeping with a changing perception of the industry.

Why couldn't other fuels, such as nuclear, wind, biomass and solar energy, be introduced so that they become part of the RRR mix? If serious investments were made in these fuels, then they could indeed prove to seriously increase RRR numbers.

In the past, the Athabasca Tar Sands was not seen as a fuel which could be added to a company's reserve. Yet over time the consensus changed and is now seen as a

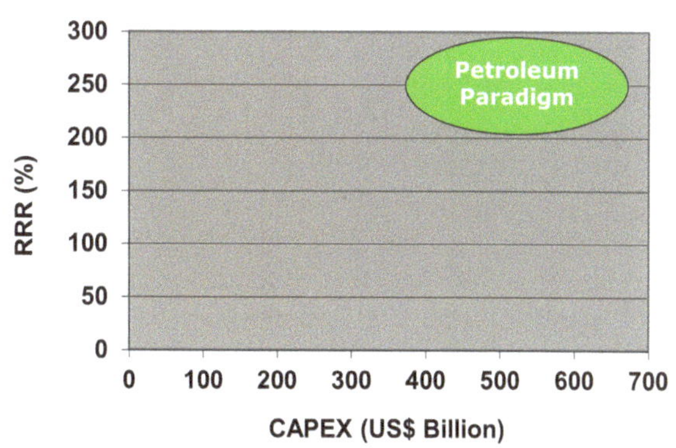

Fig. 15 Petroleum paradigm revisited. *Source* Gerard Kreeft EnergyWise, 2015

legitimate reserve base. The same type of reporting could also be done for sustainable development/alternative fuels.

We should not be surprised that alternative fuels are not taken seriously. All of the majors commit symbolic funding to move this part of the industry forward. What is used to develop renewable energy is not reported separately (Fig. 15).

If sustainable investments represented say 10% or $US 70 billion, a signal would be sent to the investor community that given the serious level of these investments, they should be included as part of their RRR calculation. This serious increase in sustainable/alternative fuel budgets, if continued for the rest of the decade, would no doubt increase RRR rates and get the industry out of the 'Petroleum Paradigm'. Why couldn't RRR rise to 200 or 300%?

Instead of talking about RRR, it is more feasible to talk in terms of Energy Units: which are CO_2 neutral, and which would have as simple policy the user pay for any environmental damage. The inertia of CO_2 emissions (buying and selling) could be greatly stimulated if Energy Units were introduced. Much work remains to be done:

- Getting the issue of RRR redefined in the energy sector;
- Educating policy-makers and regulators on this issue;
- Defining how such a road map should take place.

Certainly this should be the issue for activist shareholders around the globe! And it would solve the strategic fault line that the sector is facing.

6 Back to the Future

While predicting energy trends is risky at the best of times, there are certain scenarios that would seem plausible.

Gas imports to the EU will continue to be an important fuel. The question is the level of gas imports. It could well be that gas imports have reached their apex, and renewable fuels become a substitute fuel. Given the political dimension that gas imports have reached, Europe could well decide that enough is enough, and see this as an incentive to further develop their renewable fuels, thus voting for security of supply.

The current muddle of cheap coal, the lack of a CO_2 policy and subsidized renewables will continue to play out for the rest of the decade. The EU and the energy industry must develop a midterm strategy.

Energy companies will continue to write down more assets. A contributing factor is the decentralized role of storage, energy mobility and creation of energy communities. And the oil companies...if they continue their pursuit of only hydrocarbons, their sustainable future will be very bleak.

Certainly for the energy companies who want to survive and prosper in the twenty-first century, the transition to Energy Units from RRR is a given; developing sustainable energy systems both at the micro- and macro-level is a daunting task. This is the challenge.

References

Auer J, Slomba L (2014) The changing energy mix in Germany, 26 June 2014. Deutsche Bank Research

BP's Annual Statistical Review (2014)

Bünger U (2014) Moving power to gas from demonstration project to realization: signposting the economics. Global Energy Village

Clingendael International Energy Programme, CIEP, Fact Sheet 2014

Cornot S, Senior Economist, Cedigaz, Presentation at Global Energy Village 2014, entitled "How do we move from the Cheap Coal Era to One Where Green Energy Storage starts to make economic sense?"

Energy Policies of IEA Countries: European Union 2014

EON.com (Falkenhagen)

EY (2013) Benchmarking European power and utility asset impairments; Lessons from 2012. www.EY.com

Grama Y (2012) Analysis of Russian oil and gas reserves. Int J Energy Econ Policy 2(2):82–91. ISSN 2146-4553. Department of Diplomacy, National Chengchi University Taiwan. www.econjournals.com

Guidelines for Application of the Petroleum Resources Management System (2011) Society of Petroleum Engineers (SPE 2011)

ITM Power.com (Thuga Group's P-2-G plant exceeds expectations) 2015

Kreeft G (2015) EnergyWise, 2015. Internal documents

Mearns E (2015) Energy matters, presentation global energy village 2015

Petroleum Resources Management System (2007) Society of Petroleum Engineers (SPE 2007)

Power-to Gas Wikipedia

Schorn, Patrick President, Operations and Integration, Schlumberger, Cowan and Company, 4th annual ultimate energy conference, New York, December 2014

www.atimes.com

www.Gazprom.com

www.rian.ru

www.shell.com

www.statoil.com

2009–2012 Trennium work report, June 2012, Working Committee, Underground Gas Storage, International Gas Union

Kolibri 360 GmbH
Ferdinand-Jühlke-Straße 7
99095 Erfurt, Deutschland
produktsicherheit@kolibri360.de